T0350150

Silicon Carbide 2008— Materials, Processing and Devices

MATERIALS RESEARCH SOCIETY
SYMPOSIUM PROCEEDINGS VOLUME 1069

Silicon Carbide 2008— Materials, Processing and Devices

Symposium held March 25–27, 2008, San Francisco, California, U.S.A.

EDITORS:

Michael Dudley
Stony Brook University
Stony Brook, New York, U.S.A.

C. Mark Johnson
University of Nottingham
Nottingham, United Kingdom

Adrian R. Powell
Cree, Inc.
Durham, North Carolina, U.S.A.

Sei-Hyung Ryu
Cree, Inc.
Durham, North Carolina, U.S.A.

Materials Research Society
Warrendale, Pennsylvania

CAMBRIDGE
UNIVERSITY PRESS

University Printing House, Cambridge CB2 8BS, United Kingdom

One Liberty Plaza, 20th Floor, New York, NY 10006, USA

477 Williamstown Road, Port Melbourne, VIC 3207, Australia

314-321, 3rd Floor, Plot 3, Splendor Forum, Jasola District Centre, New Delhi - 110025, India

79 Anson Road, #06-04/06, Singapore 079906

Cambridge University Press is part of the University of Cambridge.

It furthers the University's mission by disseminating knowledge in the pursuit of education, learning and research at the highest international levels of excellence.

www.cambridge.org
Information on this title: www.cambridge.org/9781605110394

Materials Research Society
506 Keystone Drive, Warrendale, PA 15086
http://www.mrs.org

© Materials Research Society 2008

First published 2008
First paperback edition 2012

Single article reprints from this publication are available through University Microfilms Inc., 300 North Zeeb Road, Ann Arbor, MI 48106

CODEN: MRSPDH

A catalogue record for this publication is available from the British Library

ISBN 978-1-605-11039-4 Hardback
ISBN 978-1-107-40855-5 Paperback

CONTENTS

Preface xi

Materials Research Society Symposium Proceedings.......................................xii

BULK MATERIAL AND CHARACTERIZATION

* Bulk Growth of SiC...3
  Peter Wellmann, Ralf Müller, Sakwe Aloysius Sakwe,
  Ulrike Künecke, Philip Hens, Mathias Stockmeier,
  Katja Konias, Rainer Hock, Andreas Magerl, and
  Michel Pons

Alternative Routes to Porous Silicon Carbide...15
  Bettina Friedel and Siegmund Greulich-Weber

* Influence of Crystal Growth Conditions on Nitrogen
Incorporation During PVT Growth of SiC...23
  Darren Hansen and Mark Loboda

* High-Resolution Photoinduced Transient Spectroscopy
of Defect Centers in Undoped Semi-Insulating 6H-SiC...............................33
  Pawel Kaminski, Roman Kozlowski,
  Marcin Miczuga, Michal Pawlowski,
  Michal Kozubal, and Jaroslaw Zelazko

Studies of c-Axis Threading Screw Dislocations in
Hexagonal SiC..45
  Yi Chen, Ning Zhang, Xianrong Huang,
  Govindhan Dhanaraj, Edward Sanchez,
  Michael F. MacMillan, and Michael Dudley

* Links Between Etching Grooves of Partial Dislocations
and Their Characteristics Determined by TEM in 4H SiC............................53
  Jean-Pierre Ayoub, Michael Texier, Gabrielle Regula,
  Bernard Pichaud, and Maryse Lancin

* Point Defects in SiC..65
  Ádám Gali, Michel Bockstedte, Ngyen Tien Son,
  and Erik Janzén

*Invited Paper

Determination of the Core-Structure of Shockley Partial Dislocations in 4H-SiC 77
> Yi Chen, Ning Zhang, Xianrong Huang,
> Joshua D. Caldwell, Kendrick X. Liu,
> Robert E. Stahlbush, and Michael Dudley

Propagation and Density Reduction of Threading Dislocations in SiC Crystals During Sublimation Growth 83
> Ping Wu, Xueping Xu, Varatharajan Rengarajan,
> and Ilya Zwieback

Deep-Level Defects in Nitrogen-Doped 6H-SiC Grown by PVT Method 89
> Pawel Kaminski, Michal Kozubal, Krzysztof Grasza,
> and Emil Tymicki

Stress Mapping of SiC Wafers by Synchrotron White Beam X-ray Reticulography 95
> Ning Zhang, Yi Chen, Edward Sanchez,
> Michael F. MacMillan, and Michael Dudley

Effect of Annealing Temperature on SiC Wafer Bow and Warp 101
> Xueping Xu and Chris Martin

EPITAXIAL MATERIAL AND CHARACTERIZATION

Residual Stress in CVD-Grown 3C-SiC Films on Si Substrates 109
> Alex A. Volinsky, Grygoriy Kravchenko,
> Patrick Waters, Jayadeep Deva Reddy,
> Chris Locke, Christopher Frewin, and
> Stephen E. Saddow

* **Silicon Carbide Hot-Wall Epitaxy for Large-Area, High-Voltage Devices** 115
> M.J. O'Loughlin, K.G. Irvine, J.J. Sumakeris,
> M.H. Armentrout, B.A. Hull, C. Hallin, and
> A.A. Burk Jr.

*Invited Paper

Influence of Growth Conditions and Substrate Properties on Formation of Interfacial Dislocations and Dislocation Half-Loop Arrays in 4H-SiC(0001) and (000-1) Epitaxy............123
Hidekazu Tsuchida, Isaho Kamata, Kazutoshi Kojima,
Kenji Momose, Michiya Odawara, Tetsuo Takahashi,
Yuuki Ishida, and Keiichi Matsuzawa

* Suitability of 4H-SiC Homoepitaxy for the Production and Development of Power Devices............129
Christian Hecht, Bernd Thomas, Rene Stein, and
Peter Friedrichs

* Improved SiC Epitaxial Material for Bipolar Applications............141
Peder Bergman, Jawad ul Hassan, Alex Ellison,
Anne Henry, Philippe Godignon, Pierre Brosselard,
and Erik Janzén

Lateral/Vertical Homoepitaxial Growth on 4H-SiC Surfaces Controlled by Dislocations............151
Yoosuf N. Picard, Andrew J. Trunek,
Philip G. Neudeck, and Mark E. Twigg

Factors Influencing the Growth Rate, Doping, and Surface Morphology of the Low-Temperature Halo-Carbon Homoepitaxial Growth of 4H SiC with HCl Additive............157
Galyna Melnychuk, Huang De Lin,
Siva Prasad Kotamraju, and Yaroslav Koshka

Observation of Asymmetric Wafer Bending for 3C-SiC Thin Films Grown on Misoriented Silicon Substrates............163
Marcin Zielinski, Marc Portail, Thierry Chassagne,
Slawomir Kret, Maud Nemoz, and Yvon Cordier

Epitaxial Growth on 2° Off-axis 4H SiC Substrates with Addition of HCl............169
Jie Zhang, Swapna Sunkari, Janice Mazzola,
Becky Tyrrell, Gray Stewart, R. Stahlbush,
J. Caldwell, P. Klein, Michael Mazzola, and
Janna Casady

*Invited Paper

High-Frequency Electron Paramagnetic Resonance Study of the As-Deposited and Annealed Carbon-Rich Hydrogenated Amorphous Silicon-Carbon Films.....................175
A.V. Vasin, E.N. Kalabukhova, S.N. Lukin,
D.V. Savchenko, V.S. Lysenko, A.N. Nazarov,
A.V. Rusavsky, and Y. Koshka

Characterization and Growth Mechanism of $B_{12}As_2$ Epitaxial Layers Grown on (1-100) 15R-SiC.....................181
Hui Chen, Guan Wang, Michael Dudley, Zhou Xu,
James H. Edgar, Tim Batten, Martin Kuball,
Lihua Zhang, and Yimei Zhu

DEVICE PROCESSING AND CHARACTERIZATION

Simultaneous Formation of n- and p-Type Ohmic Contacts to 4H-SiC Using the Binary Ni/Al System 189
Kazuhiro Ito, Toshitake Onishi, Hidehisa Takeda,
Susumu Tsukimoto, Mitsuru Konno, Yuya Suzuki,
and Masanori Murakami

Influence of Shockley Stacking Fault Expansion and Contraction on the Electrical Behavior of 4H-SiC DMOSFETs and MPS Diodes.....................195
Joshua David Caldwell, Robert E. Stahlbush,
Eugene A. Imhoff, Orest J. Glembocki,
Karl D. Hobart, Marko J. Tadjer, Qingchun Zhang,
Mrinal Das, and Anant Agarwal

Performance of SiC Microwave Transistors in Power Amplifiers.....................203
S. Azam, R. Jonsson, E. Janzen, and Q. Wahab

Long-Term Characterization of 6H-SiC Transistor Integrated Circuit Technology Operating at 500°C 209
Philip G. Neudeck, David J. Spry, Liang-Yu Chen,
Carl W. Chang, Glenn M. Beheim, Robert S. Okojie,
Laura J. Evans, Roger D. Meredith, Terry L. Ferrier,
Michael J. Krasowski, and Norman F. Prokop

Effect of SiC Power DMOSFET Threshold-Voltage Instability.....................215
A.J. Lelis, D. Habersat, R. Green, A. Ogunniyi,
M. Gurfinkel, J. Suehle, and N. Goldsman

* SiC-Based Power Converters 221
    Leon Tolbert, Hui Zhang, Burak Ozpineci, and
    Madhu S. Chinthavali

3D Thermal Stress Models for Single Chip SiC Power
Sub-Modules...233
    Bang-Hung Tsao, Jacob Lawson, and James Scofield

MOS Capacitor Characteristics of 3C-SiC Films Deposited
on Si Substrates at 1270°C 239
    Li Wang, Sima Dimitrijev, Leonie Hold,
    Frederick Kong, Philip Tanner, Jisheng Han,
    and Gunter Wagner

Impact Ionization in Ion Implanted 4H-SiC Photodiodes 245
    Wei Sun Loh, Eric Z.J. Goh, Konstantin Vassilevski,
    Irina Nikitina, John P.R. David, Nick G. Wright, and
    C. Mark Johnson

Degradation of Majority Carrier Conductions and
Blocking Capabilities in 4H-SiC High Voltage Devices
Due to Basal Plane Dislocations...251
    Sei-Hyung Ryu, Qingchun Zhang, Husna Fatima,
    Sarah Haney, Robert Stahlbush, and Anant Agarwal

Effective Channel Mobility in Epitaxial and Implanted
4H-SiC Lateral MOSFETs..257
    Sarah Kay Haney, Sei-Hyung Ryu, Sarit Dhar,
    Anant Agarwal, and Mark Johnson

Spontaneous and Piezoelectric Polarization Effects on
the Frequency Response of Wurtzite Aluminium Gallium
Nitride/Silicon Carbide Heterojunction Bipolar Transistors....................263
    Choudhury Praharaj

Effect of Base Impurity Concentration on DC Characteristics
of Double Ion Implanted 4H-SiC BJTs...269
    Taku Tajima, Satoshi Uchiumi, Kenta Tsukamoto,
    Kazumasa Takenaka, Masataka Satoh, and
    Tohru Nakamura

*Invited Paper

Correlation Between the V-I Characteristics of (0001) 4H-SiC PN Junctions Having Different Structural Features and Synchrotron X-ray Topography.........................273
Ryoji Kosugi, Toyokazu Sakata, Yuuki Sakuma,
Tsutomu Yatsuo, Hirofumi Matsuhata,
Hirotaka Yamaguchi, Ichiro Nagai, Kenji Fukuda,
Hajime Okumura, and Kazuo Arai

Author Index.........................279

Subject Index.........................283

PREFACE

This volume contains 37 papers presented at Symposium D, "Silicon Carbide—Materials, Processing and Devices," held March 25–27 at the 2008 MRS Spring Meeting in San Francisco, California. Silicon carbide (SiC) is a very robust semiconductor material being actively developed for high power and high temperature applications, especially in the field of power electronics and sensors for harsh environments. Symposium D was the fifth in a continuing series of symposia that covered SiC growth, defects, and devices. The symposium consisted of 17 invited papers, 37 oral presentations, and 23 poster presentations given during the three days of the symposium. The scope covers most of the current research efforts focused on the development of SiC as a viable materials system for high power device structures. This includes the development of bulk material growth processes, along with the epitaxial processes required to grow SiC on both SiC substrates and on Si substrates, and results on the latest characterization techniques used to provide insight and understanding of the processes involved. Device fabrication, processing, and characterization are also covered with the results of continued discovery and optimization of processes, leading to continually improving final device characteristics. This symposium highlighted many of the advancements that have been achieved since the 12th International Conference on Silicon Carbide and Related Materials 2007 (ICSCRM2007) held in Kyoto, Japan.

We wish to thank all of the invited speakers for their outstanding presentations. We also thank all of the contributors and participants who made this symposium successful. We gratefully acknowledge the financial support provided by Cree, Inc., SULA Technologies, and the Materials Research Society

Michael Dudley
C. Mark Johnson
Adrian R. Powell
Sei-Hyung Ryu

June 2008

xi

MATERIALS RESEARCH SOCIETY SYMPOSIUM PROCEEDINGS

Volume 1066 — Amorphous and Polycrystalline Thin-Film Silicon Science and Technology—2008, A. Nathan, J. Yang, S. Miyazaki, H. Hou, A. Flewitt, 2008, ISBN 978-1-60511-036-3

Volume 1067E —Materials and Devices for "Beyond CMOS" Scaling, S. Ramanathan, 2008, ISBN 978-1-60511-037-0

Volume 1068 — Advances in GaN, GaAs, SiC and Related Alloys on Silicon Substrates, T. Li, J. Redwing, M. Mastro, E.L. Piner, A. Dadgar, 2008, ISBN 978-1-60511-038-7

Volume 1069 — Silicon Carbide 2008—Materials, Processing and Devices, A. Powell, M. Dudley, C.M. Johnson, S-H. Ryu, 2008, ISBN 978-1-60511-039-4

Volume 1070 — Doping Engineering for Front-End Processing, B.J. Pawlak, M. Law, K. Suguro, M.L. Pelaz, 2008, ISBN 978-1-60511-040-0

Volume 1071 — Materials Science and Technology for Nonvolatile Memories, O. Auciello, D. Wouters, S. Soss, S. Hong, 2008, ISBN 978-1-60511-041-7

Volume 1072E —Phase-Change Materials for Reconfigurable Electronics and Memory Applications, S. Raoux, A.H. Edwards, M. Wuttig, P.J. Fons, P.C. Taylor, 2008, ISBN 978-1-60511-042-4

Volume 1073E —Materials Science of High-k Dielectric Stacks—From Fundamentals to Technology, L. Pantisano, E. Gusev, M. Green, M. Niwa, 2008, ISBN 978-1-60511-043-1

Volume 1074E —Synthesis and Metrology of Nanoscale Oxides and Thin Films, V. Craciun, D. Kumar, S.J. Pennycook, K.K. Singh, 2008, ISBN 978-1-60511-044-8

Volume 1075E —Passive and Electromechanical Materials and Integration, Y.S. Cho, H.A.C. Tilmans, T. Tsurumi, G.K. Fedder, 2008, ISBN 978-1-60511-045-5

Volume 1076 — Materials and Devices for Laser Remote Sensing and Optical Communication, A. Aksnes, F. Amzajerdian, N. Peyghambarian, 2008, ISBN 978-1-60511-046-2

Volume 1077E —Functional Plasmonics and Nanophotonics, S. Maier, 2008, ISBN 978-1-60511-047-9

Volume 1078E —Materials and Technology for Flexible, Conformable and Stretchable Sensors and Transistors, 2008, ISBN 978-1-60511-048-6

Volume 1079E —Materials and Processes for Advanced Interconnects for Microelectronics, J. Gambino, S. Ogawa, C.L. Gan, Z. Tokei, 2008, ISBN 978-1-60511-049-3

Volume 1080E —Semiconductor Nanowires—Growth, Physics, Devices and Applications, H. Riel, T. Kamins, H. Fan, S. Fischer, C. Thelander, 2008, ISBN 978-1-60511-050-9

Volume 1081E —Carbon Nanotubes and Related Low-Dimensional Materials, L-C. Chen, J. Robertson, Z.L. Wang, D.B. Geohegan, 2008, ISBN 978-1-60511-051-6

Volume 1082E —Ionic Liquids in Materials Synthesis and Application, H. Yang, G.A. Baker, J.S Wilkes, 2008, ISBN 978-1-60511-052-3

Volume 1083E —Coupled Mechanical, Electrical and Thermal Behaviors of Nanomaterials, L. Shi, M. Zhou, M-F. Yu, V. Tomar, 2008, ISBN 978-1-60511-053-0

Volume 1084E —Weak Interaction Phenomena—Modeling and Simulation from First Principles, E. Schwegler, 2008, ISBN 978-1-60511-054-7

Volume 1085E —Nanoscale Tribology—Impact for Materials and Devices, Y. Ando, R.W. Carpick, R. Bennewitz, W.G. Sawyer, 2008, ISBN 978-1-60511-055-4

Volume 1086E —Mechanics of Nanoscale Materials, C. Friesen, R.C. Cammarata, A. Hodge, O.L. Warren, 2008, ISBN 978-1-60511-056-1

MATERIALS RESEARCH SOCIETY SYMPOSIUM PROCEEDINGS

Volume 1087E —Crystal-Shape Control and Shape-Dependent Properties—Methods, Mechanism, Theory and Simulation, K-S. Choi, A.S. Barnard, D.J. Srolovitz, H. Xu, 2008, ISBN 978-1-60511-057-8

Volume 1088E —Advances and Applications of Surface Electron Microscopy, D.L. Adler, E. Bauer, G.L. Kellogg, A. Scholl, 2008, ISBN 978-1-60511-058-5

Volume 1089E —Focused Ion Beams for Materials Characterization and Micromachining, L. Holzer, M.D. Uchic, C. Volkert, A. Minor, 2008, ISBN 978-1-60511-059-2

Volume 1090E —Materials Structures—The Nabarro Legacy, P. Müllner, S. Sant, 2008, ISBN 978-1-60511-060-8

Volume 1091E —Conjugated Organic Materials—Synthesis, Structure, Device and Applications, Z. Bao, J. Locklin, W. You, J. Li, 2008, ISBN 978-1-60511-061-5

Volume 1092E —Signal Transduction Across the Biology-Technology Interface, K. Plaxco, T. Tarasow, M. Berggren, A. Dodabalapur, 2008, ISBN 978-1-60511-062-2

Volume 1093E —Designer Biointerfaces, E. Chaikof, A. Chilkoti, J. Elisseeff, J. Lahann, 2008, ISBN 978-1-60511-063-9

Volume 1094E —From Biological Materials to Biomimetic Material Synthesis, N. Kröger, R. Qiu, R. Naik, D. Kaplan, 2008, ISBN 978-1-60511-064-6

Volume 1095E —Responsive Biomaterials for Biomedical Applications, J. Cheng, A. Khademhosseini, H-Q. Mao, M. Stevens, C. Wang, 2008, ISBN 978-1-60511-065-3

Volume 1096E —Molecular Motors, Nanomachines and Active Nanostructures, H. Hess, A. Flood, H. Linke, A.J. Turberfield, 2008, ISBN 978-1-60511-066-0

Volume 1097E —Mechanical Behavior of Biological Materials and Biomaterials, J. Zhou, A.G. Checa, O.O. Popoola, E.D. Rekow, 2008, ISBN 978-1-60511-067-7

Volume 1098E —The Hydrogen Economy, A. Dillon, C. Moen, B. Choudhury, J. Keller, 2008, ISBN 978-1-60511-068-4

Volume 1099E —Heterostructures, Functionalization and Nanoscale Optimization in Superconductivity, T. Aytug, V. Maroni, B. Holzapfel, T. Kiss, X. Li, 2008, ISBN 978-1-60511-069-1

Volume 1100E —Materials Research for Electrical Energy Storage, J.B. Goodenough, H.D. Abruña, M.V. Buchanan, 2008, ISBN 978-1-60511-070-7

Volume 1101E —Light Management in Photovoltaic Devices—Theory and Practice, C. Ballif, R. Ellingson, M. Topic, M. Zeman, 2008, ISBN 978-1-60511-071-4

Volume 1102E —Energy Harvesting—From Fundamentals to Devices, H. Radousky, J. Holbery, B. O'Handley, N. Kioussis, 2008, ISBN 978-1-60511-072-1

Volume 1103E —Health and Environmental Impacts of Nanoscale Materials—Safety by Design, S. Tinkle, 2008, ISBN 978-1-60511-073-8

Volume 1104 — Actinides 2008—Basic Science, Applications and Technology, B. Chung, J. Thompson, D. Shuh, T. Albrecht-Schmitt, T. Gouder, 2008, ISBN 978-1-60511-074-5

Volume 1105E —The Role of Lifelong Education in Nanoscience and Engineering, D. Palma, L. Bell, R. Chang, R. Tomellini, 2008, ISBN 978-1-60511-075-2

Volume 1106E —The Business of Nanotechnology, L. Merhari, A. Gandhi, S. Giordani, L. Tsakalakos, C. Tsamis, 2008, ISBN 978-1-60511-076-9

Volume 1107 — Scientific Basis for Nuclear Waste Management XXXI, W.E. Lee, J.W. Roberts, N.C. Hyatt, R.W. Grimes, 2008, ISBN 978-1-60511-079-0

Prior Materials Research Society Symposium Proceedings available by contacting Materials Research Society

Bulk Material and Characterization

Mater. Res. Soc. Symp. Proc. Vol. 1069 © 2008 Materials Research Society

Bulk Growth of SiC

Peter Wellmann[1], Ralf Müller[1], Sakwe Aloysius Sakwe[1], Ulrike Künecke[1], Philip Hens[1], Mathias Stockmeier[2], Katja Konias[2], Rainer Hock[2], Andreas Magerl[2], Michel Pons[3]
[1]Institute for Materials Department 6, Electronic Materials, University of Erlangen-Nürnberg, GERMANY.
[2]Institute for Crystallography, University of Erlangen-Nürnberg, GERMANY.
[3]LTPCM, INPG, Grenoble, FRANCE

corresponding author: peter.wellmann@ww.uni-erlangen.de

ABSTRACT

The paper reviews the basics of SiC bulk growth by the physical vapor transport (PVT) method and discusses current and possible future concepts to improve crystalline quality. In-situ process visualization using x-rays, numerical modeling and advanced doping techniques will be briefly presented which support growth process optimization. The „pure" PVT technique will be compared with related developments like the so called Modified-PVT, Continuous-Feeding-PVT, High-Temperature-CVD and Halide-CVD. Special emphasis will be put on dislocation generation and annihilation and concepts to reduce dislocation density during SiC bulk crystal growth. The dislocation study is based on a statistical approach. Rather than following the evolution of a single defect, statistic data which reflect a more global dislocation density evolution are interpreted. In this context a new approach will be presented which relates thermally induced strain during growth and dislocation patterning into networks.

INTRODUCTION

Commercial silicon carbide (SiC) substrates for electronic device applications are generally grown by the so-called PVT (physical vapor transport) growth, also called seeded sublimation technique. The growth process takes place in a quasi-closed graphite crucible system at elevated temperatures above 2000°C. Today 2", 3" and 4" substrates of the two polytypes 4H-SiC and 6H-SiC are commercially available with low defect density. Depending on the application, nitrogen doped n-type and semi-insulating materials are available. In the case of p-type doping only medium conducting material is on the market. 3C-SiC is currently not commercially available in big volume.

VAPOR GROWTH OF BULK SiC CRYSTALS

Fundamental Growth Process

Other than the widely used semiconductor materials silicon and gallium arsenide, silicon carbide single crystals cannot be grown from the melt. SiC decomposes peritectically at approx. 2800°C (figure 1a) into a Si melt with approx. 13% of dissolved C and into pure C which forms graphite clusters. At around 2000°C to 2400°C, however, sublimation of silicon carbide and defined re-crystallization may be carried out. This so called physical vapor transport (PVT) or seeded sublimation technique was first developed by Tairov and Tsvetkov [1] in 1978, who

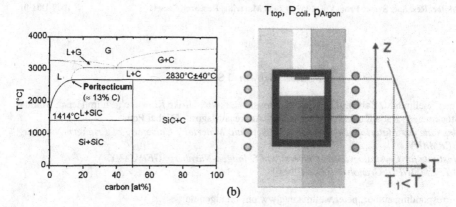

Figure 1. (a) Phase diagram of SiC. (b) Sketch of physical vapor transport (PVT) growth setup.

modified the self seeded Lely [2] process. The growth cell contains a SiC source material (often SiC powder) and SiC seed which is placed at a slightly colder position (figure 1b). The SiC source material sublimes and re-crystallizes at the colder seed as single crystal.

Although in general the growth process is called to be a physical phenomena, in reality the SiC source material sublimes, mainly forming the molecules Si, Si_2C and SiC_2 which are transported from the hot source zone to the colder SiC crystal growth interface, where they again form a SiC crystal by a chemical reaction. The latter circumstance is one of the reasons, why the C/Si ratio, which is a well known key parameter in classical CVD of SiC using silane and propane (see e.g.) [3], is also of particular interest for SiC bulk crystal growth. Although well established and optimized, the PVT process allows only an indirect control of the sublimation and re-crystallization process. Growth velocity may be controlled by heating power (P_{coil} in figure 1b) and, hence, growth setup temperature (T_{top} in figure 1b) and inert gas pressure (p_{Argon} in figure 1b). Temperature gradient may be set by crucible and isolation material design, as well as inductive heating coil position. The control of parameters like the above mentioned C/Si ratio is only indirectly possible.

Numerical Modeling

To improve and optimize the growth process, numerical modeling of the temperature field as well as SiC sublimation and mass transport has become state-of-the-art [4-7]. Using this tool, especially the radial temperature gradient can be minimized, which is a main source for thermally induced strain and, hence, dislocation generation due to strain relaxation.

X-ray In-Situ Visualization and SiC Mass Transport Labeling

Unlike the well known Czochralski growth process of the semiconductor materials Si and GaAs, the PVT setup offers no optical access to the growth cell interior. For process visualization Wellmann et al. [8] presented a digital x-ray setup that allowed to follow SiC powder evolution

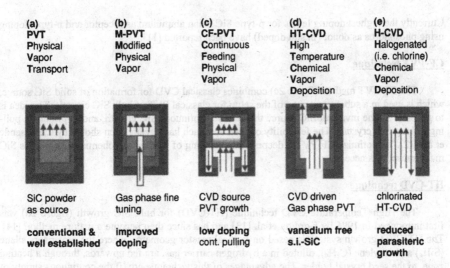

(a) PVT Physical Vapor Transport	(b) M-PVT Modified Physical Vapor	(c) CF-PVT Continuous Feeding Physical Vapor	(d) HT-CVD High Temperature Chemical Vapor Deposition	(e) H-CVD Halogenated (i.e. chlorine) Chemical Vapor Deposition
SiC powder as source	Gas phase fine tuning	CVD source PVT growth	CVD driven Gas phase PVT	chlorinated HT-CVD
conventional & well established	**improved doping**	**low doping cont. pulling**	**vanadium free s.i.-SiC**	**reduced parasiteric growth**

Figure 2. Derivates of the well established PVT process, which allow improvements for selected issues, like improved doping or high purity in the case of M-PVT or HT-CVD, respectively.

(grain re-growth and graphitization) and SiC crystal growth (shape of growth interface, growth velocity) in-situ during growth. Using ^{13}C as marker, Hero et al. [9] in addition addressed the SiC-species related mass transport through the growth cell.

Open Issues

Today, growth of 4" SiC single crystalline material has become state-of-the-art (see e.g. [10]). Major emphasis is now put on dislocation density reduction. In the case of the sever micropipe defect, reduction to almost zero defect density has been accomplished [10]. Further developments focus on the reduction of threading edge, threading screw and basal plane dislocations.

ADVANCED SiC VAPOR GROWTH APPROACHES

In the following four growth techniques will be addressed – M-PVT (Modified PVT), CF-PVT (Continuous Feeding PVT), HT-CVD (High Temperature CVD) and HCVD (halide CVD using chlorinated gases) – which allow a more flexible control of gas phase composition than classical PVT. These techniques allow overcoming the limitation of PVT. In the latter technique, the deposition chemistry is fixed by temperature and can be mainly guided by the design of the crucible. Consequently, the degrees of freedom are very limited.

M-PVT technique

The idea of the M-PVT (figure 2b) technique is to use an additional gas pipe for fine tuning of the gas phase composition in the growth cell. Major advantage is the improved doping.

Currently the highest doping levels for p-type SiC using aluminum as acceptor and n-type doping using phosphorus as donor (in situ doped) have been reported [11].

CF-PVT technique

The CF-PVT method (figure 2c) combines classical CVD for formation of solid SiC source, which is used in a subsequent zone of the setup for classical PVT of bulk SiC crystals. The idea is to generate a time invariant SiC source that allows continuous SiC growth, and in principle, pulling of long SiC crystals. The feasibility of such approach has already been shown by Chaussende et al. [12]. To optimize CF-PVT a deeper understanding of feeding gas chemistry as well as SiC mass transport is necessary.

HT-CVD technique

The High Temperature CVD technique (HT-CVD) for bulk SiC growth (figure 2d) was first introduced in 1996 by Kordina et al. [13] and has since then become a mature method [14]. The HT-CVD growth system is based on vertical reactor geometry, where the precursors silane (SiH_4) and ethylene (C_2H_4), diluted in a hydrogen carrier gas, are fed upwards, through a heating zone, to the seed crystal holder. The advantages of this technique are (i) the continuous supply of the source material, (ii) the relatively economical availability of high purity gases, (iii) the direct control of the C/Si ratio and (iv) the principle ability of pulling the growing crystal.

HCVD technique

The Halide CVD technique and its special reactor geometry (figure 2e) were designed with the aim to maximize the source gas efficiency conversion into single crystalline SiC and minimize parasitic deposition. As in all other SiC bulk growth techniques, a vertical crucible design is used. The silicon source gas (e.g. $SiCl_4$ in Ar carrier gas) and the carbon source gas (e.g. C_3H_8 in Ar:H_2 (1:1) carrier gas) are fed through separate injector tubes into the hot furnace zone (typical gas flux of both species: 1slm) where crystal growth takes place at elevated temperatures (typical T: 2020 °C) [15]. A typical ambient pressure is 200 Torr. The arrangement of separate gas injection allows the precursor gas to be pre-heated to temperatures near growth temperature without cross-reaction and hence parasitic SiC formation and deposition. Due to their thermal stability, the role of the halogenated silicon gas source is to eliminate reactions with the graphite furnace parts prior the crystal growth process in front of the SiC seed surface. The HCVD growth furnace as introduced in [15] is capable of producing crystals with diameters up to 100 mm.

DISLOCATION EVOLUTION DRUING BULK SiC CRYSTAL GROWTH

Experimental Details – Growth Process

The study of the influence of the seed temperature (or growth interface temperature) on dislocation evolution during SiC crystal growth was conducted systematically by applying novel approaches to vary the temperature during a single growth run.

Three sets of experiments were conducted: (i) Crystal growth was carried out using standard growth conditions, i.e. constant growth temperature and pressure (crystal A). The temperature was 2300°C. This growth experiment served as a reference to those with continuous

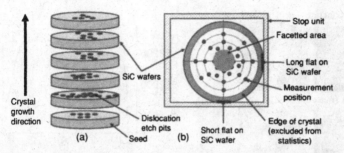

Figure 3. Statistical approach method for the study of dislocation evolution in SiC single crystals: (a) scheme of etch pit tracking method, (b) positioning of wafer on microscope stage.

temperature change described in points (ii) and (iii) below. (ii) Crystal growth was carried out in such a way that the growth temperature was increased continuously from about 2150°C at the beginning of growth until about 2350°C at the end of growth during a single growth run, while all other conditions remained constant (crystal B). Finally, in a third set of experiments, a reverse temperature mode to that in (ii) was implemented to serve as verification to the latter during dislocation analysis. The growth temperature was decreased continuously from about 2350°C to about 1900°C over the entire growth period (crystal C).

In-situ observation of the growth cell using a digital X-ray imaging technique [8] during the growth process allowed us to follow the crystal growth progress with respect to the growth time and growth temperature making it possible to assign each wafer from the crystals grown with continuous temperature change to the temperature regime in which it was produced with a high accuracy.

Experimental Details – Characterization

Delineation of the dislocations in the wafers was performed by means of defect selective etching in molten KOH. This was achieved in a specially designed and calibrated etching furnace with *in-situ* temperature measurement configuration; the temperature measurements were carried out directly in the KOH melt. Details about the KOH defect etching furnace and optimization of the etching process for dislocation analysis are given in reference [16].

The study of dislocation evolution was carried out by applying a statistical approach. In this method the dislocations are tracked along the length of the crystal from wafer to wafer. The dislocations are counted and their density evaluated in the same crystal or wafer position. In this respect it is not a single dislocation that is tracked but rather the dislocation density obtained statistically is used to interpret dislocation evolution. Each dislocation type is analyzed separately.

Further elaboration of the structural properties of the materials with respect to the applied parameters of crystal growth was performed using high energy X-ray diffraction (HE-XRD) measurements in a triple-axis configuration [3]. The energy of the X-rays was 62.5keV and penetration depths of up to 6cm could be achieved.

Atomic force microscopy (AFM) measurements were carried out on the growth fronts of the as-grown crystals to study their step structure. From the step structure, in relation to crystal growth temperature, we could obtain substantial information about the kinetic processes at the growth front and their relation to dislocation evolution. Figure 3 illustrates this principle.

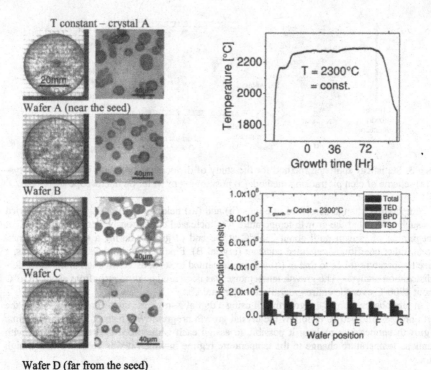

T constant – crystal A

Wafer A (near the seed)

Wafer B

Wafer C

Wafer D (far from the seed)

Figure 4. Dislocation analysis in a crystal (crystal A) grown with a constant temperature.

Results and Discussion - Dislocation Density versus Crystal Growth Temperature

We have observed a distinct relationship between the crystal growth or seed temperature and the overall dislocation density during PVT bulk growth of SiC crystals. On investigating a number of crystals with similar doping levels grown with different temperatures, we have seen a reduction in the dislocation density with increase in the growth temperature. A systematic investigation using crystals grown with continuously changing process temperature confirms these results. In these crystals the run-to-run variations that may occur during growths with different process temperatures in separate runs can be ruled out. By continuously changing the process temperature in a single run, different portions of the crystal are produced at different temperatures such that the effect of high temperature versus low temperature crystal growth can be compared even more accurately by evaluating the overall dislocation density along the crystal growth axis. By comparing the trend in the dislocation density along the crystal growth axis of these crystals with the reference crystal grown at a constant temperature, we were able to establish a solid relationship of the density to growth temperature.

T rises - crystal B

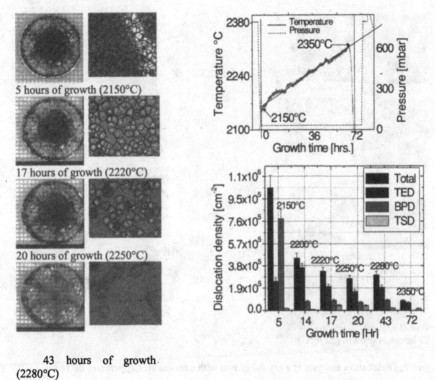

5 hours of growth (2150°C)

17 hours of growth (2220°C)

20 hours of growth (2250°C)

43 hours of growth (2280°C)

Figure 5. Dislocation analysis in a crystal grown with continuous temperature increase

The results of dislocation analysis in these crystals are shown in figures 4-6. On the left of the figures (first two columns) are KOH etching images, i.e. dislocation mapping over the whole wafer surface at a lower magnification (column 1) and etching images taken at a higher magnification from the center of the wafer (column 2). The darker regions in the mapping images correspond to a high dislocation density. On the right of the images are the applied temperature profiles during the growth processes and dislocation statistics evaluated over the wafer surfaces in several wafer positions.

Crystal A, grown with a constant temperature of 2300°C (figure 4) shows a constant dislocation density of about 2×10^5cm^{-2} along the crystal axis. This trend changes conspicuously in crystals grown with a continuous change of the growth temperature during a single growth run. In crystal B, grown with continuous temperature increase (figure 5), the dislocation density decreases dramatically with the growth temperature. It is high at early growth stages (approx. 10^6cm^{-2}), in low-temperature growth regimes, and then decreases abruptly in high growth

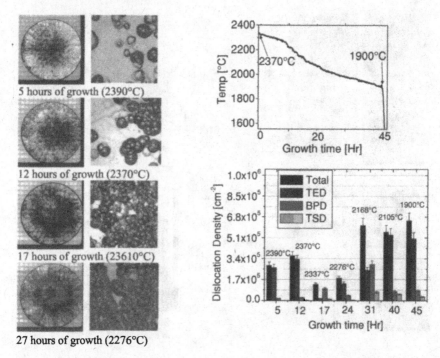

T decreases – crystal C

5 hours of growth (2390°C)

12 hours of growth (2370°C)

17 hours of growth (23610°C)

27 hours of growth (2276°C)

Figure 6. Dislocation analysis in a crystal grown with continuous temperature decrease.

temperature regimes to a value of about $8 \times 10^4 \text{cm}^{-2}$. The dislocation density at the initial growth stage is clearly higher than that of the seed used for this growth. KOH etching and optical microscopy revealed the density in the seed to be about $1.2 \times 10^5 \text{cm}^{-2}$. In the middle of the seed, excluding the edge, it was found to be about $6.8 \times 10^4 \text{cm}^{-2}$. We attribute the high dislocation density at the initial point of growth to low growth temperature because we did not observe such a high density in, e.g., crystal A in which the temperature was high in the early growth stages. In crystal C, grown with continuous temperature decrease (figure 6), the trend in the dislocation density follows the reverse order to that of crystal B, i.e. low dislocation density ($1–3 \times 10^5 \text{cm}^{-2}$) at early growth stages (high growth temperature regime) and high dislocation density (about $6.8 \times 10^5 \text{cm}^{-2}$) at later stages (low growth temperature regime). Generally, the total dislocation density in the high-temperature growth regimes of crystals B and C, grown by continuously changing the process temperature is comparable to that in crystal A, grown with a constant high temperature.

Evidently, from these results, the dislocation density of PVT-grown SiC crystals is inversely proportional to the temperature of crystal growth. This is a clear indication that high quality crystals can be achieved by high temperature growth.

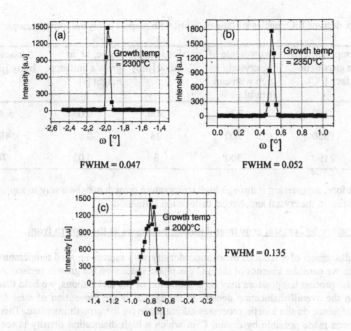

Figure 7. ($\overline{2}110$) rocking curves of the phase boundary regions of crystals (a) with constant growth temperature (crystal A), (b) with increasing growth temperature (crystal B) and (c) with decreasing growth temperature (crystal C).

High energy X-ray diffraction (HE-XRD)

To further elaborate the influence of crystal growth temperature on the quality of the crystals we performed HE-XRD measurements in a triple axis configuration [17] on the crystal portions encompassing the growth interfaces (caps). These were produced under different temperature regimes. The measurements were confined in the facetted areas of the crystals and the ($\overline{2}110$) reflection was measured. The energy of the X-rays was 62.5keV and penetration depths of up to 6cm could be achieved.

The high-temperature-grown crystals exhibit narrow rocking curves (figures 7a and 7b) as compared to the low-temperature-grown counterpart (figure 7c). The FWHM is by a factor 3 greater in the low temperature grown crystal than the high temperature grown ones. This implies that in the low temperature grown sample a high mosaicity and many sub-grain boundaries are present, which correlate with a high dislocation density as compared to those grown at high temperatures. This is in good agreement with the results of defect selective-etching and dislocation statistics.

In addition, HE-XRD measurements provided evidence that lattice plane bending in high temperature-grown SiC crystals is lesser than in low temperature grown ones. Lattice plane bending may, on its own, induce dislocation formation due to the accompanying mis-orientation in the

Table 1: Exp. derived SiC species surface mobility with respect to crystal growth temperature.

Crystal	Temperature of the growth interface [°C]	Growth rate of the crystal / region with cap [μm/h]	Step terrace, w [μm]	Time, Δt, to grow a unit step height [s]	Surface mobility [μm/s]
A	2300	220	300	0.03	5000
B	2350	200	28	0.03	486
D	2190	300	3	0.02	78

crystal. Therefore, suppressing it through high temperature growth may be a way to suppress misoriented domains in the crystal and, hence, dislocation formation.

High temperature SiC crystal growth and kinetic processes at the growth front

In the discussion of a reduced dislocation density with increasing seed temperature during crystal growth we consider kinetics of the SiC gas phase species on the growth surface. Although an effect of the process temperature may be annihilation of the dislocations, we hold that its major impact on the overall dislocation density in the crystal is the prevention of their formation through its influence on the kinetic processes taking place on the growth interface. This consideration has been made possible by crystal C in which a high dislocation density is seen as the temperature reduces. Therefore, in the context of dislocation reduction through kinetic effects we regard the temperature-enhanced surface mobility of the gas phase species.

At high growth temperatures the SiC gas phase species possess a high surface mobility at the growth interface and, therefore, a high kinetic energy. As such they are able to migrate to the kinks where they are best incorporated into the growing crystal. The migration rate depends on the temperature and in such a growth configuration a step flow growth mode is favored. At low temperatures, on the other hand, the species will possess a low surface mobility and the migration rate will be low. If a sufficient number of species are close together or coalesce a stable growth center may arise because some species will be adequately bonded together and will not be driven back into the gas phase. This results in the formation of growth centers or islands on the growth surface resulting to a non-step flow growth mode. The interaction between the growth centers results to strain and, hence, to dislocations. The dislocation density will be proportional to the density of growth islands.

Using atomic force microscopy (AFM) to study the step structure of the growth fronts of the crystals we were able to estimate the surface mobility of the SiC gas phase species as a function of the process temperature. We found a strong relationship between the step structure and the growth temperature; the terrace width of the steps increases with temperature. From this, it is possible to calculate the surface mobility from the terrace width and the time it takes for species to reach the kink, assuming that it landed on the middle of the terrace. This time is equal to the time it takes to grow a unit step (1.6nm for 6H-SiC) and can be estimated from the growth rate of the crystal. The surface mobility can be estimated from the expression: $v_x = \Delta x / \Delta t = w/2\Delta t$, where Δx is the distance the species cover to reach the kink and is equal to half the terrace width w. Δt is the time it takes to reach the kink and is equal to the time to grow a unit step height.

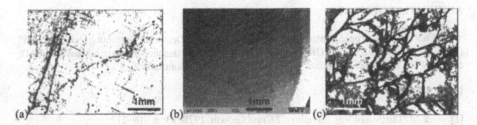

Figure 8. Dislocation cell structures in 6H-SiC: (a) p-type doped, optical microscopy, (b) p-type doped, SEM, and (c) n-type doped optical microscopy.

For crystal C, grown with continuous temperature decrease the step structure could not be determined because it was very irregular. Table 1 gives the results of the surface mobility of the gas phase species with respect to crystal growth temperature; another crystal D grown at a constant but low temperature is included. As expected, the surface mobility of the SiC gas phase species increases with temperature.

Dislocation alignment in SiC single crystals

Optical microscopy on the vicinal (0001)Si faces of etched SiC wafers reveals a peculiar alignment of the dislocations in cell structures, independent of the doping or polytype (4H- and 6H-SiC). In figure 8 optical micrographs are shown to depict dislocation networks in 6H-SiC; (a-b) p-type 6H-SiC and (c) n-type 6H-SiC. The dislocation density is very low inside the cell as compared to the cell boundaries which are actually low angle grain boundaries. The cellular nature of dislocations can best be observed by low magnification microscopy (optical or scanning electron microscopy).

Patterning of dislocations into cells has been observed in other compound semiconductors, e.g., GaAs, PbTe, CdTe and, also, in metals and metallic alloys [18, 19]. A possible mechanism for their formation may be energy-related, i.e. the reduction of strain energy through clustering into cells and grain boundaries. The overlapping strain energies of individual dislocations reduce the overall strain energy in the crystal. However, according to Rudolph [19], the patterning may also be driven by self-organizing processes in the framework of equilibrium or non-equilibrium thermodynamics by considering that the rate of entropy produced within the crystal at high temperatures evokes self-ordered patterning of the already present or currently generated dislocations.

CONCLUSIONS

Bulk SiC single crystal growth of up to 4" in diameter by the PVT technique can be called state-of-the-art. Numerical modeling of the growth process has become one of the most import tools to further improve crystal quality. Special techniques like M-PVT, HT-CVD, CF-PVT, H-CVD are of interest, if specific crystal properties are of importance. Today, further reduction of dislocation density may be called the main challenge in bulk SiC crystal growth. Approaches that relate growth kinetics, thermally induced strain and dislocation interaction through networking from a fundamental materials science point of view are expected to be most promising.

ACKNOWLEDGEMENTS

This work has been financially supported by the "Deutsche Forschungsgemeinschaft" (DFG, contract WE2107/3 – FOR476), the "Deutscher Akademischer Austauschdienst" (DAAD, contract PKZ D-0502202) and the "STAEDTLER Stiftung" (contract Hf/eh 11/07).

REFERENCES

[1]. Y.M. Tairov and V.F. Tsvetkov, J.Cryst.Growth, 1978. 43: p. 209-212.

[2]. J.A. Lely, Ber.Dt.Keram.Ges., 1955. 32(8): p. 299-231.

[3]. H. Fujiwara, K. Danno, T. Kimoto, T. Tojo, and H. Matsunami, J.Cryst.Growth, 2005. 281: p. 370-376.

[4]. D. Hofmann, M. Heinze, A. Winnacker, F. Durst, L. Kadinski, P. Kaufmann, Y. Makarov, and M. Schäfer, J.Cryst.Growth, 1995. 146: p. 214-219.

[5]. M. Pons, E. Blanquet, J.M. Dedulle, I. Garcon, R. Madar, and C. Bernard, J.Electrochem.Soc., 1996. 143(11): p. 3727-3735.

[6]. P.J. Wellmann, Z. Herro, S.A. Sakwe, P. Masri, M. Bogdanov, S. Karpov, A. Kulik, M. Ramm, and Y. Makarov, Mat.Sci.Forum, 2004. 457-460: p. 55-58.

[7]. M.V. Bogdanov, A.O. Galyukov, S.Y. Karpov, A.V. Kulik, S.K. Kochuguev, D.K. Ofengeim, A.V. Tsiryulnikov, M.S. Ramm, A.I. Zhmakin, and Y.N. Makarov, J.Cryst.Growth, 2001. 225: p. 307-311.

[8]. P.J. Wellmann, M. Bickermann, D. Hofmann, L. Kadinski, M. Selder, T.L. Straubinger, and A. Winnacker, J.Cryst.Growth, 2000. 216: p. 263-272.

[9]. Z.G. Herro, P.J. Wellmann, R. Püsche, M. Hundhausen, L. Ley, M. Maier, P. Masri, and A. Winnacker, J.Cryst.Growth, 2003. 258(1-3): p. 261-267.

[10]. S.G. Müller, M.F. Brady, A.A. Burk, H.M.D. Hobgood, J.R. Jenny, R.T. Leonard, D.P. Malta, A.R. Powell, J.J. Sumakeris, V.F. Tsvetkov, and C.H.C. Jr., Superlattices and Microstructures, 2006. 40: p. 195–200.

[11]. P. Wellmann, P. Desperrier, R. Mueller, T. Straubinger, A. Winnacker, F. Baillet, E. Blanquet, J.M. Dedulle, and M. Pons, J.Cryst.Growth, 2005. 275(1-2): p. e555-e560.

[12]. D. Chaussende, F. Baillet, L. Charpentier, E. Pernot, M. Pons, and R. Madar, J.Electrochem.Soc, 2003. 150: p. G653.

[13]. O. Kordina, C. Hallin, A. Ellison, A.S. Bakin, I.G. Ivanov, A. Henry, R. Yakimova, M. Touminen, A. Vehanen, and E. Janzen, Appl.Phys.Lett., 1996. 69(10): p. 1456-1458.

[14]. A. Ellison, B. Magnusson, B. Sundqvist, G. Pozina, J.P. Bergman, E. Janzén, and A. Vehanen, Mater.Sci.Forum, 2004. 457-460: p. 9-14.

[15]. H.J. Chung, A.Y. Polyakov, S.W. Huh, S. Nigam, M. Skowronski, M.A. Fanton, B.E. Weiland, and D.W. Snyder, J.Appl.Phys., 2005. 97: p. 084913.

[16]. S.A. Sakwe, R. Müller, and P.J. Wellmann, J.Cryst.Growth, 2006. 289(2): p. 520-526.

[17]. C. Seitz, A. Rempel, A. Magerl, M. Gomm, W. Sprengel, and H.E. Schaefer, Mater.Sci.Forum, 2003. 433-436: p. 289-292.

[18]. P. Rudolph, C. Frank-Rotsch, U. Juda, M. Naumann, and M. Neubert, J.Cryst.Growth, 2004. 265: p. 331-340.

[19]. P. Rudolph, Cryst.Res.Technol., 2005. 40(1-2): p. 7-20.

Mater. Res. Soc. Symp. Proc. Vol. 1069 © 2008 Materials Research Society 1069-D01-03

Alternative Routes to Porous Silicon Carbide

Bettina Friedel[1], and Siegmund Greulich-Weber[2]

[1]Physics, University of Cambridge, Cavendish Laboratory, JJ Thomson Avenue, Cambridge, CB30HE, United Kingdom

[2]Physics, University of Paderborn, Warburger Strasse 100, Paderborn, Germany

ABSTRACT

A low-cost alternative route for large-scale fabrication of high purity porous silicon carbide is reported. This allows a three-dimensional arrangement of pores with adjustable pore diameters from several 10 nanometers to several microns. The growth of SiC is here based on a combined sol-gel and carbothermal reduction process. Therein tetraethoxysilane is used as the primary silicon and sucrose as the carbon source. We provide two different sol-gel based ways for preparation of porous SiC, obtaining either a regular porous or a random porous type. Regular porous SiC with monodisperse ordered spherical pores of predefined size is obtained via liquid infiltration of a removable opal matrix. Whereas random porous material with polydisperse pores of an adjustable size distribution range, but without order, can be achieved via free gas phase growth. This is performed by degradation of granulated sol-gel prepared material inside a sealed reaction chamber, resulting in a SiO/CO/SiC rich gas atmosphere, which causes SiC growth inside the granulate itself. For both types doping of the initially semi-insulating porous SiC is possible either during the sol-gel preparation or via the gas phase during the following annealing procedure. As probing dopants we have used P, N, B and Al, which are well known from 'conventional' SiC. Composition and structure of the obtained material was investigated using scanning electron microscopy, X-ray diffraction, nuclear magnetic resonance and Fourier transform infrared spectroscopy.

INTRODUCTION

SiC is not only a powerful material for electronic and optoelectronic applications but also for advanced photonic applications such as photovoltaic devices, taking advantage of the large electronic bandgap. Due to its excellent properties in particular its hardness and its chemical inertness it is a challenge to use SiC devices also in harsh environment as e.g. filters or catalytic converters at high temperatures [1]. Main drawbacks using SiC for such applications are the expensive production and difficulties to process the material. Especially porous SiC is currently under discussion for various applications. For sophisticated applications porous SiC is usually fabricated destructive by labour intensive electro-chemical etching of electronic quality wafers [2]. Although notable progress has been achieved in the recent past, high quality porous structures as needed for photonic applications in silicon carbide have not been achieved so far. However, it depends on the intended application, which kind of porosity is sufficient. There are also methods starting with SiC powder [3] leading to ceramic porous SiC, which however, is inappropriate for electronic or photonic applications. We are reporting here on rather inexpensive constructive fabrication methods providing a large range of products of high purity porous SiC.

All methods to be presented are based on a sol-gel process combined with carbothermal reduction [4]. At around 1800°C we obtain 3C-SiC single crystal growth. Porous SiC can be grown on various substrates like commercial SiC wavers or sapphire and as free standing bulk material [5]. Pre-structured SiC with well defined monodisperse periodic pores is achieved by using a close packed two or three dimensional carbon spheres template, which is infiltrated with sol-gel SiC precursor [6,7]. The carbon template is removed afterwards, thus a SiC skeleton is left. The pores diameter can be adjusted between several 100nm and 2μm due to an appropriate choice of carbon spheres. The SiC finally consists of polycrystalline microcrystals surrounding the empty spheres, or in the case of thin SiC layers consisting of a monocrystalline 3C-SiC film. Using a modified sol-gel process we are able to convert any graphite part into 3C-SiC [8]. Also this material is consisting of 3C-SiC micro crystals, while the porosity depends on the graphite raw material. The main advantage of this process is that a given graphite part is preserved in shape and size after conversation of graphite into silicon carbide. Thus any porous structure easily prepared from carbon will be converted into SiC with our method.

EXPERIMENTAL DETAILS

Base Material Synthesis.

Precursor solution: Starting material either for template infiltration or for further processing towards the precursor granulate, is a sucrose containing silica sol. This was prepared via hydrolysis and condensation of tetraethoxysilane (TEOS) in ethanolic solution with deionized water in the presence of hydrochloric acid. A specific amount of sucrose, leading to a Si:C ratio of 1:4 was added via prior dissolution in deionized water. The obtained sol is either used immediately (coating, infiltration) or is kept for further processing.

Carbon charged silica granulate: The sucrose silica sol is allowed to gel fast (depending on the HCl content), followed by a slow drying process under ambient conditions. Finally the solid carbohydrate charged silica gel pieces are annealed at 1000°C under inert atmosphere (argon or nitrogen). The obtained granules are the basic material for all our gas phase supported SiC preparations.

Porous Silicon Carbide Preparation.

SiC with ordered uniform spherical pores: In this case porous SiC is grown by liquid sol-gel precursor infiltration in a removable template. Thereby a colloidal crystal of monodisperse spherical carbon particles is used as the template [6,7]. This regular structure is infiltrated by filling the voids between the spheres with the above described SiC precursor solution (see Fig 1a). To cure the liquid material, it is annealed at 1000°C in argon gas atmosphere for a few seconds and cooled down to room temperature. Due to shrinking processes of the precursor material, this infiltration cycle has to be repeated until a sufficient high fill factor in the structure is achieved.

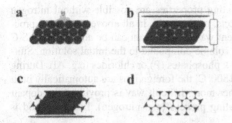

Figure 1: Schematical the preparation of uniform porous SiC, by (a) Infiltration of a colloidal crystal with liquid SiC precursor, (b) high temperature annealing of the composite in argon gas atmosphere at 1800°C, (c) removing of the colloidal crystal template, by burning in air and finally (d) the inverted structure of silicon carbide

Afterwards the obtained compound structure of spherical carbon particles and carbon rich silica glass is annealed to 1800°C for a few minutes in argon gas atmosphere (see Fig. 1b) to convert the precursor to silicon carbide. Finally the spherical carbon particles are removed by heating the whole structure short time at 1000°C in air (see Fig 1c), leaving a silicon carbide scaffold consisting of uniform spherical interconnected pores (see Fig 1d). Therein the final pore size can be adjusted by the initially diameter of the carbon spheres used to grow the template. These can be synthesized with a size from 100nm to 2µm.

Free gas phase growth: The special composition of the solid SiC precursor, described above makes it possible to coat any kind of heat resistant (up to 1800°C) substrate with a polycrystalline porous SiC film. Therefore the granulated precursor material and substrate are placed in a gradient furnace under static argon atmosphere. As soon as the precursor reaches 1700°C, it starts to degrade, causing a SiC rich gas atmosphere, which causes the growth of joined SiC micro crystals on the slightly cooler substrate. Depending on time and the used amount of precursor, films with a thickness between 100µm and several millimetres can be grown. Especially thicker films, e.g. grown on graphite can be used freestanding after removal of the substrate.

Conversion of carbonaceous parts in porous SiC: In this technique carbon bodies (e.g. from graphite or glassy carbon) are fully or partially converted into porous silicon carbide without deformation or shrinkage. This method is also gas phase based and though uses the same experimental setup and conditions as the previous one, just the substrate is replaced by a carbon body of any shape or size. Due to the nature of carbonaceous materials, the forthcoming process differs to free gas phase growth of films. Gaseous SiO and CO, created along with the SiC vapour during degradation of the precursor granulate, increase porosity of the carbon material due to oxidation. On the other hand by SiO interaction also first silicon carbide seed crystals are generated inside the pores. Further SiC vapour condensates on these SiC seeds, crystals grow inside the structure, covering or even closing the pores. Both of these processes continue until the precursor is used up. Depending on the size of the body to be converted, the exchanging process has to be repeated with renewed precursor. The porosity of the obtained SiC body depends mainly on the initial porosity of the carbon body. Shape and size of the body are not changed.

Introduction of dopants in porous SiC.

As probing dopants we have used phosphorous, nitrogen, boron and aluminium, which are the most wanted ones from the technological side in 'conventional' semiconductor SiC. In contrast to customary SiC, sol-gel SiC is not initially nitrogen doped. In contrast to commercially available SiC here p-type doping and semi-insulating properties are possible without nitrogen donor compensation or activation of intrinsic defects, respectively. In all above mentioned processes foreign ions can be introduced in two different ways. The dopant can be added during SiC precursor preparation, by adding either itself or a soluble compound to the initial solution. Suitable candidates are nitrates (N doping), borates (B), phosphates (P) or chlorides (e.g. Al). During degradation of the precursor, during annealing at 1800°C, the foreign ions are automatically built in into the growing SiC crystal structure. The other more difficult way is providing the dopant via gas phase during the high temperature annealing process (e.g. nitrogen). Latter method is only suitable for selected elements.

Characterization.

The grown and the converted materials respectively have been verified as SiC by nuclear magnetic resonance spectroscopy (NMR), Fourier transform infrared spectroscopy (FTIR) and X-ray diffraction (XRD). The pore size and structure of the porous samples has been determined by scanning electron microscopy (SEM). The nature of introduced dopants could be observed via electron paramagnetic resonance spectroscopy (EPR).

RESULTS AND DISCUSSION

Uniform porous SiC.

Although there are some techniques known to grow porous silicon carbide with ordered structures, our method is unique for the growth of SiC with monodisperse spherical pores. Therefore well ordered carbon opals (as shown in Fig. 2) are needed, their fabrication has been reported elsewhere REF. After repeated infiltration with the liquid precursor sol and the high temperature annealing step, the material is turned in a closed SiC-carbon compound structure. Thereby the SiC shows a strong tendency to grow monocrystalline nearly without crystal defects, although the carbon spheres imply an enormous intrusion in the growing area.

 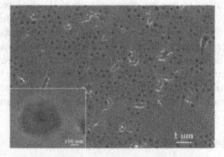

Figure 2: Colloidal crystal from monodisperse carbon spheres, prepared by a sound supported sedimentation process [7].

Figure 3: SEM image of an inverted colloidal crystal on a monocrystalline SiC substrate. Inlay shows magnification of a single pore.

Only when the surface is polished, it becomes possible to remove the remaining carbon spheres from the structure by annealing in oxygen gas atmosphere. An example of such a laid open structure, here prepared on a monocrystalline 6H-SiC substrate, is presented in Fig. 3. The inlay shows the magnification of a single pore, a slight tendency of the pores to form hexagonally shaped can be seen.

Gas phase grown films.

With this method porous polycrystalline SiC films have been grown on various substrates, as for example on silicon carbide, on sapphire or on glassy carbon. Thicknesses of the films could be obtained from few microns to several millimetres easily. The suspicion, silica (quartz or cristobalite) could be present in the grown films as well, caused by SiO gas in the reaction chamber, could be cleared out by characterization of the material with NMR and XRD.

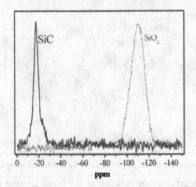
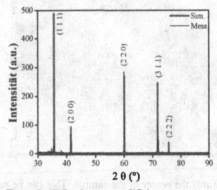

Figure 4: ^{29}Si-MAS-NMR spectrum of a freestanding mesoporous SiC layer (dark solid curve). No silica has been found (light grey line indicates the position of a usual SiO$_2$ signal).

Figure 5: XRD of a porous SiC layer, grown on a glassy carbon substrate (solid black line). The simulated reflexes indicate the typical intensity distribution of a 3C-SiC powder diagram.

Even with very sensitive ^{29}Si solid state NMR spectroscopy no silica traces, usually found at a chemical shift of -110ppm, could be detected (Fig. 4). Just the signal at -17ppm, typical for silicon carbide has been observed. XRD reflexes from porous films (see example Fig. 5, measured at a porous SiC layer grown on a glassy carbon substrate) show concerning the appearance angle good accordance to calculated 3C-SiC powder spectra [9]. Although the intensities are not comparable, because of preferred crystal orientation of the grown films, this shows clearly the exclusive growth of 3C-SiC in all samples. Have films been grown with an adequate thickness on a removable substrate, as for example graphite or glassy carbon, it is possible to achieve a freestanding porous film. Are 3 dimensional parts covered, even SiC shells of these bodies can be obtained. An example therefore is pre-

Figure 6: Porous polycrystalline SiC shell of a former graphite substrate.

sented in Fig. 6, showing the porous SiC shell of a former completely covered glassy carbon substrate, which has been burned off afterwards (by annealing in oxidative atmosphere).

Converted carbon.

To discover the properties of the conversion process and the obtained porous silicon carbide samples, a large range of probe bodies has been either completely or at least partially converted. This method, based on molecular exchange processes, offers several advantages. To achieve a requested silicon carbide body, a carbon copy is processed to the requested shape and dimensions and is easily converted into porous silicon carbide. No direct mechanical processing of SiC is necessary this way. Thereby the carbon part is transformed to SiC from the outside

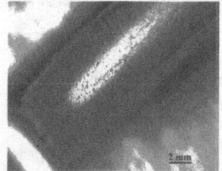 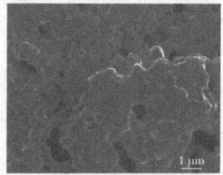

Figure 7: Cross section of a half way converted sample disc.

Figure 8: SEM image showing the porous polycrystalline structure of the cross section of a converted sample

towards the center of the sample. That can be seen clearly, when looking at the cross section of an incompletely converted sample disc (Fig. 7), whose remaining carbon has been removed by annealing in air. The white center of the disc is very soft and shows the beginning conversion. Another advantage of our conversion method is the mostly higher density compared to the samples made from other cheap pressing methods and additionally much more clean.

Doped porous SiC.

The sol-gel derived samples have been prepared with several dopants of interest for electronic applications of SiC, as e.g. N, Al, P and B. Also the introduction of rare earth ions has been tested. The presence of foreign ions has been investigated with EPR, a very powerful tool for verification of dopants in semiconductors.

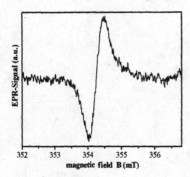

Figure 9: EPR spectrum of nitrogen doped porous 3C-SiC.

Is has to be mentioned, that pure undoped silicon carbide, obtained from our sol-gel based method, shows initially no EPR signal, different to most commercially available SiC. Thereby nearly all of the tested introduced dopants could be detected positively in our samples. One example is shown in Fig. 9, a free gas phase grown sample doped with nitrogen, which has been introduced by adding nitric acid to the initial precursor solution. The measurement has been performed at 11K, at a frequency of ν_{EPR} = 9.9416 GHz. The g-factor could be determined with g = 2.0050 and is therefore according very well to the value known from monocrystalline 3C-SiC. The ^{14}N hyperfine interaction with 3.5MHz is not resolved.

CONCLUSIONS

The different reported sol-gel based growth processes for porous SiC deliver very clean materials of selectable pore size and pore arrangements, which are therefore suitable for photonic as well as for electronic applications. The conversion of carbonaceous bodies into porous light-weight and stable silicon carbide, offers a useful easy method for the production of requested SiC parts, without having to process the hard SiC directly. Two different doping procedures allow n- and p-type doping with shallow donors such as nitrogen and phosphorous and shallow acceptors such as boron and aluminium. Overall, our described methods are easy in production at notably low costs. Numerous applications of doped and undoped porous SiC are imaginable, for instance in photonic, electronic, but also for machine parts or filtering media.

REFERENCES

1. J.M. Tulliani, L. Montanaro, T.J. Bell, V. Swain, J. Amer. Ceram. Soc. 1999, 82, 961.
2. Y. Shishkin, W.J. Choyke, R.P. Devaty, J. Appl. Phys. 2004, 96, 2311.
3. Brevier - Technische Keramik, Hrsg.: Informationszentrum Technische Keramik (IZTK) und Technische Kommission der Fachgruppe Technische Keramik, Fahner Verlag, Lauf 1998.
4. B. Friedel, S. Greulich-Weber, Materials Science Forum 2006, 527-529, 759.
5. B. Friedel, Thesis, University of Paderborn 2007.
6. B. Friedel, S. Greulich-Weber, Small 2006, 2, 859.
7. B. Friedel, S. Greulich-Weber, Mater. Res. Soc. Symp. Proc. 2007, 951, 0951-E06-27.
8. B. Friedel, S. Greulich-Weber, Patent pending DE102006055469.8.
9. K. Yvon, W. Jeitschko, E. Parthé, J. Appl. Cryst. 1977, 10, 73.

Mater. Res. Soc. Symp. Proc. Vol. 1069 © 2008 Materials Research Society   1069-D01-04

Influence of Crystal Growth Conditions on Nitrogen Incorporation During PVT Growth of SiC

Darren Hansen, and Mark Loboda
Science and Technology, Dow Corning Compound Semiconductor Solutions, AUB1007, P.O. Box 994, Midland, MI 48686-0994

ABSTRACT

The control and understanding of the incorporation of nitrogen during SiC PVT continues to play an important role in SiC crystal growth. Nitrogen acts both as a dopant and an impurity depending on the growth conditions and desired resistivity. Epitaxial growth by CVD provides some insight into N incorporation in terms of the face effects, temperature, and impact of the chemical species in terms of the C/Si ratio. This paper will present experimental results showing trends regarding nitrogen incorporation during SiC PVT. Various crystal growth processes operated under constant nitrogen partial pressures were found to produce wide ranges of SiC resistivity. These effects will be analyzed in light of the process impact on gas phase elemental composition (1), crystal stress (2), dopant activation (3) and crystal defectivity (4). The goal of this paper is to provide additional insights regarding nitrogen incorporation during SiC PVT, and in turn drive towards a more holistic approach to control the resistivity of 4H n+ SiC material, based on the understanding established from SiC epitaxy technology.

INTRODUCTION

SiC is a promising material because of its mechanical, electrical, and physical properties. 4H material in particular has a favorable bandgap, thermal conductivity, and carrier mobility relative to the other common SiC polytypes. In addition to the large amount of effort that has been placed on improving the overall defect densities of SiC materials, for example the reduction of micropipes, thermal decomposition voids, and polytype control, the electrical properties of the underlying SiC substrate are a critical aspect to the SiC substrate application in devices. For devices where the substrate actively conducts current as in some high power electronics the goal is to drive the resistivity of the SiC materials ever lower so that the SiC substrate does not negatively impact the overall device performance through introduction of parasitic resistances [2-12]. For high frequency devices, it is necessary to have the SiC material have minimum free carriers so as not to introduce loss into the overall device performance [13-20]. As a result of these two important applications it is becoming increasingly important to be able to manipulate the resistivity of the 4H substrate materials over a wide range of resistivities.

The primary dopant for n-type 4H materials is N [2,4-12], although consideration has been given to P dopants as well [3,12]. To that end work to understand nitrogen incorporation in bulk growth by physical vapor transport (PVT) and epitaxial growth has occurred [1-12]. In epitaxial growth of SiC layers the dopant incorporation can be well understood in terms of a site competition model [1]. In this model dopants such as B, Al, P, and N have been well studied in a variety of growth applications [1]. For example, on the Si face N is known to sit on the C site and, therefore, competes with C for a place in the SiC growing lattice. As a result of this it possible to actively control the dopant incorporation in growing SiC layers by controlling the C/Si ratio. Similar arguments have been demonstrated for other impurities such as B and Al which compete with Si sites on the growing lattice. The C face for growing epitaxial layers provides a slightly more complicated picture [1]. B, Al and others for example still demonstrate site competition with Si sites. However, N has actually been reported to possess a somewhat dual competition with both Si and C depending on the level of the C/Si ratio [1].

This complexity of the site competition model that has been studied in detail for epitaxial layers has also been somewhat studied in bulk growth applications [2-12]. For example, it has been reported that the C face, most commonly used for 4H crystal growth, incorporates nitrogen more rapidly than the Si face and that the N incorporation is a function of the C/Si ratio [1,2,4,8,9,11]. In addition, the N incorporation is not a straightforward function of the C/Si ratio. A typical resistivity distribution for SiC materials is one that shows concentric resistivity contours as shown in Figure 1. A central region with typically the lowest resistivity material is associated with the crystal facet. This region incorporates the most nitrogen reportedly due to the nature of the step structure on the growing facet of the SiC crystal [9,11]. Specifically, the steps are more widely spaced on the (0001) orientation than on the curved, stepped portion of the remainder of the growing crystal. The resistivity profile in the crystal is controlled by the distribution of dopants within the reaction vessel as well due to spatial variations in the reactive species, i.e. C/Si ratio, near the ,growing crystal surface,

$$\rho(r,\theta,z,t) \propto \frac{[N(t)](r,\theta,z(t)) * [C(t)](r,\theta,z(t))}{[Si(t)](r,\theta,z(t))} \quad (1)$$

Where r, θ, and z are the coordinates within the growth cell, t is time, [N] is the concentration of nitrogen, [C] is the concentration of C-containing species, and [Si] is the concentration of Si containing species.

Figure 1. Resistivity distribution in a typical 4H n+ SiC wafer. Units for resistivity are in Ω-cm.

Reducing resistivity in SiC crystals is a challenge not only due to the process extremes of growth conditions [2], but also due to the potential introduction of defects associated with increasing the dopant concentrations [5-7,12]. Therefore, resistivity improvements in the SiC materials system are not simply a matter of introducing more dopants into the growing crystal. First, too much nitrogen into the growing SiC crystal may lead to severe brittleness and increased wafer warp [5-7,12]. In addition, SiC materials can readily change polytype during growth which on a local level is associated with the formation of stacking faults (SF's). SF's have been a key focus for SiC material improvements during the past several years [20]. Studies have demonstrated that the amount of nitrogen in the material can influence the formation of SFs in the substrate which can degrade device performance [5-7].

Obtaining highly resistive SiC materials is also a significant technical challenge [13-18]. Obtaining semi-insulating materials is accomplished by introducing compensating centers at a level comparable to or greater than the residual dopants in the material which are typically N or B. One method for introducing these centers has involved the use of vanadium [14,15]. Another approach is to drive the impurities low enough to the point where the material deep centers, for example, carbon vacancy, can compensate for residual carriers [13,14,16-18]. Reducing the unintentional impurities in SiC to levels near or below 10^{15} cm^{-3} remains a significant challenge and requires a detailed understanding of dopant incorporation chemistry.

Typical wafer resistivity specifications from commercial vendors show a potential for variation of 40% or more. Within wafer variations are typically in the neighborhood of 25% based on a maximum to minimum resistivity variation. These resistivity variations are quite large and need to be improved to minimize the SiC wafer impact on device performance. A significant body of work remains to be completed to improve the resistivity target across the entire dopant spectrum at the same time as driving towards to a much more uniformly conducting or resistive material system on a repeatable basis. The goal of this paper is to provide some insight into factors affecting the resistivity of 4H n+ SiC materials.

EXPERIMENT

4H n+ SiC crystals were grown using physical vapor transport (PVT) on a 4H SiC, C-face seed. A standard graphite crucible was used to grow both 3" diameter 4H n+ SiC crystals. The growth temperature was between 2000 and 2300°C at sub-atmospheric pressures. N_2 and Ar gas were added to the growth chamber to control resistivity. The controlled partial pressure of N_2 is a function of the total gas flows and total reactor pressure external to the crucible environment. The partial pressure of N_2 internal to the crucible is not known in this work. Growth rates for all investigations were between 0.1 and 1 mm/hr. The gas phase [N] concentration did not impact the crystal growth rate nor did growth rate significantly impact resistivity compared to normal process variability.

The electrical properties of the resultant 4H n+ SiC crystals were investigated using eddy current resistivity mapping, capacitance voltage (CV), and Hall effect tests. Chemical determination of N_2, B, Al, Ti, and V was determined using secondary ion mass spectroscopy (SIMS). In this paper individual wafer resistivity profiles are characterized by two variables: maximum wafer resistivity measured in Ω-cm and uniformity defined as:

$$Uniformity(\%) = \frac{(\rho_{max} - \rho_{min})}{\rho_{min}} * 100 \qquad (2)$$

Where ρ_i is the resistivity indicated by the subscript.

Modeling of the sublimation growth of 4H SiC was carried out using the Virtual Reactor^{TM} code produced by STR, Inc. A full thermal and mass flux quasi-steady state model was employed for the sublimation growth process.

DISCUSSION

Discussion of Resistivity Trends Based on Large Data Sets

High power electronic devices require that the 4H n+ SiC substrate be as benign as possible. This means that the wafer defects must be low enough so as not too negatively impact device performance. In addition, wafer resistivity should be as low and as uniform as possible to provide predictable device yield and operating characteristics for devices made within a single wafer and across multiple wafers.

The typical resistivity profile within the same SiC crystal is shown in Figure 2. The observed behavior for the resistivity distributions can be consistently understood in terms of the changing C/Si ratio in the PVT process internal to the crucible [1,2,4,8]. Thermal modeling of the PVT growth environment reveals radial and axial distributions of the C/Si species similar to the resistivity maps, suggesting that this parameter influences the resistivity profile [21]. The source becomes depleted of Si and the vapor becomes more C rich. As a result, the C/Si ratio during PVT growth continuously increases and leads to a continuous suppression of N incorporation as the crystal grows..

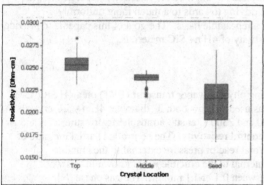

Figure 2. Typical wafer resistivity distribution within 4H SiC n+ doped crystal locations.

Several studies have focused on methods to improve the wafer resistivity uniformity. A key variable that has been found to impact the resistivity profile has been in control of the growing crystal shape through control of the radial thermal gradients [8,9,19]. As shown in Figure 3, for crystals grown with 15% lower radial gradients the resistivity uniformity variation is reduced. The method to control the crystal shape revolved around controlling the thermal radial gradients employed in the PVT growth environment as guided by numerical modeling. These changes are consistent not only with a method to improve wafer resistivity uniformity, but have also been directly tied to improving crystallinity as well [19].

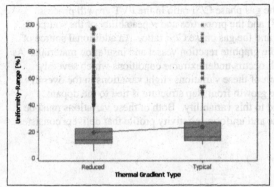

Figure 3. Demonstration of 4H n+ SiC uniformity improvement based on thermal gradient control. The crystal crystal growth fron shape was changed by 15% through control of the radial thermal gradient.

Application of gradient reduction to crystal growths producing large volumes of wafers has confirmed a significant and statistically valid improvement in wafer uniformity as shown in Figure 4. The median wafer uniformity improved 22% to 12% with a shift to reduced thermal gradients in the PVT growth environment.

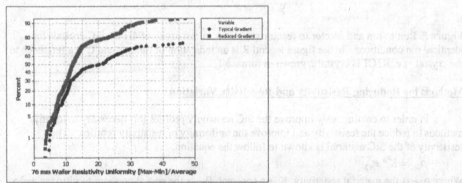

Figure 4. Large data set demonstration of within wafer resistivity uniformity improvements related to changes in thermal gradient in 4H n+ SiC crystal growth.

A more challenging problem to consistent wafer resistivity in PVT is shown in Figure 5. This figure shows the maximum wafer resistivity and the uniformity obtained for a sequence of runs on the same PVT reactor and on identical furnaces. These charts, therefore, represent the run-to-run and reactor-to-reactor repeatability of the resistivity for this PVT process. As can be seen from these charts there still a considerable range of variability within the growth process. The potential origin of this resistivity variation must be addressed for a long term stable and repeatable growth process. There are several sources of error that may be considered, but perhaps the most difficult to address is the nature of the vaporization of the SiC source material.

As has been reported in several articles the gas phase C/Si ratio in the PVT growth process impacts the dopant incorporation [1,2,4,8] and the properties and repeatability of the source material impacts the source vaporization and the gas phase C/Si ratio. An additional source of error may in the repeatable assembly of the graphite reaction vessel and insulation materials. As has been discussed previously SiC growth occurs under extreme conditions which severely complicate the process control. As a result of these variations slight variations in the overall crystal growth shape are noted. Since the growth front step structure is tied to the dopant incorporation [8,9] the resistivity is linked to this variability. Both of these variations must be addressed in order to develop a repeatable and uniform resistivity profile that delivers consistent wafer properties to device makers.

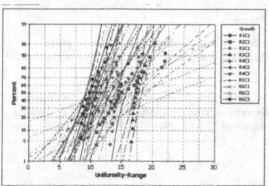

Figure 5. Run to run and reactor to reactor variations in resistivity of 4H n+ SiC crystals for identical run conditions. In the figure legend R is an indication of furnace and C is a reference to the crystal. *i.e.* R1C1 is crystal 1 grown in furnace 1.

Methods for Reducing Resistivity and Resistivity Variation

In order to continuously improve the SiC resistivity profiles it is necessary to identify methods to reduce the resistivity and improve the uniformity of resistivity profiles. The resistivity of the SiC material is known to follow the equation:

$$\rho_{SiC} = K * P_{N2}^{x} \tag{3}$$

Where ρ_{SiC} is the material resistivity, K is a constant, P_{N2} is the partial pressure of nitrogen and x is a power layer typically less than 1. Equation (1) implies that the most straightforward method to adjust the SiC material resistivity is to adjust the partial pressure of N in the crucible. However, there are other factors influencing the material resistivity. The other factors that impact material resistivity are located in the parameter, K. For example, the source vaporization and resulting C/Si ratio can impact the value of K. Source material preparation and subsequent evolution is a key factor to SiC growth and in particular in dopant incorporation as this phenomenon can modify the gas phase composition. In addition, parameters such as crystal stress could impact the parameter, particularly if the residual stress can negatively impact the carrier mobility. Carrier mobility can be impacted in the SiC material if the source or nature of the residual stress can impact parameters such electron scattering. Therefore, source material and crystal defects that limit electron mobility are two potential contributors to resistivity through the parameter, K.

A test was performed comparing two types of SiC source material with the same molar ratio of C/Si, but a significantly different morphology. For PVT growth under nominally the same condition the resistivity profile for select wafers is provided in Figure 6. It is important to note for these experiments that the growth rate for the experiment was effectively the same with the resulting crystals being within the typical variation of the process. First, it has been observed that the material resistivity decreased significantly. The resistivity data suggests the modified source material effectively provided a significantly lowered C/Si ratio relative to the standard source material. As a result of the modified C/Si ratio it is reasonable to expect a significant change in N incorporation based on the site competition model [1]. Another interesting observation from this experiment, however, is a general improvement in the resistivity uniformity. As can be seen from Figure 6, the wafer-to-wafer resistivity variation is significantly reduced with the modified source material. This indicates that the source material evolves in a manner that leads to a more consistent C/Si ratio through the duration of PVT growth. As a result, this modified source chemistry provides a method to not only reduce the resistivity of the 4H n+ SiC material but also reduces its variation within the crystal.

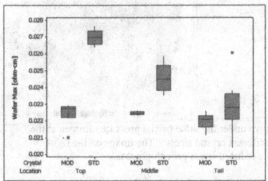

Figure 6. Impact of source material on 4H n+ SiC crystal resistivity and uniformity. MOD represents the modified source morphology and STD represents the standard source morphology

A variable that has been identified to impact SiC wafer quality is the stress level found in the growing SiC crystal [19]. There are several sources of stress in the SiC growth process. For example the nature of the PVT growth process requires that axial and radial temperature gradients be established in the growth cell in order to support and encourage SiC vapor transport from the source to seed. The degree and nature of these gradients and the absolute temperature of the reaction itself can create a significant amount of stress in the crystal. Figure 7 shows that resistivity is also impacted by the PVT process that results in lower crystal stress. These materials were grown under similar partial pressures of nitrogen with similar crystal growth temperatures as validated by numerical modeling. The SIMS values for these materials show that the N concentration for the typical process is 6.1×10^{18} cm^{-3} while for the low stress process is 7.0×10^{18} cm^{-3}. What is particularly intriguing, however, is that the overall material resistivity for the standard material resistivity is 0.024 Ω-cm while the low stress approach reaches 0.020

Ω-cm. To achieve comparable resistivity in the standard process a 100% N_2 ambient is required during PVT. For some crystals produced under reduced stress growth conditions the resistivity is below the system detection limit of 0.007 Ω-cm. This trend can be understood by several possible scenarios. First, the lower stress growth process results in the N incorporation being on more electrically active sites. Previous studies have suggested that the inactive N in a typical SiC crystal may be as low as 50% [10] although there is not much agreement in the literature [12]. The other possibility is that the mobility of the two materials is significantly different. In fact, uniaxial strain is known to impact the resistivity of SiC materials [20]. The basic interpretation, therefore, is that value of K suggested equation (3) is also impacted with the reduced stress level in growth. The reduced stress approach for growing SiC crystal provides not only an improved crystal quality with lower residual stress and improved crystal quality, but also a lower resistivity material for the same partial pressure of nitrogen.

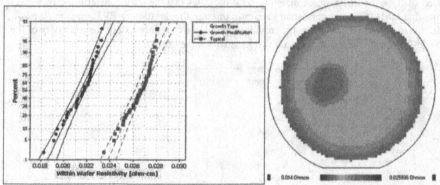

Figure 7. Resistivity levels for wafers grown under the same partial pressure nitrogen at the same temperature, but with significantly different crystal stress. The image on the right provides resistivity across the wafer for the reduced stress growth process.

CONCLUSIONS

Manufacturing 4H n+ SiC substrates with uniform resistivity is still a significant technical challenge. Due to the nature of the PVT growth environment and the dependence of impurity incorporation on the gas phase composition, that is the C/Si ratio, dopant incorporation in 4H n+ substrates continues to be non-uniform. These non-uniformities can manifest not only within wafers from the same crystal growth run, but also from wafers across several growth runs. This is largely due to the difficult nature of the controlling the SiC source material and growth processes effects that leads to changes in the repeatability of the C/Si ratio of the source vapor during PVT growth. Several mechanisms influencing the 4H n+ resistivity levels and uniformities were discussed in terms of altering the nature of the SiC source, controlling the thermal gradients, and also reducing the residual material stresses during the growth of 4H n+ SiC materials. Future work on resistivity improvements are still required in order to controllably produce a uniformly doped 4H n+ SiC substrate that meets the current and future demands of SiC high power and high frequency devices.

ACKNOWLEDGMENTS

This work was supported in part by ONR Contract #N00014-05-C-0324 (Program Officer: Dr. Paul Maki).

REFERENCES

1. D.J. Larkin, Phys. Stat. Sol. (B) 202, 305 (1997)
2. R.C. Glass, D. Henshall, V.F. Tsvetkov, and C.H. Carter, Phys. Stat. Sol. (B) 202, 149 (1997).
3. R. Muller, U. Kunecke, D. Queren, S.A. Sakwe, and P. Wellman, Chem. Vap. Deposition 12, 557 (2006).
4. D. Schulz, K. Irmscher, J. Dolle, W. Eiserbeck, T. Muller, H.J. Rost, D. Siche, G. Wagner, and J. Wollweber, Mat. Sci. Forum 339-342, 87 (2000).
5. K. Irmscher, M. Albrecht, M. Rossberg, H.J. Rost, D. Siche, and G. Wagner, Physica B 376-377, 338 (2006).
6. H.J. Rost, J. Doerschel, K. Irmscher, M. Rossberg, D. Schulz, and D. Siche, J. Cryst. Growth 275, e451 (2005).
7. H.J. Rost, J. Doerschel, K. Irmscher, D. Schulz, and D. Siche, J. Crys. Growth 257, 75 (2003).
8. N. Sugiyama, A. Okamoto, and T. Tani, Inst. Phys. Ser. 142, 489 (1996).
9. K. Onoue, T. Nishikawa, M. Katsuno, N. Ohtani, H. Yashiro, and M. Kanaya, Phys. Ser. 142, 65 (1996).
10. N. Schulze, D.L. Barrett, G. Pensl, S. Rohmfeld, and M. Hundhausen, Mat. Sci. and Eng. B61-62, 44 (1999).
11. K. Onoue, T. Nishikawa, M. Katsuno, N. Ohtani, H. Yahiro, and M. Kanaya, Jpn. J. Appl. Phys. 35, 2240 (1996).
12. K. Semmelroth, N. Schulze, and G. Pensl, J. Phys.: Condens. Matter 16, S1597 (2004).
13. V.F. Tsvetkov, S.T. Allen, H.S. Kong, and C.H. Carter, Inst. Phys. Ser. 142, 17 (1996).
14. H. McD. Hobgood, R.C. Glass, G. Augustine, R. Hopkins, J. Jenny, M. Skowronski, W.C. Mitchell, and M. Roth, Appl. Phys. Lett. 66, 1364 (1995).
15. M. Bickermann, D. Hofmann, T.L. Straubinger, R. Weingartner, P.J. Wellman, and A. Winnacker, Appl. Surface Science 184, 84 (2001).
16. D.L. Barrett, J.P. McHugh, H.M. Hobgood, R.H. Hopkins, P.G. McMullin, and R.C. Clarke 128, 358 (1993)
17. M. Naitoh, K. Hara, F. Hirose, S. Onda, J. Cryst. Growth 237-239, 1192 (2002).
18. E. Schmitt, T. Straubinger, M. Rasp, and A.D. Weber, Superlattices and Microstructures 40, 320 (2006).
19. V.M. Lyubimskii, Sov. Phys. Semcond. 11, 437 (1977).
20. See for example works published as part of International Conference on SiC and Related Materials 2007 held in Otsu, Japan.
21. Q. Li, A. Polyakov, M. Skowronski, E. Sanchez, M. J. Loboda, M. A. Fanton, Timothy Bogart and R. D. Gamble, Mat. Sci. Forum 527-529, 51 (2006).

Mater. Res. Soc. Symp. Proc. Vol. 1069 © 2008 Materials Research Society   1069-D02-01

High-Resolution Photoinduced Transient Spectroscopy of Defect Centers in Undoped Semi-Insulating 6H-SiC

Pawel Kaminski[1], Roman Kozlowski[1], Marcin Miczuga[2], Michal Pawlowski[1,2], Michal Kozubal[1], and Jaroslaw Zelazko[1]

[1]Institute of Electronic Materials Technology, ul. Wólczyńska 133, Warszawa, 01-919, Poland
[2]Military University of Technology, ul. Kaliskiego 2, Warszawa, 00-908, Poland

ABSTRACT

High-resolution photoinduced transient spectroscopy (HRPITS) has been applied to studying defect centers controlling the charge compensation in semi-insulating (SI), vanadium-free, bulk 6H- SiC. The photocurrent relaxation waveforms were digitally recorded in the temperature range of 50 – 750 K and a new approach to extract the parameters of defect centers from the temperature-induced changes in the time constants of the waveforms has been implemented. It is based on a two-dimensional analysis using the numerical inversion of the Laplace transform. As a result, the images of spectral fringes depicting the temperature dependences of the emission rate of charge carriers for defect centers are created. Using the new procedure for the analysis of the photocurrent relaxation waveforms and the new way of the visualization of the thermal emission rate dependences, a number of shallow and deep defect levels ranging from 80 to 1900 meV have been detected. The obtained results indicate that defect structure of undoped SI bulk 6H-SiC is very complex and the material properties are affected by various point defects occupying the hexagonal and quasi-cubic lattice sites.

INTRODUCTION

Semi-insulating (SI) silicon carbide is an important material in terms of manufacturing substrates for high power microwave devices and integrated circuits with GaN HEMT structures as active elements. The SI properties of SiC are achieved by compensating residual shallow level impurities such as nitrogen and boron with deep-level defects pinned the Fermi level position near the middle of the band gap. Recently, vanadium-free (undoped) SI SiC crystals with the Fermi level pinned to deep levels introduced by native defects have been widely investigated. However, the existing knowledge on the electrical activation of native defects and residual impurities in these materials is insufficient for complete controlling performance of electronic devices. The defect structure of SiC crystals is fairly complex due to the occurrence of inequivalent lattice sites for impurity atoms and intrinsic point defects. In SI 6H-SiC, the point defects can be located in three inequivalent lattice sites – the hexagonal site h and two quasi-cubic sites k_1 and k_2. Since the defect centers have different activation energies for hexagonal and cubic sites, there are a large number of energy levels in the bangap of these materials and high-resolution spectroscopic techniques are necessary to fully understand the compensation mechanism.

In this paper, we present new experimental results, obtained by the high-resolution photoinduced transient spectroscopy (HRPITS), on the electronic properties of grown-in defect centers in high-purity 6H-SiC. The photocurrent transients were measured in the temperature range of 50-750 K. For the analysis of the temperature induced changes in the photocurrent relaxation waveforms, the experimental data are transformed into the images using the two-dimensional procedure based on the numerical inversion of the Laplace transform. As a result, the sharp spectral fringes depicting the temperature dependences of the emission rate of charge carriers for detected defect centers are visualized.

EXPERIMENT

The samples used in this work were prepared from the commercially available wafers with a thickness of ~ 388 μm cut out perpendicularly to the c axis (<0001> direction) from SI vanadium-free 6H-SiC boules. The (0001) Si-face surface of the wafers was polished according to the requirements for epitaxial substrates and the backside was of optical quality. The resistivity at room temperature was ~3.0x10^7 Ωcm, respectively. In the temperature range of 600 – 700 K, the activation energy of dark conductivity was 1.06 eV. The nitrogen concentration in the material, determined by Secondary Ion Mass Spectroscopy (SIMS) was ~ 3x10^{16} cm^{-3}. For the HRPITS measurements, arrays of two co-planar Ohmic contacts were made by evaporating a 20-nm layer of Cr and a 300-nm layer of Au on the front surface (Si-face) of the wafers followed by annealing at 500 °C. The width of the gap between two co-planar contacts was 0.7 mm. The samples with dimensions of 4 x 9 mm^2 were cut from the central parts of the wafers. The measurements of photocurrent transients were carried out in the temperature range of 50 – 700 K with steps ranging from 2 to 5 K. The excess charge carriers were generated by means of a semiconductor laser emitting the beam with the wavelength of 375 nm (3.31 eV). The photon flux was 1.3x10^{18} cm^{-2}s^{-1}. For the measurements in the temperature range 50-320 K, the width of the UV excitation pulses was 2 s and the repetition period was 10 s. The voltage applied between the two co-planar contacts was 150 V. The measurements in the range 320 – 700 K were carried out at the width of excitation pulses equal to 0.1 s; the repetition period was 1 s and the applied voltage was 300 V. The voltage was supplied from a Keithley 2410 high-voltage source meter. The photocurrent transients were amplified using a Keithley 428 fast current amplifier (conductance-voltage converter) and then digitized with a 12-bit amplitude resolution and a 1-μs time resolution. In order to improve the signal to noise ratio the digital data were averaged usually taking 50 – 500 waveforms. For further processing each photocurrent relaxation waveform was normalized with respect to the photocurrent amplitude at the end of the UV pulse. The detailed description of the experimental setup dedicated to characterization of defect centers in SI wide bangap materials has been presented elsewhere [1].

When the time constant of the relaxation waveform at a given temperature is much longer than the carrier lifetime ($\tau_{rel} \gg \tau_n$) and the retrapping of the charge carriers is neglected, the relaxation can be described by the sum of the exponential components whose time constants are equal to the reciprocals of the thermal emission rate of charge carriers for defect centers participating in the thermal emission process [1, 2]. So, by analyzing the waveforms recorded in a temperature range, the parameters of defect centers can be determined directly from the temperature dependences of the emission rate fitted with the Arrhenius formula [2]

$$e_T(T) = AT^2 \exp(-E_a/k_BT) \tag{1}$$

where e_T is the thermal emission rate of electrons or holes, T is the temperature, E_a is the activation energy, k_B is the Boltzmann constant and $A=\gamma\sigma$ is the product of the material constant γ and the apparent capture cross-section σ for electrons or holes. As a result, the activation energy E_a and the pre-exponential factor A, related to the capture cross-section, are calculated for the detected defect centers. The temperature dependences of the emission rate are obtained by means of a numerical procedure involving the two-dimensional spectral analysis of the recorded photocurrent relaxation waveforms using the inverse Laplace transform algorithm [2, 3]. As a result, the temperature changes in the time constants of the photocurrent relaxation waveforms are visualized in the 3D space as the spectral surface being a function of two variables: the temperature (T) and emission rate (e_T). The processes of thermal emission of charge carriers from detected defect levels are seen as the sharp folds and in order to extract the parameters of defect centers, the ridgelines of the folds are found and approximated using the Arrhenius equation [4]. The projections of the folds and ridgelines on the plane given by the axes (T, e_T) create the image of the temperature dependences of emission rate for the detected traps.

RESULTS AND DISCUSSION

The image of Laplace spectral fringes obtained by the two-dimensional analysis of the photocurrent relaxation waveforms recorded for a sample of undoped SI 6H-SiC at temperatures ranging from 50 to 150 K is presented in figure 1.

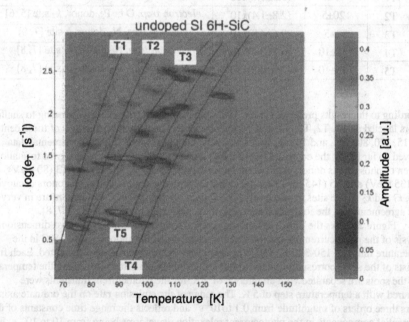

Figure 1. Image of Laplace spectral fringes obtained by the two-dimensional analysis of the photocurrent relaxation waveforms measured in the temperature range of 50-150 K for a sample of undoped SI 6H-SiC. The solid lines illustrate the temperature dependences of emission rate for detected defect centers.

The amplitude of the folds corresponding to the fringes is also shown. Five spectral fringes for defect centers labeled as T1 – T5 were resolved. The solid lines are the defect signatures illustrating the temperature dependences of the emission rate of charge carriers for the defect centers. The lines were obtained directly from the approximation of the Laplace spectral folds ridgelines with the Arrhenius equation using least-squares fitting. So, each solid line is described by equation (1) and its position on the plane depends on the values of the activation energy E_a and the pre-exponential factor A. The values of the activation energy and pre-exponential factor determined for the defect centers detected in the temperature range of 50-150 K are summarized in Table I.

Table I. Summary of parameters determined from the Laplace spectral fringes for defect centers in undoped SI 6H-SiC observed at temperatures 50-150 K.

Trap label	Activation energy E_a (meV)	Pre-exponential Factor A $(s^{-1}K^{-2})$	Tentative identification
T1	115±5	$(1.3\pm0.7)\times10^5$	*electron trap*, O or P_{Si} donor , k_1 site [5, 6]
T2	120±5	$(2.8\pm1.4)\times10^4$	*electron trap*, O or P_{Si} donor, k_2 site [5, 6]
T3	80±5	$(9.6\pm4.8)\times10^1$	*electron trap*, N_C donor, *h* site [7, 8]
T4	135±10	$(6.2\pm3.1)\times10^3$	*electron trap*, N_C donor, k_1 site [7, 8]
T5	145±10	$(1.0\pm0.5)\times10^4$	*electron trap*, N_C donor, k_2 site [7, 8]

According to the results presented in the Table I, five spectral fringes corresponding to shallow donors labeled as T1, T2, T3, T4, and T5 are observed. The activation energies of these centers are 115, 120, 80, 135, and 145 meV, respectively. On the grounds of the experimental data received so far [5, 6], the centers T1 (115 meV) and T2 (120 meV) can be assigned to shallow oxygen or phosphorus donors at different lattice sites (k_1 and k_2). The centers T3 (80 meV), T4 (135 meV) and T5 (145 meV) can be identified with the shallow nitrogen donors occupying the h, k_1 and k_2 lattice sites, respectively. The activation energies of these centers are in very good agreement with the ionization energies of the particular nitrogen donors [7, 8].

Figure 2 shows the image of Laplace spectral fringes obtained by the two-dimensional analysis of the photocurrent relaxation waveforms recorded for undoped SI 6H-SiC in the temperature range of 150-320 K. The fringes are very distinctive and well separated. Each fringe consists of the spots corresponding to various values of the emission rate. Along the temperature axis, the spots are separated with an interval of 5 K, for the photocurrent transients were measured with a temperature step of 5 K. The range of the emission rate on the ordinate axis covers three orders of magnitude from 0.1 to 10^2 s^{-1} and reflects the range time constants of the exponential components in the photocurrent relaxation waveforms being from 10 to 10^{-2} s. In the image shown in figure 2, we managed to distinguish 16 closely spaced spectral fringes related to the thermal emission of charge carriers from defect centers labeled as T6 – T21. The values of the activation energy E_a and the pre-exponential factor A, extracted from the photocurrent relaxation waveforms measured at temperatures ranging from 150 to 320 K, are listed in Table II. The presented results indicate that the activation energies of the detected traps are in the range of 225 – 840 meV. The defect centers were tentatively identified by comparing their parameters with the electronic properties of known point defects in 6H-SiC established by electron paramagnetic resonance (EPR), optical absorption, photoluminescence (PL), deep-level transient capacitance spectroscopy (DLTS) and optical admittance spectroscopy (OAS), as well as by thermally stimulated current (TSC) and temperature dependent Hall effect (TDH) measurements [8-23]. In view of the results reported in [8], the centers T6 (225 meV), T7 (245 meV) and

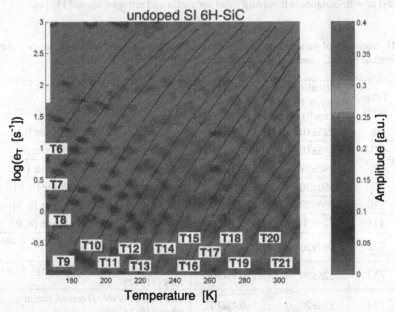

Figure 2. Image of Laplace spectral fringes obtained by the two-dimensional analysis of the photocurrent relaxation waveforms measured in the temperature range of 150-320 K for a sample of undoped SI 6H-SiC. The solid lines illustrate the temperature dependences of emission rate for detected defect centers.

T8 (290 meV) are likely to be isolated aluminum acceptors occupying the hexagonal (h) and quasi-cubic k_1 and k_2 lattice sites, respectively. Moreover, the centers T9 (360 meV), T10 (400 meV) and T11 (420 meV) can be assigned to boron acceptors [8, 9]. The former is presumably related to the boron acceptor at the h site and the two latter are likely to be attributed to the impurity at the k_1 and k_2 lattice sites, respectively. In other words we resolved the levels associated with the most common acceptors in 6H-SiC at the hexagonal and two cubic sites. It should be noted, that both the aluminum and boron concentrations, determined in the samples of undoped SI 6H-SiC by the Glow Discharge Mass Spectroscopy (GDMS), were of the order of 1×10^{16} cm^{-3}. The centers T12 (490 meV) and T13 (525 meV) are likely to be related to a native point defect observed by DLTS as an electron trap E_i (E_c- 0.50 eV) in n-type 6H-SiC irradiated with electrons or neutrons [10, 11]. On the grounds of the annealing behavior and the EPR results, this defect was identified with the $EI5$ center attributed to the positively charged carbon vacancy V_C^+ [11, 12]. So, there is strong experimental evidence pointing towards E_i being a carbon vacancy. Thus, we assume that the centers T12 (490 meV) and T13 (525 meV) can be attributed to the carbon vacancy located at the hexagonal and quasi-cubic lattice sites, respectively. Similarly, the centers T17 (620 meV) and T18 (625 meV) can be identified with the known electron traps Z_1 (E_c- 0.62 eV) and Z_2 (E_c- 0.64 eV), respectively, which are characteristic of both 4H- and 6H-SiC [11, 13]. Moreover, there are strong experimental facts indicating that the so called Z_1/Z_2 center can be associated with silicon vacancies V_{Si} at various lattice sites (h and k_1, k_2) or with complexes involving these vacancies and nitrogen atoms [13, 14].

Table II. Summary of parameters determined from the Laplace spectral fringes for defect centers in undoped SI 6H-SiC observed at temperatures 150-320 K.

Trap label	Activation energy E_a (meV)	Pre-exponential factor A ($s^{-1}K^{-2}$)	Tentative identification
T6	225±10	$(6.5\pm3.3)\times10^1$	*hole trap*, Al acceptor, *h* site [8]
T7	245±10	$(4.9\pm2.5)\times10^2$	*hole trap*, Al acceptor, k_1 site [8]
T8	290±20	$(6.6\pm3.3)\times10^3$	*hole trap*, Al acceptor, k_2 site [8]
T9	360±20	$(9.0\pm4.5)\times10^4$	*hole trap*, B acceptor, *h* site [8, 9]
T10	400±20	$(2.1\pm1.1)\times10^5$	*hole trap*, B acceptor, k_1 site [8, 9]
T11	420±20	$(1.4\pm0.7)\times10^5$	*hole trap*, B acceptor, k_2 site [8, 9]
T12	490±20	$(2.0\pm1.0)\times10^6$	*electron trap*, E_i (0.50 eV), V_C at *h* site [11, 12]
T13	525±20	$(3.1\pm0.9)\times10^6$	*electron trap*, E_i (0.50 eV), V_C at k_1, k_2 sites [11, 12]
T14	575±20	$(1.3\pm0.4)\times10^7$	*hole trap* 0.58 eV, *D* center, boron related [8, 15, 16]
T15	725±20	$(2.0\pm0.6)\times10^9$	unknown nature
T16	670±20	$(9.8\pm2.9)\times10^7$	*hole trap* 0.65 eV, boron –native defect complex [15, 17]
T17	620±20	$(4.5\pm1.4)\times10^6$	*electron trap*, Z_1/Z_2 center, V_{Si} (*h* site) related, V_{Si}-N [11, 14]
T18	625±20	$(1.6\pm0.5)\times10^6$	*electron trap*, Z_1/Z_2 center, $V_{Si}(k_1, k_2$ sites) related, V_{Si}-N [11, 14]
T19	725±20	$(3.3\pm1.0)\times10^7$	unknown nature
T20	780±20	$(8.0\pm2.4)\times10^7$	*electron trap*, residual vanadium acceptor at k_2 site [18, 19]
T21	840±20	$(2.6\pm0.8)\times10^8$	*electron trap*, residual vanadium acceptor at k_1 site [18, 19]

So, the centers T17 (620 meV) and T18 (625 meV) are likely to be related to silicon vacancies occupying the hexagonal and quasi-cubic lattice sites, respectively. The trap T14 (575 meV) can be identified with the level at $E_v + 0.58$ eV reported for the so called D center [8]. Apart from the shallow boron level whose ionization energy was about 0.39 eV, the D center as a deep acceptor was for the first time observed in the boron-doped epitaxial 6H-SiC. Although the microscopic structure of this defect is still an open question, in view of the reported results the D center is a complex involving boron and a native defect, presumably the carbon vacancy [15, 16]. The trap T16 (670 meV) can be identified with the level at $E_v + 0.65$ eV observed by optical DLTS in 6H-SiC bulk crystals grown by Halide Chemical-Vapor Deposition (HCVD). According to the reported suggestions, this level can be assigned to a deep acceptor involving boron and silicon antisite [17]. The traps T15 and T19 have the same activation energy but the capture cross-section for charge carriers for the trap T15 is by two orders of magnitude higher than that of trap T19. So, in the image shown in figure 2, the fringe corresponding to trap T15 is much shifted towards lower temperatures compared to that for the trap T19. The origin of the both traps remains unknown.

The image of Laplace spectral fringes obtained by the two-dimensional analysis of the photocurrent relaxation waveforms recorded for undoped SI 6H-SiC at temperatures ranging from 320 to 700 K is presented in figure 3.

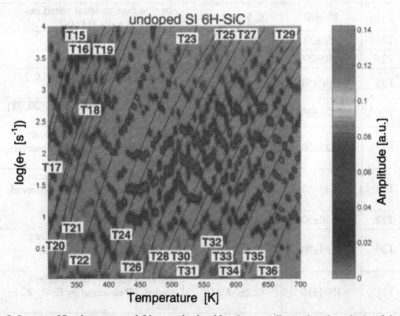

Figure 3. Image of Laplace spectral fringes obtained by the two-dimensional analysis of the photocurrent relaxation waveforms measured in the temperature range of 320-700 K for a sample

of undoped SI 6H-SiC. The solid lines illustrate the temperature dependences of emission rate for detected defect centers.

The density of the spots in the image shown in figure 3 is much higher than that in figure 2 because the photocurrent transients in the range 320-700 K were measured with a temperature interval of 2 K. The range of the emission rate on the ordinate axis covers nearly four orders of magnitude from 1 to 10^4 s^{-1} and reflects the range of the exponential components time constants in the photocurrent relaxation waveforms being from 1 to 10^{-4} s. Therefore, the fringes for the traps T15 –T21, observed in figure 2 at lower values of the emission rate, are seen in figure 3 at the higher emission rate values. The values of the activation energy E_a and the pre-exponential factor A, extracted from the photocurrent relaxation waveforms measured at temperatures ranging from 320 to 700 K, are listed in Table III.

Table III. Summary of parameters determined from the Laplace spectral fringes for defect centers in undoped SI 6H-SiC observed at temperatures 320-700 K.

Trap label	Activation energy E_a (meV)	Pre-exponential factor A $(s^{-1}K^{-2})$	Tentative identification
T22	865±20	$(2.9±0.9)×10^7$	electron trap, residual vanadium acceptor at h site [18, 19]
T23	950±20	$(2.4±0.6)×10^7$	unknown nature
T24	910±20	$(2.6±0.7)×10^6$	unknown nature
T25	1030±30	$(1.8±0.5)×10^7$	UD1 line at 1.0 eV, LT PL [20, 21], a native defect at h site
T26	1120±30	$(5.3±1.3)×10^7$	UD2 line at 1.13 eV, LT PL [20, 21], a native defect at k_1, k_2 sites
T27	1225±30	$(3.3±0.8)×10^8$	hole trap 1.24 eV, carbon vacancy related defect labeled XX observed by EPR, h site [22]
T28	1240±30	$(1.1±0.3)×10^8$	hole trap 1.24 eV, carbon vacancy related defect labeled XX observed by EPR, k_1, k_2 sites [22]
T29	1130±30	$(4.1±1.0)×10^6$	unknown nature
T30	1265±30	$(2.3±0.6)×10^7$	hole trap 1.29 eV, native defect labeled PP observed by EPR, h site [22]
T31	1295±30	$(2.2±0.6)×10^7$	hole trap 1.29 eV, native defect labeled PP observed by EPR, k_1, k_2 sites [22]
T32	1480±40	$(2.6±0.7)×10^8$	electron trap, residual vanadium donor at h site [18, 19]
T33	1655±40	$(2.4±0.6)×10^9$	electron trap, residual vanadium donor at k_2 site [18, 19]

T34	1635±40	$(4.4±1.1)×10^8$	*electron trap*, residual vanadium donor at k_1 site [18, 19]
T35	1885±40	$(6.8±2)×10^9$	PL line at 1.8 eV, a native defect at h site [23]
T36	1895±40	$(2.0±0.5)×10^9$	PL line at 1.8 eV, a native defect at k_1, k_2 sites [23]

According to the results shown in Tables II and III, the centers T20 (780 meV), T21 (840 meV) and T22 (865 meV) are identified with the vanadium acceptors located at the quasi-cubic sites k_2, k_1 and the hexagonal site, respectively. The identification is based on the recently reported experimental data for the vanadium acceptor levels in vanadium-doped SI 6H-SiC [18, 19]. Taking into account the results reported in [20, 21], the centers T25 (1030 meV) and T26 (1120 meV) can be considered to be related to the same native defect observed in undoped SI 6H-SiC through the Low Temperature (LT) photoluminescence lines known as UD1 and UD2. The small differences in the values of the activation energy and pre-exponential factor in the Arrhenius equation indicate that the centers can originate from the same defect located at various lattice sites. It should be added that the lowest ionization energy corresponds to the hexagonal site, and the highest to the quasi-cubic site k_2 [7]. Similarly, the centers T27 (1225 meV) and T28 (1240 meV), observed as the hole trap at E_v +1.24 eV by photo EPR measurements, can be assigned to one native defect occupying the h site and the k_1 or k_2 site, respectively [22]. This type of arguments allows attributing the centers T30 (1265 meV) and T31 (1295 meV) to the same native defect labeled as PP, as well as assigning the very deep centers T35 (1885 meV) and T36 (1895 meV) to another native defect located at the hexagonal and cubic lattice sites, respectively [22, 23]. In view of the results reported in [18, 19], the centers T32 (1480 meV), T33 (1655 meV) and T34 (1635 meV) can be assumed to be residual vanadium donors occupying the h, k_2 and k_1 sites, respectively.

Figure 4 represents a visualization of the values of the activation energy E_a and pre-exponential factor A extracted from the photocurrent relaxation waveforms recorded for undoped SI 6H-SiC in the whole temperature range 50-700 K. The standard error bars are also shown.

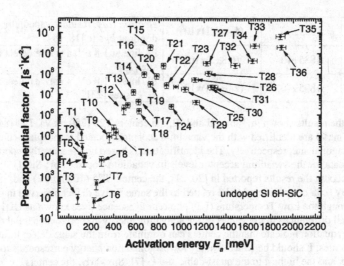

Figure 4. Visualization of the activation energy and pre-exponential factor values for all the traps detected in a sample of undoped SI 6H-SiC single crystal.

The presented results indicate that through the implementation of the new numerical algorithm to the two-dimensional analysis of the temperature-induced changes in the photocurrent relaxation waveforms measured in the temperature range of 50-700 K, we have managed to scan the bandgap of SI 6H-SiC with a very high resolution. As a result, thirty six traps, labeled as T1 – T36, with activation energies ranging from 80 to 1900 meV were found. According to the identification given in Tables I, II and III, the traps are related to the residual impurities and native defects located at the hexagonal and the two cubic lattice sites. So, the traps T1 (115 meV) and T2 (120 meV) are presumably oxygen or phosphorus-related shallow donors at the k_1 and k_2 sites, respectively. The traps T3 (80 meV), T4 (135 meV) and T5 (145 meV) are the nitrogen-related shallow donors at the h, k_1 and k_2 sites, respectively. On the other hand, the traps T6 (225 meV), T7 (245 meV) and T8 (290 meV) are the aluminum-related shallow acceptors at the h, k_1 and k_2 sites, respectively. Moreover, the traps T9 (360 meV), T10 (400 meV) and T11 (420 meV) are the boron-related shallow acceptors at the h, k_1 and k_2 sites, respectively. The traps T12 (490 meV), T13 (525 meV) and T17 (620 meV), T18 (625 meV) are produced by carbon and silicon vacancies at various lattice sites, respectively. The deep traps T20 (780 meV), T21 (840 meV) and T22 (865 meV) seem to be the residual vanadium-related acceptors at the k_2, k_1 and h sites, respectively. On the other hand, the midgap traps T32 (1480 meV), T33 (1655 meV) and T34 (1635 meV) are the residual vanadium-related donors at the h, k_2 and k_1 sites, respectively. In addition to that, we can distinguish the following pairs of traps: T25 (1030 meV) and T26 (1120), T27 (1225 meV) and T28 (1240 meV), T30 (1265 meV) and T31 (1295 meV), as well as T35 (1885 meV) and T36 (1895 meV). Each pair is presumably related to the specific native defect of unknown origin located at various lattice sites.

CONCLUSIONS

For the first time, the high-resolution photoinduced transient spectroscopy has been applied to studying defect levels in a single crystal of undoped SI 6H- SiC. A new approach to the analysis of the temperature-induced changes in the photocurrent relaxation waveforms measured in the temperature range of 50-700 K, as well as to the visualization of the experimental results and extracting the parameters of defect centers has been presented. It is based on creating images of the sharp spectral fringes using the two-dimensional numerical procedure employing the inverse Laplace transform. The potentialities of the new approach are exemplified by resolving 36 traps with activation energies ranging from 80 to 1900 meV. The traps were found to be related to residual impurities and native defects occupying the hexagonal and two cubic lattice sites. The mechanism responsible for semi-insulating properties of undoped 6H-SiC seems to be complex. In the charge compensation important roles play the residual nitrogen donors and boron acceptors, the residual vanadium donors and acceptors, as well as the deep-level native defects whose atomic configurations still remain unknown.

ACKNOWLEDGMENTS

This work was supported by the Polish Ministry of Science and Higher Education under Grant No. 3 T10C 028 30.

REFERENCES

1. P. Kamiński, R. Kozłowski, M. Kozubal, J. Żelazko, M. Miczuga, M. Pawłowski, *Semiconductors* 41, 414 (2007).
2. M. Pawłowski, P. Kamiński, R. Kozłowski, S. Jankowski, M. Wierzbowski, *Metrology and Measurement Systems* XII, 207 (2005).
3. S. Provencher, *Comp. Phys. Comm.* 27, 229 (1982).
4. P. Kamiński, S. Jankowski, R. Kozlowski, and J. Bedkowski, *Mater. Res. Soc. Symp. Proc.* 994, F03-14, 2007.
5. T. Dalibor, H. Trageser, G. Pensl, T. Kimoto, H. Matsunami, D. Nizhner, O. Shigiltchoff, and W. J. Choyke, *Mater. Sci. Eng.* B61-62, 454 (1999).
6. N. T. Son, A. Henry, J. Isoya, M. Katagiri, T. Umeda, A. Gali, and E. Janzén, *Phys. Rev. B* 73, 075201-1-16 (2006).
7. J. Schneider and K. Maier, *Physica B* 185, 199 (1993).

8. G.L. Harris (ed.), *Properties of Silicon Carbide*, INSPEC-IEE, London 1995.
9. S.R. Smith, O. Evwaraye, W.C. Mitchel, and M.A. Capano, *J. Electron. Mater.* **28**, 190 (1999).
10. C. Hemmingsson, N.T. Son, O. Kordina, E. Janzen, and J.L. Lindström, *J. Appl. Phys.* **84**, 704 (1998).
11. X. D. Chen, S. Fung, C. C. Ling, C. D. Beling, and M. Gong, *J. Appl. Phys.* **94**, 3004 (2003).
12. V.Ya. Bratus, T.T. Petrenko, S.M. Okulov, and T. L. Petrenko, *Phys. Rev. B* **71**, 125202-1-22 (2005).
13. I. Pintilie, L. Pintilie, K. Irmscher, and B. Thomas, *Appl. Phys. Lett.* **81**, 4841 (2002).
14. U. Gerstman, E. Rauls, Th. Fraunheim, and H. Overhof, *Phys. Rev. B* **67**, 205202 (2003).
15. A. O. Evwaraye, S. R. Smith, W. C. Mitchel, and H. McD. Hobgood, *Appl. Phys. Lett.* **71**, 1186 (1997).
16. A. v. Duijn-Arnold, T. Ikoma, O. G. Poluektov, P. G. Baranov, E. N. Mokhov, and J. Schmidt, *Phys. Rev. B* **57**, 1607 (1998).
17. S. W. Huh, H. J. Chung, S. Nigam, A. Y. Polyakov, Q. Li, M. Skowronski, E. R. Glaser, W. E. Carlos,B. V. Shanabrook, M. A. Fanton and N. B. Smirnov, *J. Appl. Phys.* **99**, 013508 (2006).
18. P. Kamiński, R. Kozłowski, M. Miczuga, M. Pawłowski, M. Kozubal, and M. Pawłowski, *J. Mater. Sci.: Mater. Electron.* (2008) (in press).
19. M. E. Zvanut, V. V. Konovalov, H. Wang, W. C. Mitchel, W. D. Mitchell, G. Landis, *J. Appl. Phys.* **96**, 5484 (2004).
20. Z.-Q. Fang, B. Claflin, D. C. Look, and G. C. Farlow, *J. Electron. Mater.* **36**, 307 (2007).
21. W.C. Mitchel, W.D. Mitchell, Z.Q. Fang, D.C. Look, S. R. Smith, H. E. Smith, Igor Khlebnikov, Y.I. Khlebnikov, C. Basceri, and C. Balkas, *J. Appl. Phys.* **100**, 043706 (2006).
22. D.V. Savchenko, E.N. Kalabukhova, S.N. Lukin, Tangali S. Sudarshan, Yuri I. Khlebnikov, W.C. Mitchel, S. Greulich-Weber, Mater. Res. Soc. Symp. Proc. **911**, B0507 (2006).
23. N. E. Korsunska, I. Tarasov, V. Kushnirenko, and S. Ostapenko, " High-temperature photoluminescence spectroscopy in p-type SiC", *Semicond. Sci. Technol.* **19**, 833 (2004).

Mater. Res. Soc. Symp. Proc. Vol. 1069 © 2008 Materials Research Society　　　　1069-D02-03

Studies of c-Axis Threading Screw Dislocations in Hexagonal SiC

Yi Chen[1], Ning Zhang[1], Xianrong Huang[2], Govindhan Dhanaraj[1], Edward Sanchez[3], Michael F. MacMillan[3], and Michael Dudley[1]

[1]Department of Materials Science and Engineering, Stony Brook University, Stony Brook, NY, 11794-2275

[2]National Synchrotron Light Source II, Brookhaven National Laboratory, Upton, NY, 11794-2275

[3]Dow Corning Compound Semiconductor Solutions, Midland, MI, 48611

ABSTRACT

In our study, closed-core threading screw dislocations (TSDs) and micropipes were studied using synchrotron x-ray topography of various geometries. The Burgers vector magnitude of TSDs can be quantitatively determined from their dimensions in back-reflection x-ray topography, based on ray-tracing simulation. This has been verified by the images of elementary TSDs. Dislocation senses of closed-core threading screw dislocations and micropipes can be revealed by grazing-incidence x-ray topography. The threading screw dislocations can be converted into Frank partial dislocations on the basal planes which has been confirmed by transmission synchrotron x-ray topography.

INTRODUCTION

Devices made from silicon carbide (SiC) possess promising advantages in high temperature, high power and high frequency applications due to its unique combination of properties (e.g., high thermal conductivity, high breakdown voltage, chemical stability). However, various defects, e.g., closed-core threading screw dislocations (TSDs), micropipes (MPs), triangular defects (TDs), stacking faults (SFs) and low angle grain boundaries (LAGBs), can degrade the performance of SiC devices significantly. MP densities have recently been significantly lowered. However, TSDs remain in densities ranging from $10^3 - 10^4/cm^2$. TSDs in SiC are of great interest from several different points of view. First, it should be noted that TSDs play a critical role in maintaining the desired polytype by promoting spiral step-flow growth. Second, they have been found to degrade device performance, reducing the breakdown voltage by 5-35% [1]. Third, the forest of TSDs in the as-grown crystal poses a significant barrier which creates a pinning effect on the glide of the basal plane dislocations (BPDs) either during growth or post-growth cooling, leading to localized higher densities concentration of BPDs. Fourth, TSDs have been found to interact with SFs in SiC epitaxial films expanding under forward bias, creating prismatic SFs [2].

EXPERIMENTAL

SiC crystals used in the study are commercial 3-inch 4H-SiC wafers, grown by the physical vapor transport (PVT) technique, with 8° off-cut towards [11-20]. Grazing-incidence topographs using the (11-28) plane reflections and back-reflection topographs using basal plane reflection were taken from the Si-face side. A fine-scale copper grating with periodicity of 50 μm was placed in the incident X-ray beam. The setting up in reticulography is similar and can be referred to Lang's report.[3] The imaging was carried out at the Stony Brook Synchrotron Topography Station, Beamline X-19C, at the National Synchrotron Light Source at Brookhaven National Laboratory, and XOR-33BM at Advanced Photon Source, at Argonne National Laboratory using VRP-M holographic films at a specimen-to-film distance of 10~15 cm for back-reflection geometry and 25~35 cm for grazing-incidence geometry, respectively. The viewpoint is from behind the x-ray film throughout our discussion.

DISCUSSION

Direct determination of Burgers vector magnitude of elementary TSDs

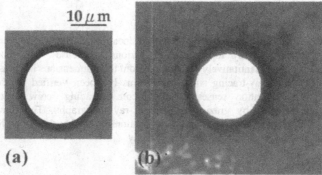

Figure 1. (a) Ray-tracing simulation of a 1c TSD at specimen-to-film distance of 12.5 cm. (b) The monochromatic topograph showing a TSD with smallest Burgers vector.

The length of smallest unit vector along the c-axis in 4H-SiC is 10.05 Å and the magnitude of Burgers vectors of elementary TSDs has been assumed to be the length of the unit cell along c-axis. Monochromatic topography with ultra-high geometrical resolution makes it possible for us to quantitatively verify the magnitude of the Burgers vectors of elementary TSDs. Figure 1(b) shows a highly magnified monochromatic topograph of a 1c TSD recorded using basal plane reflection at specimen-to-film distance of 12.5 cm. The diameter of the white circle is ~ 19.5 μm. By simulating the topographic image of 1c TSD using different Burgers vector magnitudes (integral number of Si-C bilayer thickness, 2.5125 Å) as a fitting parameter, the image of a dislocation with Burgers vector 10.05 Å [Figure 1(a)] exactly matches the experimental observation. This indicates that the Burgers vector magnitude of elementary TSD is equal to the length of one unit cell along the c-axis.

Dislocation sense of micropipes

The nucleation mechanism of micropipes (MPs) still remains obscure. They can be nucleated at hexagonal voids where the Burgers vector of same-sign closed-core TSDs aggregate to form MPs [4]. Furthermore, Dudley et al. [5] proposed the nucleation of opposite-sense MPs at inclusions. Therefore, the dislocation sense of MPs contains critical information regarding their formation mechanisms. We have developed a novel technique based on grazing-incidence x-ray topograph using pyramidal reflections, which can be used to unambiguously and non-destructively map the dislocation sense of MPs in commercial SiC wafers [6,7]. Figure 2 shows a typical (11-28) grazing-incidence x-ray topograph containing many MPs. The MP images appear as nearly white ovals of various dimensions surrounded dark contrast. They can be grouped into two categories according to their canting features: the white ovals are canted to one or the other side of the g-vector. The observations in topograph can be well interpreted by computer simulation based on ray-tracing principles. Detailed simulation process and discussion can be found in Ref. 8 and the simulated images of 8c MPs are illustrated in the inset of Figure 2. The sense of MPs can be subsequently be determined according to the canting feature of the white oval contrast. These are marked in the figure. Back-reflection reticulography has been used to confirm this prediction [6].

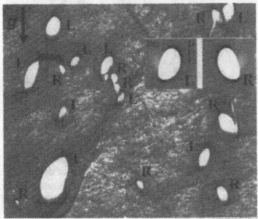

Figure 2. (11-28) grazing-incidence topograph showing the canted elliptical images of MPs. They are canted to one or the other side of the g-vector. Their sense can be unambiguously determined by the canting feature based on simulation (see inset).

Dislocation sense of closed-core TSDs

Figure 3 is a monochromatic topograph showing 13 elementary TSDs. The small white dots correspond to the threading edge dislocations. Different from MPs, it is very difficult to determine the sense of closed-core TSDs from the canting feature of the white contrast because of two reasons: 1) the dislocation line direction can be tilted away from c-axis so that shape of the roughly elliptical white contrast changes and 2) the white contrast in the topographic images of TSDs is small so that it makes the observation obscure. Clearly, the TSDs visible can be divided into two groups, according to the position of the enhanced perimeter contrast relative to the g-vector (each type is indicated by an arrow with one of two orientations). They are marked by "L" or "R" in Figure 3. The two TSDs near the top edge of the view are quite close and form a TSD pair. In order to determine the sense of the TSDs, the ray-tracing method, taking into account surface relaxation effects, has been used to simulate the grazing-incidence topographic images of closed-core TSDs. Figure 3 shows simulated images of left-handed ("L") and right-handed ("R") TSDs. They appear as asymmetric, roughly elliptical white features with perimeters of dark contrast which thicken along one side and at both ends. For TSDs with line directions slightly inclined to the c-axis, both the eccentricity and the inclination angle of the roughly elliptical features change slightly but the enhancement of perimeter contrast along one side persists, acting as an indicator of sign. By comparing the topographic images of TSDs in Figure 3 and the simulation (inset in Figure 3), one can find that the TSDs marked by "L" are left-handed while TSDs "R" are right-handed. "Small Bragg angle" back-reflection topography has been carried out to confirm the sense determination by grazing-incidence topography [7].

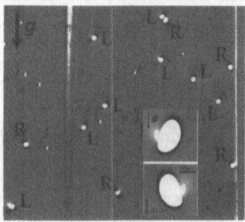

Figure 3. A monochromatic x-ray topograph containing TSDs of difference sign. Their sense can be determined based on the position of the thickened dark contrast in their images (marked by "L" or "R"). Inset shows the simulated images of left-handed ("L") and right-handed ("R") TSDs, respectively.

Conversion of TSDs into SFs in Chemical Vapor Deposition (CVD) epilayers

During CVD homo-epitaxy of SiC, the TSDs in the substrate are usually observed to propagate into the epilayer without any change. However, recent studies indicated that the TSDs roughly parallel to c-axis can be converted into defects in the basal plane during CVD growth [9]. Figure 4 shows a back-reflection topograph from a 100 μm epilayer grown on 8° off-cut substrate, where linear features are observed lying parallel to the off-cut direction. The length of these defects recorded in transmission mode correlates to the dimension of the basal plane in the epilayer along the off-cut direction, indicating that they are nucleated close to the substrate/epilayer interface. They appear as either a single linear defect or two closely-spaced linear defects with spacing gradually increasing toward down-step direction (rightward). They behave differently from the BPDs observed in back-reflection XRT, since the BPDs appear as either white stripes with dark contrast bands at both edges, or dark lines. For further studies of the characters of the linear defects, the substrate was thinned down to ~50 μm. Back-reflection topographs have been recorded from both the epilayer surface and substrate side for further investigations. Figure 5 gives two (00012) topographs recorded from the epilayer surface (a) and substrate side (b), respectively. Due to the limited penetration depth of x-ray, only information within certain thickness close to the x-ray incident surface was recorded. The linear defect shows faint contrast in the topograph recorded from the substrate side and such faint contrast is probably due to the polishing marks which obscure its image. Almost all the TSDs have one-to-one correspondence between the two topographs, indicating the TSDs simply propagate into the epilayer during CVD growth, although their location may shift slightly due to the tilt of dislocation lines. However, we notice that there does exist one TSD in Figure 5(b) [marked by triangle], which disappears in Figure 5(a). It is connected to the left end of the linear defect. Therefore, the linear defect is probably converted from a TSD existing in the substrate.

Figure 4. Back-reflection x-ray topograph of a 100 μm epilayer showing linear defects indicated by triangles on the figure formed during CVD growth, lying roughly parallel to the off-cut direction.

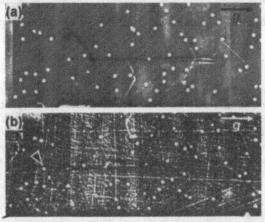

Figure 5. Back-reflection topographs recorded from the epilayer side (a) and substrate side (b). A TSD marked in (b) disappears after epilayer growth and it is connected to the end of a linear defect.

These linear defects are probably extrinsic SFs bounded by Frank partial dislocations in order to conserve the Burgers vectors. The Frank partials are so closely-spaced that they can not be separated in the topograph and appear a single linear defect. In order to further investigate such defect, an example containing two well-spaced dislocations has been selected for transmission topograph. Figure 6 shows a series of transmission topographs for quantitative studies of the linear defect. The SF area is always out of contrast in three {1-100} reflections, indicating that it is not a Shockley-type SF. It is visible in (0-111) and (-1011) reflections, but invisible in [-13-22] reflection, indicating that the fault vector is ½[0001] (not ¼[0001]). Therefore, the SF is bounded two Frank partials with Burgers vectors ½[0001] and the Burgers vector is conserved during the conversion. Please note that the SF might contain an in-plane shift, which also satisfies the situation in Figure 6.

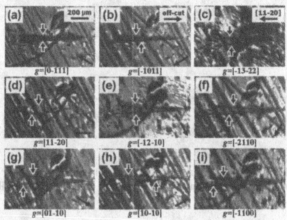

Figure 6. Transmission topographs of a linear defect containing two well-separated Frank partial dislocations.

CONCLUSIONS

The dislocation character of MPs and TSDs is discussed. The magnitude of the Burgers vector of elementary TSD has been quantitatively confirmed by using ultra-high geometrical resolution monochromatic topography, which is the length of the unit cell along c-axis. The sense of MPs/TSDs can be unambiguously determined by using grazing-incidence topography. The TSDs can be converted into Frank partial dislocations during CVD growth and the TSD-converted Frank-type SF has been quantitatively verified using transmission x-ray topography.

ACKNOWLEDGMENTS

This work is supported in part by ONR grants N00011400010348, N000140110302 and N000140211014 and by Dow Corning Corporation under contract numbers N0001405C0324 and DAAD1701C0081. Thank Dr. H. Tsuchida and I. Kamata for supplying sample for Fig. 5. Topography experiments were carried out at the Stony Brook Synchrotron Topography Facility, Beamline X-19C, at the NSLS (Contract no. DE-AC02-76CH00016) and Beamline XOR-33BM, Advance Photon Source, Argonne National Laboratory. The XOR/UNI-CAT facility is supported by the U.S. Department of Energy under Award No. DEFG02-91ER45439. The APS is supported by the U.S. Department of Energy, Office of Science, Office of Basic Energy Sciences, under Contract No. DE-AC02-06CH11357.

REFERENCES

[1] P. G. Neudeck, W. Huang, and M. Dudley, Solid-State Electron. 42, 2157 (1998)

[2] Y. Chen, M. Dudley, K. X. Liu and R. E. Stahlbush, Appl. Phys. Lett. 90, 171930 (2007)

[3] A. R. Lang and A. P. W. Makepeace, J. Phys. D: Appl. Phys. 32, A97 (1999)

[4] T. A. Kuhr, E. K. Sanchez, M. Skowronski, W. M. Vetter and M. Dudley, J. Appl. Phys. 89, 4625 (2001)

[5] M. Dudley, X. R. Huang, W. Huang, A. Powell, S. Wang, P. Neudeck and M. Skowronski, Appl. Phys. Lett. 75, 784 (1999)

[6] Y. Chen, G. Dhanaraj, M. Dudley, E. K. Sanchez, and M. F. MacMillan, Appl. Phys. Lett. 91, 071917 (2007)

[7] Y. Chen and M. Dudley, Appl. Phys. Lett. Vol. 91, 141918 (2007)

[8] Y. Chen, M. Dudley, E. K. Sanchez and M. F. MacMillan, J. Electron. Mater. 2007 (in press)

[9] H. Tsuchida, I. Kamata and M. Nagano, J. Cryst. Growth 306, 254 (2007)

Mater. Res. Soc. Symp. Proc. Vol. 1069 © 2008 Materials Research Society   1069-D02-04

Links Between Etching Grooves of Partial Dislocations and Their Characteristics Determined by TEM in 4H SiC

Jean-Pierre Ayoub, Michael Texier, Gabrielle Regula, Bernard Pichaud, and Maryse Lancin
IM2NP, CNRS, Aix-Marseille Université, av. Escadrille Normandie Niemen, Marseille, 13397, France

ABSTRACT

We introduce defects into $(11\bar{2}0)$ oriented highly N-doped 4H-SiC by surface scratching, bending and annealing in the brittle regime. Emerging defects at the sample surface are revealed by chemical etching of the deformed samples. The etch patterns are constituted of straight bulges exhibiting various topographical features. These etch figures correspond to the emergence of double stacking faults dragged by a pair of partial dislocations. In this paper, we discuss the links between the etch figure characteristics and the defect nature. Results obtained by optical and atomic force microscopy are completed by structural analysis of defects performed by transmission electron microscopy. Mobility of partial dislocations in 4H-SiC is discussed and correlated to their core composition and to the effect of the applied mechanical stress.

INTRODUCTION

The exceptional properties of SiC which is both a ceramic and a wide band gap semiconductor have stimulated the rapid development of growth techniques of single-crystal, single-polytype 4H and 6H-SiC. As a result, commercial wafers of excellent crystalline perfection and controlled purity are now available. However, the performances of SiC based electronic devices are controlled by the long range defects –dislocations or stacking faults- which may develop during manufacturing. Despite numerous studies, their nucleation and propagation mechanisms are still matter of concern.

So far, the dislocation dynamics was indirectly determined by plasticity experiments [1-5]. As expected, the plasticity mechanisms involve dissociated perfect dislocations above the brittle-ductile temperature and partial dislocations (PD) dragging stacking faults (SF) below. However, the multiplicity of the created SFs in N-doped 4H-SiC is a matter of debate. Pirouz and co-workers have found single SFs and multiple SFs after indentation or compression tests [1,6-7]. By surface grinding and annealing at 1100°C, Skowronski and co-workers have only created double SFs (DSF), which are faults created by two PDs gliding in two adjacent glide planes [8]. Similarly, in samples strained by cantilever bending at 550°C or 700°C after surface scratching, we only observed DSFs [9]. Recently, Mussi et al. have found SFs of various multiplicities after uniaxial compression tests [10].

The results are also controversial concerning the core composition of the leading PDs. In 4H, 6H and 3C-SiC, Pirouz and co-workers have shown that the leading PDs have always a silicon core [1,3]. In good agreement with these authors, we have shown that in 4H-SiC all the leading PDs have a silicon core whereas C-core PDs are created but are not mobile at such temperatures [9,11]. Mussi et al. on the contrary, have found both Si-core and C-core leading PDs in 4H-SiC [10]. Surprisingly, all the calculations yet performed have given a lower energy of formation and migration for C-core PDs in SiC whatever the polytype [12-13].

On the basis of our previous TEM analyses, we have coupled a systematic study of the etch patterns by optical microscopy (OM) and atomic force microscopy (AFM) with new TEM characterisations of the PDs by weak-beam dark-field (WB-DF) imaging and large angle convergent beam electron diffraction (LACBED). This work demonstrates that it is possible to localise the (0001) and the $(000\overline{1})$ planes and to relate the orientation of the PD line versus the surface by coupling chemical etching of $\{11\overline{2}0\}$ faces, OM and AFM.

EXPERIMENTAL DETAILS

Rectangular samples were cut from N-doped (2×10^{18} cm^{-3}) 4H-SiC wafers. Their surface was parallel to the $(11\overline{2}0)$ plane and their length close to $[2\overline{2}01]$ direction. In that configuration, the (0001) glide planes are normal to the surface and at 45° from the sample length. Dislocations were created by scratching the sample surface in a direction parallel to the sample length using a diamond tip loaded with 100 g weight. The reference axes were chosen so that the {X,Y} plane is located at the half of the sample thickness. For samples under flexion this plane corresponds to the neutral plane (*cf.* Fig. 1.1) [14].

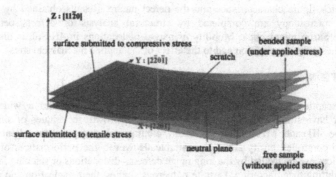

Fig. 1: Scheme of sample geometry before and after applying bending stress.

The samples were then bent at room temperature around the Y direction in cantilever mode leading, for a given X, to a maximum compressive stress on the $(11\overline{2}0)$ face and a maximum tensile stress on the $(\overline{1}\,\overline{1}20)$ face (*cf.* Fig. 1). Using this bending mode, three aspects of the average shear stress are taken into account: *(i)* the variation of the resolved shear stress (σ_r) as a function of X; *(ii)* the linear decrease and the sign inversion of σ_r beyond the neutral plane. This means that σ_r decreases along the partial dislocation segments as they penetrate deeper into the sample; *(iii)* $\sigma_r(X,Z)$ is constant along the sample width. Thus the stress is not constant in a (0001) glide plane since it is inclined at 45° from the X direction [14-15].

The bent samples were annealed under stress in Ar flux at 550°C and 700°C. The defects emerging at the sample surface were revealed by chemical etching using molten KOH (500°C, few minutes). KOH etching involves surface oxidation in the etching reaction and thus requires presence of oxygen in the molten salt. The oxygen is provided either by decomposition of the molten salt, or by the surrounding atmosphere [16].

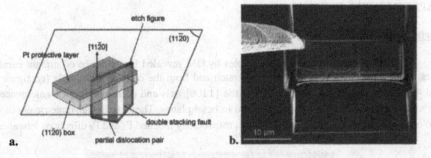

Fig. 2: a) Scheme of the first step of DBFIB (11$\bar{2}$0) thin foil preparation. Platinum protective layer is deposited on the sample surface. A slice containing the defect is cut parallel to the sample surface. b) DBFIB electronic micrograph of the box which is stuck on a micro-needle during the (11$\bar{2}$0) thin foil preparation process.

Following the KOH defect-selective etching, surface analysis was carried out by means of OM and AFM. For high resolution transmission electron microscopy (HRTEM) observations, (11$\bar{2}$0) TEM thin foils were prepared using dual-beam focused ion-beam (DBFIB) micromachining (FEI Dual Beam FIB Strata). For this, several steps are required: *(a)* Pt is deposited over one selected etch pattern; *(b)* standard FIB sectioning procedure is applied in order to remove a small box of materials (dimensions 10×6×2 μm³) as shown in figure 2; *(c)* the DBFIB box is stuck on a TEM grid; *(d)* the box is then ion milled along (11$\bar{2}$0) planes in order to remove the platinum protective layer and to thin the sample.

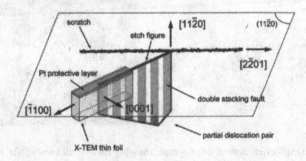

Fig. 3: Scheme of the FIB (0001) thin foil preparation. A slice is cut at the extremity of an etch figure, perpendicularly to the sample surface.

For LACBED and WB-DF analyses, (0001) TEM thin foils were prepared by FIB ion milling. The foils were prepared at the tip of etch figures, as depicted in figure 3.

EXPERIMENTAL RESULTS

Surface analysis

Overall observation of etched samples by OM revealed the presence of straight parallel lines which expand on both sides of the scratch and from the edge of the sample (*cf.* figure 4a and 4c). Those etch figures are parallel to the $[\bar{1}100]$ axis and correspond to the emergence at the sample surface of planar defects located in basal planes. They expand asymmetrically in the two opposite directions $[\bar{1}100]$ and $[1\bar{1}00]$, respectively labelled P_1 and P_2 directions hereafter.

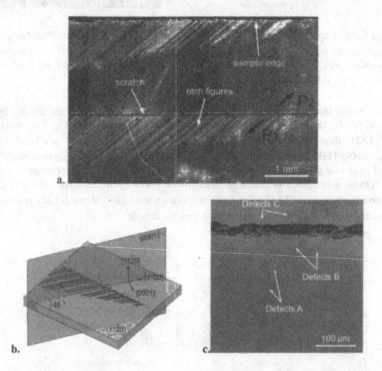

Fig. 4: a) Low magnification optical micrograph showing an overall view of the sample surface after annealing under flexion at 550°C and chemical etching. b) Schematic description of the etch pattern presenting the crystallographic orientation of the samples. c) Detail of the etch pattern close to the scratch exhibiting various families of defects: A, B and C.

Thanks to optical observations at higher magnification of areas close to the scratch, three populations of defects were discriminated as shown in figure 4c. Defects A and B extend in the P_1 direction, while defects C expand in the P_2 direction. Moreover, the length of defects A is clearly dependent on the stress variation along X while that of defects B and C is not or little influenced by it.

OM observations of the etch figure tips allowed us to distinguish five types of facieses: (i) type 1, 2, 3 and 4 facieses correspond to etch figure tips characterized by abrupt terminations, for which various geometrical shapes can be distinguished (see figure 5a); (ii) type 5 facies is very characteristic since the termination is not abrupt but presents a needle-like shape (see figure 5b).

Fig. 5: Optical micrographs showing a) the type 1, 2, 3 and 4 facieses. b) the type 5 facies.

To get further information about the topology of the sample surface, AFM analyses were carried out on several etch figures. Figure 6a shows a typical AFM image (40×40 μm^2) obtained from a region of the sample surface displaying the tips of various etch figures. Each figure appears as an extended bulge. By performing AFM observations on selected etch figures previously imaged by OM, we were able to determine that the type 1 to 4 facieses observed by OM correspond to the extremity of bulges for which the tip consist in steep faces. AFM observations performed on type 5 facieses reveal that the needle-like figures observed by OM correspond to a progressive decrease of the bulge height (see figure 6b), so that the extremity of the bulges does not appear as a steep wall.

For the sake of accuracy, several topographical profiles were extracted from AFM images performed on various bulges, along the [0001] direction. Figure 6c is representative of the bulge profiles. One can notice that all bulges present the same asymmetrical triangular print, with a maximum height of about 70 nm. The projection on the ($11\overline{2}0$) plane, of the vector normal to the steep faces is always oriented along the [000$\overline{1}$] direction, whereas the projection of the vector normal to the gently sloped faces is oriented along the opposite direction (cf. Fig. 6c).

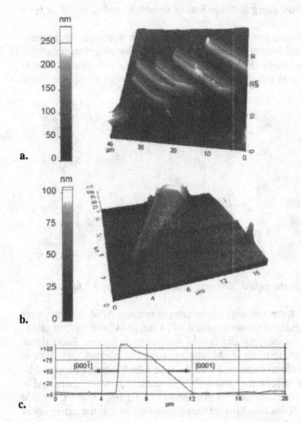

Fig. 6: a) AFM 3D-image of bulges observed at the sample surface after chemical etching. b) AFM 3D-image on type 5 etch figure tip. c) Typical topographical profile obtained on a bulge.

Defect characterization

The nature of planar defects was characterized by performing HRTEM imaging on $(11\bar{2}0)$ TEM thin foils in areas close to the scratch, where defect density is sufficient enough. These observations reveal that the observed defects systematically consist of 6 cubic bi-layers embedded in the hexagonal structure. The resulting 3C-SiC nano-bands are found parallel to the basal plane and present two distinguishable crystallographic orientations of the successive Si-C dumbbells within the nano-bands. These two cubic stackings (named a-3C or b-3C stacking in [9]) correspond to double stacking faults (DSF) which can be produced by the propagation of two Shockley partial dislocations (PD) gliding in adjacent glide-set planes. From crystallographic arguments, it is shown that only the glide of two PDs in adjacent and structurally equivalent planes (*i.e.* having the same orientation for Si-C dumbbells) may lead to the observed configurations. This implies that the shear results from the glide of two PDs either in the planes labelled G_1G_2 or G_3G_4 in [9]. Depending on the involved couple of glide planes (G_1G_2 or G_3G_4), the 3C-SiC stacking created is then found as being either of a-3C or b-3C type.

Plane view HRTEM examination was performed in a sample area containing an isolated bulge. The sample geometry and the preparation method which made use of the DBFIB device were described before. Figures 7a and 7b show the defect observed in the TEM thin box. Despite of the superficial amorphous layers produced by the FIB preparation which is crippling for HRTEM imaging, the HRTEM micrograph reveals the presence of only one defect in the box. Analysis of the image contrast also confirms that the defect, seen edge-on, consists of a DSF. The WB-DF micrograph shown in figure 7b allows us to distinguish the two bounding PDs and to estimate the equilibrium distance between them as being about 65 nanometres.

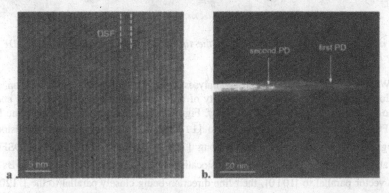

Fig. 7: a) HRTEM image showing a DSF, seen edge-on, in the DBFIB ($11\bar{2}0$) thin foil. b) WB-DF micrograph showing the contrast fringes due to the presence of the SF and the DSF left in the wake of the two bounding PDs (out of contrast with the experimental conditions used).

In order to investigate the possible correlation between the surface topography and the characteristics of the emerging dislocations bounding the DSFs, WB-DF imaging was also performed at the tip of selected etch figures. To do so, (0001) TEM thin foils were prepared by FIB as described in the section "Experimental details", either at the extremity of steep bulges (type 1) or at the extremity of needle-like bulges (type 5). Typical observations are given in figures 8a and 8b. Contrast analysis of these micrographs shows the presence of partial dislocation pairs as already observed in [9], whatever the examined etch figure extremity (*i.e.* steep or sharp bulge). They show that the presence of type 5 bulges at the sample surface is always associated with the presence of emerging dislocations whose line is almost parallel to the sample surface. On the contrary, dislocations observed at the extremity of type 1 bulges present sloped emerging segments.

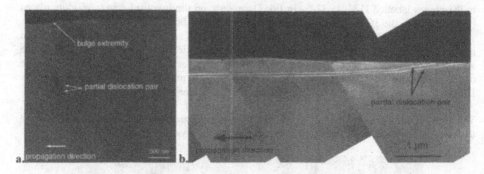

Fig. 8: a) WB-DF image showing PDs at the tip of a type 1 etch figure. b) WB-DF image showing PDs at the tip of a type 5 etch figure.

WB-DF imaging and LACBED analyses were methodically carried out on mobile dislocation segments, located at the extremity of bulges opposite to the scratch, and ended by type 1 to 4 facieses. WB-DF observations (*cf.* Fig. 9) reveal that the PDs bounding the DSFAs and DSFCs have Burgers vectors parallel to $[\bar{1}100]$. The emerging segments of dislocations bounding DSFAs are generally aligned along $[\bar{1}2\bar{1}0]$ whereas those bounding DSFCs are usually aligned along $[2\bar{1}\,\bar{1}0]$. Emerging dislocations located at the extremity of DSFBs have a Burgers vector parallel to $[10\bar{1}0]$, their line direction being closely parallel to the $[11\bar{2}0]$ axis. Thus, DSFAs, DSFBs and DSFCs are driven close to the sample surface by PD segments having a strong 30° character. In all cases, the extinction conditions are the same for both dislocations of the same pair, showing that their Burgers vectors are identical; their equilibrium distance is about (65±10) nm.

Fig. 9: WB-DF micrographs of dislocations bounding a) DSFA, b) DSFB, and c) DSFC.

LACBED patterns performed on dislocations for each defect family allowed us to fully determine the Burgers vectors characteristics (*i.e.* direction, modulus and sign) of the dislocation pairs. Taking into account the precautions necessary for the LACBED analysis of very close PDs having the same Burgers vector, specific splitting rules were applied. A detailed description of the method used can be found elsewhere [17-18].

LACBED analyses demonstrate that DSFA, DSFB and DSFC defects are bounded by dislocations with Burgers vectors respectively equal to $\frac{a}{3}[\bar{1}100]$, $\frac{a}{3}[10\bar{1}0]$ and $\frac{a}{3}[1\bar{1}00]$.

DISCUSSION

Correlation between defect characteristics and etched surface features

Examination of the surface topography of etched samples by AFM indicates that the emergence of the defects at the surface leads to bulges to appear during etching process. Bulges are associated to the straight lines observed by optical microscopy and are due to the emergence of planar defects at the sample surface. The observed relief of the etched surface is supposed to be correlated to the differences of etching rates between non-faulted and faulted areas of the crystal. As shown by HRTEM imaging, the planar defects consist of 3C-SiC bands. The presence of bulges at the emergences of DSF at the sample surface can be explained by an etching rate lower for 3C-SiC than for 4H-SiC. This hypothesis is reinforced by the WB-DF imaging performed on PDs located at the extremity of type 5 etch figures, showing that in this case, emerging dislocations are very close and almost parallel to the sample surface. Taking into account the propagation direction of the PDs, in this configuration, the 3C-SiC layer is consequently located below the PD pair. The progressive decrease of the bulge height is then due to the fast etching rate of the 4H-SiC structure situated on top of the slightly sloped emerging dislocations, and the low etching rate of the 3C-SiC structure situated below.

Topography of the bulges, perpendicularly to the defects propagation direction, shows an asymmetric profile rather triangular. This is interpreted as resulting from the direction dependence of the 4H-SiC chemical etching when etched along a polar axis. In α-SiC, the etching rate of the C-face, is several times higher than that of the Si-face [19-20], As a consequence, assuming that the interfaces between the DSFs and the 4H-SiC matrix constitute preferential sites for chemical etching, we can expect that the $(000\overline{1})$ interface (*i.e.* bounded by a C plane of the 4H-SiC phase), etches faster than the (0001) interface, in agreement with the observed profile of the bulges.

Dislocation core composition and mobility

Core reconstructions of the PDs viewed along the $[11\overline{2}0]$ direction were done for each possible Burgers vector by shearing the perfect 4H-SiC structure. An example of core reconstruction of a dislocation pair is illustrated in figure 10.

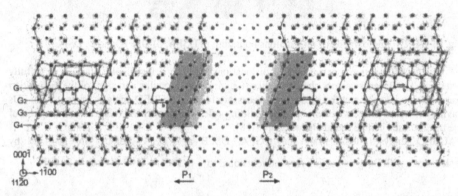

Fig. 10: Core reconstruction of a pair of adjacent dislocation loops having a Burgers vector equal to $\frac{a}{3}[\overline{1}100]$. Large and small dots respectively represent Si and C atoms. The core composition of the dislocations depends on the propagation direction.

From these reconstructions, we can then deduce the core composition of the PDs, depending on their Burgers vector, their character (*i.e.* depending on the angle between Burgers vector and dislocation line), and the P_1 or P_2 propagation direction. All the possible combinations are reported in [21]. The dislocation characteristics determined by WB-DF and LACBED analyses are consistent with PDs segments which have a Si(g) core.

Knowing that the dislocation nucleation occurred at the scratch or at its immediate vicinity, these results indicate that the mobile segments of the nucleated dislocations are, for our deformation conditions, systematically Si(g) core dislocations. This experimental evidence of a higher mobility of Si(g) segments than C(g) segments is coherent with previous observations [5,17,22] but clashes with atomistic simulations which expect the opposite tendency [12]. The lack of mobility of the C-core PD is responsible of the asymmetric expansion of the etch patterns on either sides of the scratch.

Our observations also underlined some differences of extension between families of defects, named DSFA, DSFB and DSFC. The variable DSF extensions from scratch can be correlated with the characteristics of the leading dislocations when taking into account the effect of the compressive stress on dislocation propagation. Indeed, the Peach-Köhler force acting on

62

the dislocations is dependent of the Burgers vector direction and sign. Thus, we can verify that compressive stress tends to favour the extension of dislocations having Burgers vectors equal to $\frac{a}{3}[\bar{1}100]$, and in lesser proportions, dislocations with Burgers vectors equal to $\frac{a}{3}[0\bar{1}10]$ and $\frac{a}{3}[10\bar{1}0]$. On the contrary, dislocations having a Burgers vector equal to $\frac{a}{3}[1\bar{1}00]$, $\frac{a}{3}[01\bar{1}0]$ or $\frac{a}{3}[\bar{1}010]$ must shrink under compressive stress. From core reconstructions, we note that the $\frac{a}{3}[\bar{1}100]$ and $\frac{a}{3}[10\bar{1}0]$ dislocations present 90° and 30° Si(g) segments for which propagation will be enhanced in the P_1 direction by compressive stress. In addition, some studies suggest a higher mobility of 90° PDs than 30° PDs [3,7]. We can consequently explain the high stress sensitivity of dislocations with Burgers vector equal to $\frac{a}{3}[\bar{1}100]$ (dislocations bounding DSFAs) by the fact that *(i)* they undergo the highest Peach-Kölher force and, *(ii)* they are trailed at their initial stage of development by pure edge emerging segments. The main part of applied stress being relaxed by the more mobile dislocation population, dislocations with Burgers vector equal to $\frac{a}{3}[10\bar{1}0]$ (dislocations bounding DSFBs) will be less sensitive to applied stress, as experimentally observed.

For the opposite propagation direction (*i.e.* the P_2 direction), no dislocation presents at once Si(g) core segments and Burgers vector orientation leading to a propagation favoured by compressive stress. In this direction, Si(g) core dislocations can only extend under tensile stress. The observation of $\frac{a}{3}[1\bar{1}00]$ dislocations which bond DSFCs is explained as follows: the elastic energy stored during the scratching step induces a complex stress field of limited extension. It acts locally as a compressive or a tensile stress would do. As a result, all the mobile segments are nucleated and glide over a few micrometres during the annealing. However, only the mobile segments for which propagation is favoured by compressive stress will extend further during the deformation experiment.

CONCLUSIONS

Nucleation and propagation of partial dislocations in 4H-SiC was carried out by means of scratching and bending of $(11\bar{2}0)$ wafers in the brittle domain. Optical observation of the etch pattern resulting from the chemical etching of the deformed sample surface, revealed peculiar features that permit to discriminate various populations of defects depending on their extension and propagation direction.

Topographic analyses of the etch figures associated with TEM characterization of the nucleated defects, allowed us to correlate the surface relief to the defect characteristics and the crystallographic polarity of the samples.

Finally, this work demonstrates that partial dislocation mobility in 4H-SiC is at once controlled by the dislocation core composition, the dislocation character, and the influence of the applied stress. The propagated dislocation pairs are composed by identical PDs having a Si(g) core gliding in adjacent glide-set planes, leading to the formation and extension of DSFs. A detailed analysis of the correlation between mobile dislocation characteristics and mechanical stress is reported in [21].

REFERENCES

1. X.J. Ning, N. Huvey, P. Pirouz, *J. Am. Ceram. Soc.*, **80**, (1997), p.1645.
2. A.V. Samant, W.L. Zhou, and P. Pirouz, *Phys. Stat. Sol.(a)*, **166**, (1998), p. 155.
3. A.V. Samant, M.H. Hong, P. Pirouz, *Phys. Stat. Sol. (b)*, **222**, (2000), p. 75.
4. J.L. Demenet, M.H. Hong and P. Pirouz, *Mat. Sci. Forum*, *338-342*, (2000), p. 517.
5. P. Pirouz, J.L. Demenet, M.H. Hong, *Phil. Mag. A*, **81**, (2001), p. 1207.
6. X.J. Ning and P. Pirouz, *J. Mat. Res.*, **11**, (1996), p. 884
7. P. Pirouz and J.W. Yang, *Ultramicroscopy*, **51**, (1993), p. 189.
8. H. J. Chung, J. Q. Liu, M. Skowronski, *App. Phys. Lett.*, **81**, (2002), p. 3759.
9. G. Regula, M. Lancin, H. Idrissi, B. Pichaud, J. Douin, *Phil. Mag. Lett.*, **85** (2005), p. 259.
10. A. Mussi, J. Rabier, L. Thilly, J.L. Demenet, *Phys. Stat. Sol. (c)*, **4**, (2007), p. 2929.
11. H. Idrissi, G. Regula, M. Lancin, J. Douin, B. Pichaud, *Phys. Stat. Sol. (c)*, **2**, (2005), p.1998.
12. A.T. Blumenau, C.J. Fall, R. Jones, S. Öberg, T. Frauenheim, P.R. Bridon, Phys. Rev. B, **68** (2003), p. 174108.
13. G. Savini, M.I. Heggie, S. Öberg, *Faraday Discuss.*, **134**, (2007), p. 353.
14. H. Idrissi, B. Pichaud, G. Regula, M. Lancin, *J. Appl. Phys.*, **101** (2007), p. 113533.
15. H. Idrissi, M. Lancin, G. Regula, B. Pichaud, *Mat. Sci. Forum*, **457-460**, (2004), p. 355.
16. D. Zhuang, J.H. Edgar, *Mat. Sci. Eng. R*, **48**, (2005), p.1.
17. M. Texier, G. Regula, M. Lancin, B. Pichaud, *Phil. Mag. Lett.*, **86** (2006), p. 529.
18. M. Texier, G. Regula, M. Lancin, B. Pichaud, *Phil. Mag. Lett.*, **87** (2007), p. 135.
19. R.W. Brander, A.L. Boughey, Br. *J. Appl. Phys.*, **18**, (1967), p.905.
20. M. Katsuno, N. Ohtani, J. Takahashi, HYashiro, M. Kanaya, *Jpn. J. Appl. Phys.*, **38**, (1999), p. 4661.
21. to be published.
22. M. Texier, M. Lancin, G. Regula, B. Pichaud, *proc. MSM conference*, Cambridge, UK, to be published.

Mater. Res. Soc. Symp. Proc. Vol. 1069 © 2008 Materials Research Society

Point Defects in SiC

Ádám Gali[1], Michel Bockstedte[2], Ngyen Tien Son[3], and Erik Janzén[3]

[1]Department of Atomic Physics, Budapest University of Technology and Economics, Budafoki ut 8, Budapest, H-1111, Hungary

[2]Universität Erlangen-Nürnberg, Staudtstr. 7/B2, Erlangen, D-91058, Germany

[3]Linköping University, Linköping, S-581 83, Sweden

ABSTRACT

Tight control of defects is pivotal for semiconductor technology. However, even the basic defects are not entirely understood in silicon carbide. In the recent years significant advances have been reached in the identification of defects by combining the experimental tools like electron paramagnetic resonance and photoluminescence with *ab initio* calculations. We summarize these results and their consequences in silicon carbide based technology. We show recent methodological developments making possible the accurate calculation of absorption and emission signals of defects.

INTRODUCTION

Point defects in semiconductors are pivotal for tailoring the electronic properties of materials according to needs. In silicon carbide (SiC), substitutional nitrogen or phosphorus are used for n-type doping while aluminum or boron for p-type doping. Deep defect centers compensate shallow impurities and produce semi-insulating regions in the semiconductor. These deep defect centers may be introduced un-intentionally during the technological process of doping that counteracts the efforts of the tight control of electronic conductivity of SiC. Recently, intrinsic deep defect centers have been started to apply to produce semi-insulating (SI) SIC substrate that can efficiently compensate nitrogen or boron impurities in the sample. The role of the intrinsic defects is crucial in both cases. However, even the basic intrinsic point defects are not entirely understood in SiC. The SiC electronic device problem is still a "defect engineering" problem. The first step is to identify the defects in order to control them.

Computational physics and experiments are complementary tools to attack the same problem in defect physics: computational physics can determine lots of properties of a single defect (correlation of "signals") but it cannot follow the complex processes occurring in the experiments and the accuracy of the obtained numbers are usually an order of magnitude higher than those coming from the experimental spectrum. In experiments one can characterize the defect accurately (sharp spectrum) but due to the difficulties of making correlations measurements (either due to the difficulties of sample preparation for two different experimental techniques or the complex processes occurring during the measurement) the interpretation of the signal is not straightforward and hinders the correct identification of the defect center. However, the combination of the two approaches can be very fruitful, at least, to disregard the inappropriate models for defect centers, and more: it can lead to the identification of defects. In the next sections we will summarize the computational methodologies that have been used in the defect studies and the recent results. Here, we focus our studies on *vacancy* defects. As we will show, comparison the hyperfine signals of the paramagnetic defects detected experimentally and obtained by calculations led to the identification of vacancy defects that are mostly responsible for the semi-insulating behavior of SiC samples.

COMPUTATIONAL METHODOLOGIES

We used *ab initio* supercell calculations to model the defects. We applied the traditional density functional theory within local density approximation (DFT-LDA). Depending on the actual code we used, we applied norm-conserving pseudopotentials or projector augmentation wave method to eliminate the direct calculation of the core electrons. In the case of hyperfine tensor calculations, we applied the all-electron PAW method. We used convergent basis sets and K-point sampling of the Brillouin-zone. For technical details see Refs. [1,2,3,4]. DFT-LDA supercell calculations suffer from two basic problems: i) DFT-LDA theory is a ground state theory and it is not capable of addressing excitation energies. For instance, the calculated band gap of SiC is about 1.0 eV lower than the experimental one. ii) The formation energy of the charged defects can converge slowly with the size of the supercell. Problem i) may be solved by using hybrid functionals [2] or GW calculations [5,6]. The excitonic effects in the absorption or emission processes can be taken into account by Bethe-Salpeter equation (BSE) [6]. The computational load of these methods is one order of magnitude higher than that of LDA, and they are still in development. Therefore, only few results have been obtained with these advanced methods. Problem ii) represents a technical difficulty that may be solved by using the "scaling" method [7]. Still, this involves huge computational resources even for LDA calculations. Therefore, the calculation of charge transition levels, especially, for highly charged defects needs large supercells and can be prohibitive even on supercomputers.

IDENTIFICATION OF THE VACANCY DEFECTS IN SILICON CARBIDE

In the next subsections we discuss the theory of the vacancy defects in silicon carbide and their fingerprints found experimentally. We note here that the technologically relevant polytypes of SiC are the hexagonal 4H and 6H-SiC. Since 6H-SiC contains 12 basis atoms in its primitive unit cell, computationally would be prohibitive to model the defects in the supercell constructed from this unit cell for large number of calculations. Instead, the calculations were carried out mostly in 4H-SiC supercells that contain inequivalent hexagonal (h) and cubic (k) sites. The calculational results obtained in 4H-SiC supercells may be interpolated to 6H-SiC (where there are two cubic sites near the hexagonal site) in order to compare the experimental and calculated data.

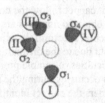

Figure 1: schematic diagram of vacancy. The vacant site is represented by a dashed circle, while the dangling bonds by the four σ functions. The labels are used in the group theory analysis.

The theory of vacancies is well-known in cubic materials. However, it might not be the case for hexagonal crystals like 4H or 6H SiC. Therefore, we analyze the vacancies in detail in 4H-SiC. The simplest vacancy defect is the isolated vacancy. The point group of the isolated vacancy is C_{3v} in 4H SiC. Taking the first neighbor atoms (I, II, III, IV) around the vacant site four dangling bonds ($\sigma_1, \sigma_2, \sigma_3, \sigma_4$; orthonormal to each other for simplicity) point to the vacancy before relaxation (see Figure 1).

These four dangling bonds form two a_1 states ($\psi_{a_1(1)} = \alpha\sigma_1 + \sqrt{\dfrac{1-\alpha^2}{3}}(\sigma_2 + \sigma_3 + \sigma_4)$ and

$\psi_{a_1(2)} = -\sqrt{1-\alpha^2}\,\sigma_1 + \dfrac{\alpha}{\sqrt{3}}(\sigma_2 + \sigma_3 + \sigma_4)$ where $0 \le \alpha \le 1$ depending on the defect) and one

double degenerate e state ($\psi_{e(3)} = \dfrac{1}{\sqrt{6}}(2\sigma_2 - \sigma_3 + \sigma_4)$ and $\psi_{e(4)} = \dfrac{1}{\sqrt{2}}(\sigma_3 - \sigma_4)$). As we shall

show the position and the order of these levels depend on the type and site of the vacancy as well as its charge state. During relaxation of the atoms the C_{3v} symmetry may be reduced to C_{1h} symmetry. In that case the a_1 states appear as a' states while the double degenerate e state splits to a' and a'' states. The occurrence of this reconstruction depends not just on the type of the vacancy and its charge state but also on the site of the vacancy in the hexagonal 4H SiC. Obviously, the group theory itself cannot predict the order and the position of the defect levels. LDA supercell calculations were applied to obtain the geometry and electronic structure of the vacancies. There are two types of isolated vacancies in SiC: carbon vacancy and silicon vacancy. We show here that the nature of these defects is quite different in SiC.

Carbon vacancy in silicon carbide

Theory predicted that one of the most abundant defects is the carbon vacancy (V_C) in SiC [8,9,1,10]. V_C can be double positively and negatively ionized, i.e., it is amphoteric. In the unrelaxed neutral state the first a_1 level falls into the valence band.

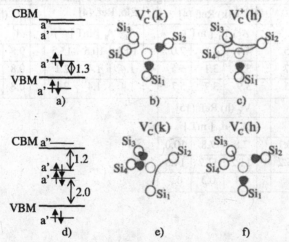

Figure 2: (a) and (d) are the calculated LDA+GW one-electron levels of V_C^0 and V_C^{2-} in eV; (b) and (c) show the spin distribution of V_C^+ at different sites; (e) and (f) show the spin distribution of V_C^- at different sites.

The second a_1 level is in the gap occupied by two electrons while the e level is empty in 4H-SiC. After relaxation the symmetry is reduced to C_{1h} and the silicon dangling bonds form long bonds pair-wise as can be seen in Figure 2(a). Interestingly, in the (+) state V_C conserves its C_{3v} symmetry at h-site while it reconstructs to C_{1h} symmetry at k-site resulting in significantly different spin density distribution (see Figure 2(b,c)). Indeed, the symmetry lowering from C_{3v} to C_{1h} was observed for V_C^+(k) [4,11], while V_C^+(h) kept its C_{3v} symmetry even at low temperature [4]. We have found that the spin density is mainly localized on one Si atom in V_C^+(h), while it is more evenly spread out on the first neighbor Si-atoms in V_C^+(k) (see Table I.) [3,4,11]. The V_C^+ signals are often observed in irradiated p-type 4H- and 6H-SiC [4,11] and are also common in high-purity semi-insulating (HPSI) 4H- and 6H-SiC substrates grown by physical vapour transport (PVT) or high-temperature chemical vapor deposition (HTCVD) [12,13]. Photo-excitation EPR (photo-EPR) showed a decrease of EI5 signal at the excitation energy of about 1.47 eV. This was previously interpreted as the (+/0) donor level of V_C in 4H-SiC [14]. However, our recent GW calculations revealed that the situation is more complex, since the transition to the split empty defect levels can be more effective, and has lower transition energy than that of the transition from the valence band edge to the defect level in the gap for V_C^+.

In the negatively charged states the atoms relax very strongly with bending to each other, and the two a' defect levels appear close to each other in the gap. This is due to a static Jahn-Teller effect because the e level will be only partially occupied. The empty a'' level goes above the conduction band minimum (CBM) (see Figure 2(d)). We have found that the geometry differs for V_C^- at different sites resulting in different spin density distribution [15] (Figure 2(e,f)). We have recently identified V_C^-(h) in electron-irradiated n-type 4H-SiC [15] as shown in Table I.

Table I. The measured hyperfine constants of V_C related EPR centers and the calculated hyperfine constants of V_C at different sites and charged states. See Figure 1 for atom labels.

	EI5, Ref. [4]			V_C^+ (k) Ref. [4]			EI6, Ref. [4]			V_C^+ (h) Ref. [4]		
	$A_\perp$, $A_\parallel$ [mT]			$A_\perp$, $A_\parallel$ [mT]			$A_\perp$, $A_\parallel$ [mT]			$A_\perp$, $A_\parallel$ [mT]		
Si$_1$	4.4	4.4	6.5	4.4	4.1	7.0	10.6	10.6	15.5	9.8	9.8	14.3
Si$_2$	3.3	3.2	4.7	3.3	3.1	5.5	1.4	1.4	2.1	0.8	0.7	1.5
Si$_{3,4}$	3.9	3.8	5.5	3.9	3.7	5.7	1.4	1.4	2.1	0.8	0.7	1.5
	HEI1, Ref. [15]			V_C^- (h) Ref. [15]								
	$A_\perp$, $A_\parallel$ [mT]			$A_\perp$, $A_\parallel$ [mT]								
Si$_1$	7.7	7.7	10.1	7.8	7.8	10.0						
Si$_2$	11.8	11.7	15.2	11.2	11.2	14.5						
Si$_{3,4}$	not observed			0.1	0.3	0.3						

The V_C^-(h) signal becomes dominating under illumination with photon energies larger than ~1.1 eV. Photo-EPR experiments [15] suggest that in irradiated n-type samples, V_C(h) may be in the double negative charge state lying at ~1.1 eV below CBM. Illumination with photo energies larger than ~1.1 eV excites electrons from the (–/2–) level to the conduction band and leaves the center in the single-negative charge state being detected by EPR. Photo-EPR of the V_C^- signal in HPSI 4H-SiC samples [13] shows similar results. This scenario is supported by theory if the self-interaction of LDA is healed by GW method: the highest occupied defect level is about 1.1 eV below CBM (see Figure 2(d)). At low temperatures, the center has C_{1h} symmetry. Above ~70 K, due to the thermal reorientation the symmetry becomes C_{3v}. An activation energy of ~20 meV was estimated for this thermal reorientation process [15].

Silicon vacancy in silicon carbide

In the silicon vacancy (V_{Si}) carbon dangling bonds point to the vacant site. These carbon dangling bonds are too localized to overlap with each other; therefore, no long bonds are created. Instead, the carbon atoms relax outwards and basically keep the C_{3v} symmetry independently on its charge state or its site in 4H SiC [8]. Our recent large supercell calculations revealed that the order of the defect levels is a_1, a_1, e in this case. The first a_1 level is resonant with the valence band not far from the top of it, while the remaining a_1 and e levels are close to each other at around 0.5-0.7 eV above the valence band edge in the neutral charge state (Figure 3(a,c)).

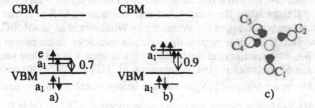

Figure 3: (a) and (b) are the calculated LDA+scissor one-electron levels of V_{Si}^0 and V_{Si}^- in eV; (c) schematic picture of the near neighbor C atoms of V_{Si} with their dangling bonds

In the neutral charge state the a_1 and e levels in the gap are occupied by two electrons. In this case the configurational interaction between these levels (and even with the first a_1 level in the valence band) may not be neglected. In cubic SiC it was shown that the configurational interaction (CI) can result in a spin-singlet ^1E ground state which can be described by two Slater-determinants, so the simple independent single-particle scheme cannot be used [16].

Taking only the defect levels in the gap the electronic configuration of V_{Si}^0 in 4H-SiC is $(a_1e)^{(2)}$. This configuration provides two triplet states and several singlet states. The $M_S=\pm1$ spin states of the ^3E and $^3A_2$ states can be described by a single Slater-determinant. In the case of ^3E

state the a_1 and e levels are occupied by a single electron with the same spin state while for $^3A_2$ state the e level is occupied by two electrons with the same spin state and the upper a_1 level is empty. Naturally, the single-particle LDA calculations resulted in S=1 spin state as the ground state of V_{Si}^0 in SiC [8,17], nevertheless, it was not reported which S=1 state was found in 4H-SiC. We found in a 576-atom supercell that ^3E state is more stable than $^3A_2$ state at both sites, but the energy difference between the ^3E and $^3A_2$ states is smaller at h-site than at k-site. We note that one of the singlet states as a possible ground state cannot be disregarded and a precise CI calculation is needed to investigate the true ground state of V_{Si}^0 in 4H-SiC. We note that V_{Si}^0 has never been unambiguously detected. Originally, T_{V2a} ODMR/EPR center was assigned to V_{Si}^0 with hyperfine interactions of four equivalent C atoms but recent pulsed-EPR and electron nuclear double resonance (ENDOR) studies [18,23] suggested the quartet spin state S=3/2 for T_{V2a} and the center is now believed to be V_{Si}^-. Also, PL centers in the near-infrared spectral region with two no-phonon lines (NPL) V_1 (1.438 eV) and V_2 (1.352 eV) in 4H-SiC and three NPLs V_1 (1.438 eV), V_2 (1.366 eV) and V_3 (1.398 eV) in the 6H polytype were observed [Sorman00]. Four and six different triplet states detected by optical detection of magnetic resonance (ODMR) by monitoring this PL band in 4H- and 6H-SiC, respectively were earlier attributed to V_{Si}^0 [21] but are now believed to be due to V_{Si}^-.

The high spin state S=3/2 has been theoretically predicted for V_{Si}^- [8,17, 1,22]. In hexagonal SiC the a_1 and e states get occupied by three electrons by parallel spins yielding $^4A_2$ state in C_{3v} symmetry. In this case the gap levels are shifted up by about 0.2 eV with respect to those of the neutral defect (Figure 3(b)). The quartet state was measured by EPR with four equivalent carbon dangling bonds possessing C_{3v} symmetry by Wimbauer et al. in 4H-SiC [22]. This center is isotropic and has a zero or negligible fine structure parameter. In the case of the spin quartet state of the T_{V2a} center, the crystal field splitting is non-zero. It is not clear yet, how the EPR/ENDOR centers of Mizuochi et al. [19], Wimbauer et al. [22] and Tv2a [20] are related to each other, since all of them are assigned to V_{Si}^-. Our calculations revealed that spin density of V_{Si}^- at different sites in 4H-SiC follow closely the center of Mizuochi et al. [23] (Table II).

Table II. Principal values of the hf-tensors of the negatively charged silicon vacancy V_{Si}^- defects in 4H-SiC as obtained from experiments and predicted by DFT-LSDA theory. See Figure 2 for atom labels.

V_{Si}^- 4H(k)			V_{Si}^- 4H(h)		
$A_\perp$, $A_\parallel$ [MHz]			$A_\perp$, $A_\parallel$ [MHz]		
36	36	82	38	38	84
32	32	76	37	37	76
V_{Si}^- (I) 4H Ref. [23]			V_{Si}^- (II) 4H Ref. [23]		
$A_\perp$, $A_\parallel$ [MHz]			$A_\perp$, $A_\parallel$ [MHz]		
33.2	33.2	80.2	33.8	33.8	80.1
28.3	28.2	76.3	31.4	31.2	79.4

We discuss now the excitation of V_{Si}^-. Since the ground state of V_{Si}^- is $^4A_2$, the only allowed transitions are promoting an electron from the lower *resonant* a_1 level to either the e level (the excited state will be ^4E) or the upper a_1 level (the excited state will be $^4A_2$). Both the initial and final states of the transitions should be ODMR active. We note that the selection rules for these transitions will be different and should be detected in polarization studies. Taking into account also the inequivalent sites it is not surprising to find several different ODMR signals associated with this PL band. The similar g-values obtained for the different ODMR signals could be the reason why no Zeeman splitting of any line in magnetic field (up to 5T) has been observed, since both initial and final state split in a similar way. Our calculated excitation energies for the resonant transitions (1.4 and 1.5 eV at h and k sites, respectively) are relatively close to the measured ones, however, this is very preliminary result because it is based on simple LDA calculations.

Our calculations predict that V_{Si} can be negatively ionized even to (4-) state in 4H SiC (where the a_1 and e levels are fully occupied). The paramagnetic (3-) and (+) states, though, have not yet been identified by EPR. By negatively ionizing the defect its formation energy decreases rapidly in n-type hexagonal SiC. From EPR and PL, it is known that the T_{V2a} center is the dominating defect in irradiated n-type SiC. It is therefore expected that in n-type material irradiated with low doses of electrons (10^{14}-10^{15} cm^{-2}) when the Fermi level still lies at the N donor level, the levels corresponding to high charge states of V_{Si} in the upper half of the bandgap should be filled. These levels work as electron traps and should be detectable by DLTS. Among the reported DLTS levels in irradiated n-type 4H-SiC, the EH2 level [23] at E_C-0.68 eV has an annealing behavior similar to that of V_{Si} [20], i.e. being annealed out at ~750°C, and is a possible candidate for the (3-/4-) level of V_{Si}.

Carbon antisite-vacancy pair in silicon carbide

In a compound semiconductor, the antisite-vacancy complex (AV) is a fundamental defect [24]. Theory predicted that the carbon AV pair ($C_{Si}V_C$), the counterpart of the silicon vacancy, is stable in p-type SiC and metastable in n-type SiC [10,25].

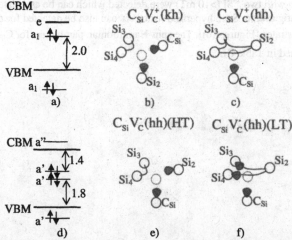

Figure 4: (a) and (d) are the calculated LDA+scissor one-electron levels of $C_{Si}V_C^0$ and $C_{Si}V_C^{2-}$ in eV (published first here); (b) and (c) show the spin distribution of $C_{Si}V_C^+$ in off-axis and on-axis configurations; (e) and (f) show the spin distribution of $C_{Si}V_C^-$ in high temperature (HT) and low temperature (LT) configurations

In $C_{Si}V_C$ defect three silicon dangling bonds come from the V_C part while one carbon dangling bond comes from C_{Si}. The unrelaxed defect possesses C_{3v} symmetry if C_{Si} is along c-axis (on-axis configurations, $C_{Si}(k)V_C(k)=C_{Si}V_C(kk)$ and $C_{Si}V_C(hh)$), while it has C_{1h} symmetry if C_{Si} is out-of c-axis (off-axis configurations, $C_{Si}V_C(kh)$ and $C_{Si}V_C(hk)$) in 4H SiC.

In the neutral charge state of the on-axis configurations the first a_1 level falls into the valence band, while the second a_1 level occupied by two electrons and the empty e level are in the band gap. The second a_1 state is mostly localized on C_{Si} while the e level is localized on the silicon dangling bonds (Figure 4(a)). In the case of off-axis configurations the e state splits to a' and a''. The former can be mixed with the lower a' state in the gap.

The (+) charge state is paramagnetic and we have recently identified by EPR in 4H SiC [26]. As expected the spin density is mostly localized on C_{Si} (Table III), and the carbon dangling bond is slightly mixed with the silicon dangling bonds in off-axis configurations (Figure 4(b,c)).

The *SI*-5 EPR center, which often is detected in HPSI 4H- and 6H-SiC substrates, has recently been identified in our joint research [24] as the carbon antisite-carbon vacancy pair in the negative charge state ($C_{Si} V_C^-$). The center can be detected after irradiation and its concentration reaches a maximum after annealing at 800-850°C. In electron irradiated n-type samples or in as-grown HPSI SiC substrates with the activation energies $E_a \sim 1.1$ eV or less, the $C_{Si} V_C$ center is in the double negative charge state and is not detected in dark. The $C_{Si} V_C^-$ signal appears after illuminating the samples with light of photon energies larger than ~1.1 eV. Once being illuminated, the signal keeps its intensity even in dark. In this case the spin density is localized in the V_C part of the defect. Since the e level is occupied by one electron in $C_{Si} V_C^-$ this system is Jahn-Teller unstable and it reconstructs to C_{1h} symmetry (Figure 3(d)). Two possible C_{1h} configurations exist with close energy at each site resulting in a low temperature and a high temperature configuration (Figure 3(e,f)). The high temperature configuration shows an effective C_{3v} symmetry due to dynamical reorientation effect resulting in a configuraitonal average for Si hyperfine signal. It was indeed shown that the symmetry of the *SI*-5 center lowers from C_{3v} to C_{1h} at temperatures below 60 K [24] and in the C_{1h} configuration, additional large hf splitting due to the interaction with two ^{29}Si (>10 mT) were detected which can be associated with the low temperature configuration. Such hyperfine structures can also be detected for the *SI*-5 center in HPSI 4H-SiC substrates [Figure 4(b)]. The spin-Hamiltionian parameters for $C_{Si} V_C^+$ and $C_{Si} V_C^-$ are summarized in Table III.

Table III. Top: HF-tensors of $C_{Si}V_C^-$ in comparison with those of SI5-center. The atoms are labeled according to Figure 3. Angular braces refer to a configurational average (cf. text). Bottom: HF-tensors of $V_C C_{Si}^+$ in comparison with those of HEI9,10 centers. The angle Θ between the main principal axis A_{zz} and the c-axis of the crystal amounts to $0°$ for the centers HEI9a/b and the axial complexes as expected from symmetry and to $108-109°$ for HEI10a/b in agreement with the finding for the non-axial centers. Three additional detectable ^{29}Si hyperfine interaction lines were found near C_{Si}, which is not shown here.

	V_C-C_{Si} (kk), LT			V_C-C_{Si} (hh), LT				SI5 30K, Ref. [24]		
	$A_\perp$, $A_\parallel$ [mT]			$A_\perp$, $A_\parallel$ [mT]				$A_\perp$, $A_\parallel$ [mT]		
Si$_2$	0.3	0.3	0.1	0.8	0.8	0.3	Si (2×)	10.16	10.04	12.9
Si$_{3,4}$	10.8	10.9	13.8	11.2	11.3	14.2	C (2×)	1.3	1.3	1.8
C$_{Si}$	0.5	0.5	0.5	0.4	0.4	0.4				
C$_{II}$	1.3	1.3	1.7	1.4	1.4	1.7				
	V_C-C_{Si} (kk), HT			V_C-C_{Si} (hh), HT				SI5 100K, Ref. [24]		
	$A_\perp$, $A_\parallel$ [mT]			$A_\perp$, $A_\parallel$ [mT]				$A_\perp$, $A_\parallel$ [mT]		
Si$_2$	15.9	15.9	20.3	15.8	15.8	20.1	Si (1×)	6.38	6.38	8.02
Si$_{3,4}$	2.7	2.7	3.6	3.2	3.3	4.3	C (1×)	1.77	1.77	2.22
$\langle$Si$_{2-4}\rangle$	7.2	7.6	8.6	7.5	7.9	8.9				
C$_{Si}$	0.1	0.1	0.1	0.2	0.2	0.2				
C$_I$	0.9	0.9	1.3	1.3	1.3	1.7				

	HEI9a, Ref. [26]			V_C-C_{Si} (hh)			HEI9b, Ref. [26]			V_C-C_{Si} (kk)		
	A_{xx},A_{yy},A_{zz} [mT]			A_{xx},A_{yy},A_{zz} [mT]			A_{xx},A_{yy},A_{zz} [mT]			A_{xx},A_{yy},A_{zz} [mT]		
C$_{Si}$	2.27	2.27	8.25	1.96	1.98	7.01	3.71	3.71	9.95	3.12	3.12	8.59

	HEI10a, Ref. [26]			V_C-C_{Si} (hk)			HEI10b, Ref. [26]			V_C-C_{Si} (kh)		
	A_{xx},A_{yy},A_{zz} [mT]			A_{xx},A_{yy},A_{zz} [mT]			A_{xx},A_{yy},A_{zz} [mT]			A_{xx},A_{yy},A_{zz} [mT]		
C$_{Si}$	2.65	2.65	8.75	1.69	1.70	6.00	2.45	2.31	8.43	1.65	1.65	6.14
Six2				2.11	2.13	2.71				2.52	2.53	3.44

Comparing to the reported DLTS data, the (1–/2–) level of $C_{Si}V_C$ is close to the EH5 level at ~E_C-1.13 eV observed in irradiated n-type 4H-SiC after annealing at ~750°C [23]. This is also in agreement with the calculations using large supercell and reliable gap correction which predicted the (1–/2–) level of $C_{Si}V_C$ to be at ~E_C-1.1 eV and ~E_C-0.9 eV for the defect at the cubic and hexagonal lattice site, respectively [24]. Annealing studies of irradiated n-type materials suggest that the center plays an important role in compensating the N donors [26].

Divacancy in silicon carbide

The closest pair of V_{Si} and V_C, the divacancy, can form during the migration of the isolated vacancies [27]. As it has been explained for $C_{Si}V_C$ four possible configurations exist for the divacancy due to the inequivalent sites in 4H SiC. We analyze the on-axis configurations in detail: it has C_{3v} symmetry with three Si dangling bonds (V_C part) and three C dangling bonds (V_{Si} part). These six dangling bonds form two a_1 and two e levels. According to the calculations the two a_1 levels are resonant with the valence band while the e levels appear in the gap (Figure 5(a,b)). The lower e level is localized on V_{Si} while the upper e level is localized on V_C. In the ground state of $V_{Si}V_C^0$ the a_1 levels are fully occupied while the lower e level is occupied by two electrons. In principle, the C_{3v} symmetry can be conserved by $^3A_2$, ^1E and ^1A multiplets by assuming $e^{(2)}$ electron configuration. The latter two S=0 states cannot be described with independent single-particle scheme, while the $^3A_2$ state can be described by putting two electrons with parallel spins on the e level. Alternatively, one can put two electrons with anti-parallel spins on the e level which is Jahn-Teller unstable and reconstructs to C_{1h} symmetry. The latter two situations can be described by independent single particle schemes (like usual LDA calculation). The LDA calculations found [28] that the high spin state (S=1) is favorable and should be the ground state. We have recently identified the P6/P7 EPR centers (see Table IV) as the neutral divacancy, $V_{Si}V_C^0$ [29].

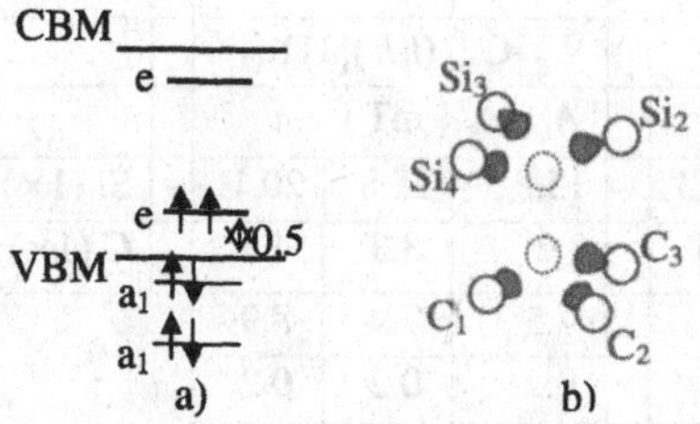

Figure 5: (a) The calculated LDA+scissor one-electron levels of $V_C V_{Si}^0$ in eV. (b) Schematic picture of the near neighbor atoms of the divacancy and their dangling bonds

Table IV. Principal values of the hf-tensors of the axial $V_C V_{Si}^0$ complexes in comparison with those of P6/P7-centers in MHz. C_I/Si_I denote the first neighor C/Si atoms near divacancy (labels of C_{1-3} and Si_{1-3} in Figure 5.) while Si_{II} atoms are the first neigbor Si atoms of C_I.

	P6b, Ref. [29]	V_C-V_{Si} (h)			P6'b, Ref. [29]	V_C (k)-V_{Si} (k)		
	$A_\perp$, $A_\parallel$ [MHz]	$A_\perp$, $A_\parallel$ [MHz]			$A_\perp$, $A_\parallel$ [MHz]	$A_\perp$, $A_\parallel$ [MHz]		
$C_I (3\times)$	53	50	110	55	56	116	47	45	104	49	49	110
$Si_{IIa} (3\times)$	12	12	12	9	9	9	13	13	13	10	10	9
$Si_{IIb} (6\times)$	9	9	9	9	9	8	10	10	10	10	10	9
$Si_I (3\times)$	3	3	3	3	4	5	3	3	3	1	1	2
	P7b, Ref. [29]	V_C (h)-V_{Si} (k)			P7'b, Ref. [29]	V_C (k)-V_{Si} (h)		
	$A_\perp$, $A_\parallel$ [MHz]	$A_\perp$, $A_\parallel$ [MHz]			$A_\perp$, $A_\parallel$ [MHz]	$A_\perp$, $A_\parallel$ [MHz]		
$C_{Ia} (1\times)$	not determined	51	52	118	52	52	110	52	52	116
$C_{Ib} (2\times)$	not determined	50	50	109	48	45	109	43	47	103

Since P6/P7 EPR centers were also detected in dark at very low temperature (1.2-8 K), the $^3A_2$ is indeed the ground state of $V_{Si} V_C^0$ [29]. The excitation scheme of $V_{Si} V_C^0$ is very complicated due to the possible configuration interaction between the e states. The *resonant* a_1 levels are not far from the lower e level. Taking this electron configuration, $a_1^{(2)} e^{(2)}$, the possible multiplets are $^3A_2$ and ^3E in the triplet state. Resonant transition from one of the a_1 level is possible forming the ^3E excited state ($^3A_2 \rightarrow {}^3E$), however, the defect level within the valence band maximum (VBM) should be filled by electrons which may hinder reemission. Further excitation is possible from the lower e level to the upper e level ($e^{(1)} e^{(1)}$ configuration of the excited state: $^3A_2 \rightarrow {}^3E$). Only careful CI calculation can predict this excitation energy. Magnetic circular dichroism of the absorption (MCDA) and MCDA-detected EPR studies in irradiated n-type 6H-SiC [30] confirmed that P6/P7 triplet centers are related to the PL band with the NPLs at 0.9887, 1.0119, 1.0300, 1.0487 and 1.0746 eV. In 4H-SiC with strong EPR signal of $V_C V_{Si}^0$, several NPLs with similar energies at 0.9975, 1.0136, 1.0507 and 1.0539 eV were also detected in the absorption spectra [31]. These strong absorption lines can also be seen in PL but are often very weak. In 4H-SiC, the relation between these NPLs and the EPR P6/P7 centers has not been confirmed. The P6/P7 centers can be detected by both ODMR [32] and EPR in 6H-SiC. In 4H-SiC, the centers have not been detected by ODMR and the reason may be due to their too weak PL.

The divacancy, like V_{Si}, can be ionized at least up to (3-) in 4H-SiC. In (3-) state the upper e defect level is occupied by an electron which is mostly localized on the carbon vacancy. The atoms reconstruct in this case reducing the symmetry to C_{1h} even for on-axis configurations. While this defect can be singly positively ionized as well, it is much more probable that it can capture a hole(h^+) possibly forming a pseudo-acceptor state, like ($V_C V_{Si}^- + h^+$) which might also result in transitions in the PL band. Our theoretical calculations predicted that the neutral divacancy can recombine with an injected hole with an excitation energy of ~1.0 eV which is close the measured transition energies [28].

CONCLUSIONS

We showed that combined theoretical and experimental studies could successfully identify vacancy defects in SiC. The hyperfine data can be relatively accurately calculated by traditional DFT methods. However, the inaccuracy in the DFT-LDA calculation of charge transition levels is too large for ambiguous identification of PL centers based on zero-phonon lines or of deep level transient spectroscopy centers. Still, our preliminary results revealed that transition from the resonant defect level to the defect level in the gap can occur for silicon vacancy and divacancy explaining the complex nature of their PL spectra. For carbon vacancy GW calculation showed that the absorption in the photo-EPR measurement is due to the transition between the defect bands. This highlights that the absorption or emission spectra should be interpreted by great care taking the defect band – defect band, or defect band – resonant band transitions into account. The calculations helped to identify the defects that are responsible the semi-insulating property of the given semi-insulating SiC samples.

ACKNOWLEDGMENTS

We like to thank Prof. Dr. Isoja and Dr. J. Steeds for valuable discussions. The identification of the vacancy complexes was a collaboration with the group of Prof. Dr. Isoja. Financial support by the Deutsche Forschungsgemeinschaft (SiC-researchgroup and grant BO1852/2-1), from the Swedish Foundation for Strategic Support program SiCMAT, the Swedish national Infrastructure for computing grant No. SNIC 011/04-8, the Swedish Foundation for International Cooperation in Research and Higher Education and the Hungarian OTKA Grant No. K-67886.

REFERENCES

1. B. Aradi et al., Phys. Rev. B 63, 245202 (2001).
2. A. Gali et al., Phys. Rev. B 68, 125201 (2003).
3. M. Bockstedte, M. Heid, and O. Pankratov, Phys. Rev. B 67, 193102 (2003).
4. T Umeda et al., Phys. Rev. B 70, 235212 (2004).
5. A. Gali et al., Appl. Phys. Lett. 86, 102108 (2005).
6. M. Bockstedte et al., Materials Science Forum, in press
7. C. W. M. Castleton, A. Höglund, and S. Mirbt, Phys. Rev. B 73, 035215 (2006).
8. A. Zywietz, J. Furthmüller, and F. Bechstedt, Phys. Rev. B 59, 15166 (1999).
9. L. Torpo and M. Marlo, J. Phys.: Condensed Matter 13, 6203 (2001).
10. M. Bockstedte, A. Mattausch, and O. Pankratov, Phys. Rev. B 68, 205201 (2003).
11. V. Ya. Bratus et al., Phys. Rev. B 71, 125202 (2005).
12. M. E. Zvanut and V. V. Konovalov, Appl. Phys. Lett. 80, 410 (2002).
13. N. T. Son et al., Mater. Sci. Forum 433-436 45 (2003); 457-460 437 (2004).
14. N. T. Son, B. Magnusson, and E. Janzén, Appl. Phys. Lett. 81, 3195 (2002).
15. T. Umeda et al., Phys. Rev. B 71, 193202 (2005).
16. P. Deák et al., Appl. Phys. Lett. 75, 2013 (1999).
17. L. Torpo, R. M. Nieminen, K. E. Laasonen, and S. Pöykkö, Appl. Phys. Lett. 74, 221 (1999)
18. N. Mizuochi et al., Phys. Rev. B 66, 235302 (2005).
19. N. Mizuochi et al., Phys. Rev. B 72, 235208 (2005).
20. E. Sörmann et al., Phys. Rev. B 61, 2613 (2000).
21. Mt. Wagner et al., Phys. Rev. B 66, 155214 (2002).
22. T. Wimbauer et al., Phys. Rev. B 56, 7384 (1997).
23. C. Hemmingsson et al., J. Appl. Phys. 81, 6155 (1997).
24. T. Umeda et al., Phys. Rev. Lett. 96, 145501 (2006).
25. E. Rauls, Th. Frauenheim, A. Gali, and P. Deák, Phys. Rev. B 68, 155208 (2003).
26. T. Umeda et al., Phys. Rev. B 75, 245202 (2007).
27. L. Torpo, T. E. M. Staab, and R. M. Nieminen, Phys. Rev. B 65, 085202 (2002).
28. A. Gali et al., Mater. Sci. Forum 527-529, 523-526 (2006).
29. N. T. Son et al., Phys. Rev. Lett. 96, 055501 (2006).
30. Th. Lingner et al., Phys. Rev. B 64, 245212 (2001).
31. B. Magnusson and E. Janzén, Mater. Sci. Forum 483-485, 341 (2005).
32. N. T. Son et al., Semicond. Sci. Technol. 14, 1141 (1999).

Mater. Res. Soc. Symp. Proc. Vol. 1069 © 2008 Materials Research Society

Determination of the Core-structure of Shockley Partial Dislocations in 4H-SiC

Yi Chen[1], Ning Zhang[1], Xianrong Huang[2], Joshua D Caldwell[3], Kendrick X Liu[3], Robert E Stahlbush[3], and Michael Dudley[1]

[1]Department of Materials Science and Engineering, Stony Brook University, Stony Brook, NY, 11794-2275

[2]National Synchrotron Light Source II, Brookhaven National Laboratory, Upton, NY, 11973-5000

[3]Naval Research Laboratory, Washington, DC, 20375-5320

ABSTRACT

Synchrotron x-ray topographs taken using basal plane reflections indicate that the electron-hole recombination activated Shockley partial dislocations in 4H silicon carbide bipolar devices appear as either white stripes with dark contrast bands at both edges or dark lines. *In situ* electroluminescence observations indicated that the mobile partial dislocations correspond to the white stripes in synchrotron x-ray topographs, while immobile partial dislocations correspond to the dark lines. Computer simulation based on ray-tracing principle indicates that the contrast variation of the partial dislocations in x-ray topography is determined by the position of the extra atomic half planes associated with the partial dislocations lying along their Peierls valley directions. The chemical structure of the Shockley partial dislocations can be subsequently determined unambiguously and non-destructively.

INTRODUCTION

Silicon carbide (SiC) has been more and more widely used in devices for high voltage and high temperature applications. However, the forward voltage drop under forward biasing in bipolar devices, due to the expansion of basal stacking faults (SFs) from the advancement of Shockley partial dislocations, degrades the SiC-based bipolar devices dramatically and eventually kills them. The advancement of Shockley partial dislocations is activated by the electron-hole recombination enhanced dislocation glide (REDG) process [1-4] and the driving force are still under discussion [5, 6, 7]. The properties of partial dislocations in semiconductors, not only their mobility but especially their possible energy states in the band gap, must depend on their structure configurations. For example, first-principles calculation indicates that the reconstructions along the dislocation lines can be electrically active by giving rise to energy states in the forbidden band gap [8]. Thus, it is of great importance to understand the chemical structures of partial dislocations so as to study their influence on energy states in the band gap. S. Ha and X. J. Ning have employed transmission electron microscopy to determine the core structure of partial dislocations in SiC [9,10]. The rigid sample preparation process in transmission electron microscopy makes it difficult and destructive to distinguish the chemical structure of the Shockley partial dislocations. We are introducing a simple, unambiguous and non-destructive way to determine the chemical structure of the Shockley partial dislocations.

EXPERIMENT

The 4H-SiC substrates used in this study were commercially available wafers grown by the physical vapor transport (PVT) technique with an 8° off-cut angle in the [11-20] direction. The epitaxial layer structure consists of a 16 μm n$^+$ buffer layer, 2×10^{18} cm^{-3}, and a 115 μm n$^-$ layer, 2×10^{14} cm^{-3}, capped by a 2 μm p$^+$ layer, 8×10^{18} cm^{-3}. Contact to the p$^+$ layer is formed by depositing aluminum and patterning a grid structure. This method has been shown to be just as effective as using fully processed PiN diodes for the purpose of electroluminescence (EL) imaging dislocations [11].Back-reflection topographs were recorded following the forward bias stressing. The specimen was then thinned down from the substrate side to ~140 μm for transmission topograph. The imaging was carried out atBeamline X-19C, NSLS and Beamline XOR-33BM, APS, using Ilford L4 nucleate plates. The ray-tracing method was used to simulate the topographic images of partial dislocations in back-reflection x-ray topography. Successful modeling has been done for the MPs/threading screw dislocations (TSDs)/basal plane dislocations (BPDs) in SiC using ray-tracing [12-14] and detailed simulation process can be referred to Ref.13.

DISCUSSION

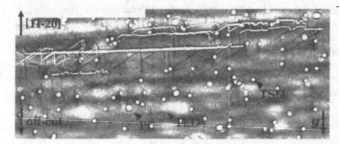

Figure 1. A (0008) back-reflection topograph showing complex webs of partial dislocations. The partial dislocations appear as either dark lines or white stripes with diffuse dark contrast bands at both edges. PC: phase contrast.

Figure 1 shows a synchrotron x-ray topograph using (0008) basal plane reflection. The circular white dots surrounded by narrow dark rings correspond to the closed-core TSDs and the small black or white dots in the grey background are threading edge dislocations (TEDs). Those white dots surrounded by diffuse dark circles are the phase contrast due to the non-uniform thickness of the Be x-ray window. The most remarkable feature visible in the topograph is the complex web of dislocations, comprising white stripes with diffuse dark contrast bands at both edges and dark lines. It has been well known that in SiC bipolar devices, the Shockley partial dislocations activated by REDG process advance under forward biasing [15]. The Shockley partial dislocations are aligned along their Periels valley directions, which are at 30° or 90° to their Burgers vectors. These dislocations other than TEDs/TSDs observed in Figure 1 are Shockley partial dislocations bounding Shockley type SFs, which are out of contrast since $g \cdot R$ is always equal to an integer in (0008) reflection (g is the reflection vector and R is the fault vector). The wider white stripes are due to the superimposition of strain fields of more than one Shockley partial dislocations. The mechanism of the contrast variation of different partial dislocation segments still remains obscure.

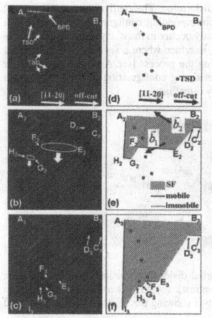

Figure 2. (a) Original dislocation configuration before forward biasing where a long BPD A_1B_1 can be observed in the epilayer. (b) One Shockley partial dislocation dissociated from the BPD advances under forward biasing and is split by a TSD into two segments E_2F_2 and G_2H_2. (c) Further expansion of SF. (d)-(f) Schematic diagrams showing the configurations in (a)-(c), respectively.

In order to understand the difference between different Shockley partial dislocations, electroluminescence has been used *in situ* to monitor the expansion process of the Shockley SF. Sequential EL images were excerpted from a large number of images, see Figure 2(a) - Figure 2(c). Figure 2(d) - Figure 2(f) are the schematics corresponding to the configurations of SF in Figure 2(a) - Figure 2(c), respectively. Figure 2(a) is the initial configuration of the dislocation structure in the epilayers before forward biasing, where a long BPD, A_1B_1, and some threading dislocations (bright dots) are visible. Some of the bright dots correspond to TSDs (marked in Figure 2(d) based on x-ray topography). This BPD A_1B_1 propagated from the substrate during the chemical vapor deposition (CVD) growth and it is screw-oriented by further Burgers vector analysis via transmission x-ray topography. The detailed Burgers vector analysis is not discussed here. This is consistent with previous report that screw-oriented BPDs are tending to propagate into the epilayer without being converted into TEDs [16,17]. During the electrical stressing, the leading partial dissociated from the original perfect BPD is a $30°$ dislocation [Burgers vector b_1 is marked in Figure 2(e)] and it advances toward the bottom side of the view. As the advancing partial dislocation encounters a TSD, the partial dislocation is split into two segments by the pinning TSD and they continue to advance albeit at different velocity. The formation of the complex web of dislocations inside the faulted area, which is resulted from the interactions between the advancing partial and TSDs, is not discussed here, though detailed interaction process can be referred to Ref. 18. The splitting of partial dislocation at the TSD can be seen in Figure 2(b) and the two mobile segments E_2F_2 and G_2H_2 continue to advance (toward the bottom, as marked by arrows). Another partial dislocation segment D_2E_2, lying along its Peierls valley direction, is expected to be deposited as they advance (invisible in EL since it emits at different spectrum) in order to maintain the continuity of the dislocation. The configuration in Figure 2(c) is observed after further expansion of SF and the corresponding configuration is illustrated in Figure 2(f). Since the electron–hole recombination only occurs in the n^- layer, the motion of the mobile partial dislocation will terminate at the p^+/n^- interface where a segment of "$60°$" partial dislocation is deposited near the p^+/n^- interface during the process [see A_2H_2 in Figure 2(b) or A_3I_3 in Figure 2(c)]. They are expected to appear as zigzag configuration in a fine scale, with each dislocation segment lying along its Peierls valley direction.

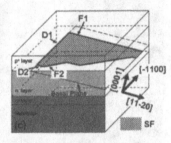

Figure 3. (a) Back-reflection topograph showing partial dislocations bounding a SF within the penetration depth of the x-ray (SF area is out of contrast). (b) (-1011) transmission topograph showing the SF area. (c) Three-dimensional schematics showing the structure of the device and configuration of the SF.

The sample was then low-temperature annealed (~500°C) for 72 hours and electrically stressed again. Back-reflection x-ray topography was subsequently recorded. Observations from EL indicate that low temperature annealing is just a reverse process of the expansion of Shockley SF [19]. Therefore, the topograph recorded after second electrical stressing process can be considered as an image taken *in situ* during the first electrical stressing. Figure 3(a) shows the topograph recorded from the same device cell as in Figure 2, using (0008) reflection. Figure 3(b) is the (-1011) transmission topograph of the same area after thinning down the device cell from the substrate side and a 3-D schematics is shown in Figure 3(c), based on topography and EL. Four partial dislocations bounding the SF are visible in the Figure 2(a) (marked by "D1", "D2", "F1" and "F2") and the SF area itself is out of contrast since $g·R$ is equal to zero. The circular white dots in the topograph correspond to elementary TSDs with Burgers vector $1c$. The complete configuration of the SF can be seen in Figure 3(b), in which the dark contrast area corresponds to the SF (some small SFs are also visible). The configuration of the SF is achieved via a process discussed in Figure 2. The dislocation segment "U" is unrelated to the SF discussed here. The configuration on the topograph is in between Figure 2(b) and Figure 2(c) and the configuration in Figure 2(c) is expected after further advancement of mobile partial segments ["D1" corresponds to A_3I_3 in Figure 2(c)]. Since the partial dislocation "D1" has not completely reached the p^+/n^- interface, it appears as zigzag configuration at certain positions on the topograph. The straight segment on "D1" is also expected to appear as zigzag at a very fine scale, with each segment lying along its Peierls valley direction. Due to the limited penetration depth of x-ray, only part of bounding partial dislocations is visible on the topograph Figure 3(a) and the invisible segments can be visualized from the SF configuration observed in Figure 3(b).

The small square shaped features observed throughout Figure 3(a) are associated with the device pattern. By comparing Figure 2 and Figure 3, we can notice that the mobile segments appear as white stripes while immobile segments appear as dark lines.

Ray-tracing method has been used to simulate the images of partial dislocations in back-reflection geometry. The Shockley partial dislocations are composed of 30° or 90° segments lying along their Peierls valley directions. The 90° segment is moving much faster [20] and it vanishes, leaving only 30° segments remaining. In our case, segments "D2", "F1" and "F2" are 30° dislocations, while "D1" is approximately a 60° segment (since it has terminated its motion upon arrival at the n^-/p^+ interface where the electron-hole recombination goes to zero). The screw component does not contribute to the topographic contrast since both $g \cdot b_s$ and $g \cdot b_s \times l$ are equal to zero (here g is the reflection vector, b_s is the Burgers vector of screw component and l is the line direction of the dislocation). Thus the contrast is exclusively arisen from the edge component of the partial dislocation. Figure 4 shows the simulated images of 60° partial dislocations with extra atomic half planes extending toward the C-face [substrate side, see (a)] and Si-face [device side, see (b)]. The simulated image in Figure 4(a) appears as a narrow dark line while the one in Figure 4(b) shows as a white stripe (~7.5 μm wide) with narrow dark bands along both edges. These dark bands along the edges of the white stripe can be seen more clearly in the inset of Figure 3(a) which shows a magnified image of the boxed region. Detailed examination shows that the simulated images show very good agreement with the recorded images, in both contrast and dimension. Therefore, the simulation indicates that the topographic contrast variation is determined by the position of the extra atomic half planes.

 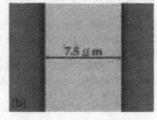

Figure 4. Ray-tracing simulated images of 60° Shockley partial dislocations with extra atomic half planes extending toward C-face (a) and Si-face (b), respectively.

It has been established that the Shockley partials in SiC are glide-set dislocations and if the extra half plane is extending toward the C-face, the corresponding Shockley partial is a C-core dislocation, otherwise a Si-core dislocation. Since the device structure is fabricated on the Si-face of a 4H-SiC substrate in our case, the partial dislocations which appears as dark lines corresponds to C-core partials [Figure 4(a)], while the partial dislocations showing as white stripes are Si-core [Figure 4(b)]. This is consistent with previous report that the mobile partials are Si-core and immobile ones are C-core [9].

CONCLUSIONS

The Shockley partial dislocations appear as either white stripes with diffuse dark contrast bands at both edges or dark lines in x-ray topography using basal plane reflection. EL observation indicated that white stripes correspond to mobile Shockley partials while dark lines correspond to immobile Shockley partials. Ray-tracing simulation indicates that such contrast variation can be subsequently used to determine the chemical structure of the Shockley partial dislocations unambiguously and non-destructively.

ACKNOWLEDGMENTS

This work is supported in part by ONR grants N0001140010348, N000140110302 and N000140211014 (contract monitor Dr. Colin Wood) and by Dow Corning Corporation under contract numbers N0001405C0324 and DAAD1701C0081. Topography experiments were carried out at the Stony Brook Synchrotron Topography Facility, Beamline X-19C, at the NSLS (Contract no. DE-AC02-76CH00016) and Beamline XOR-33BM, Advance Photon Source, Argonne National Laboratory. The XOR/UNI-CAT facility is supported by the U.S. Department of Energy under Award No. DEFG02-91ER45439. The APS is supported by the U.S. Department of Energy, Office of Science, Office of Basic Energy Sciences, under Contract No. DE-AC02-06CH11357.

REFERENCES

[1] H. Lendenmann, F. Dahlquist, N. Johansson, R. Soderholm, P. A. Nilsson, J. P. Bergman, and P. Skytt, Mater. Sci. Forum 353-356, 727 (2001)

[2] A. Galeckas, J. Linnros, and P. Pirouz, Appl. Phys. Lett. 81, 883 (2002)

[3] J. D. Weeks, J. C. Tully, and L. C. Kimerling, Phys. Rev. B 12, 3286 (1975).

[4] H. Sumi, Phys. Rev. B 29, 4616 (1984).

[5] S. Ha, M. Skowronski, J. J. Sumakeris, M. J. Paisley, and M. K. Das, Phys. Rev. Lett. 92, 175504 (2004)

[6] W. R. L. Lambrecht and M. S. Miao, Phys. Rev. B 73, 155312 (2006)

[7] M. E. Twigg, R. E. Stahlbush, M. Fatemi, S. D. Arthur, J. B. Fedison, J. B. Tucker and S Wang, Appl. Phys. Lett. 82, 2410 (2003)

[8] G. Savini, M. I. Heggie, and S. Oberg, Farad. Disc. 134, 353 (2007)

[9] S. Ha, M. Benamara, M. Skowronski, and H. Lendenmann, Appl. Phys. Lett. 83, 4957 (2003)

[10] X. J. Ning and P. Pirouz, J. Mater. Res. 11, 884 (1996)

[11] K. X. Liu, R. E. Stahlbush, K. B. Hobart, and J. J. Sumakeris, Mater. Sci. Forum 527-529, 387 (2006)

[12] W. M. Vetter, H. Tsuchida, I. Kamata, and M. Dudley, J. Appl. Cryst. 38, 442 (2005)

[13] Y. Chen, M. Dudley, E. K. Sanchez and M. F. MacMillan, Appl. Phys. Lett. 91, 071917 (2007)

[14] Y. Chen and M. Dudley, Appl. Phys. Lett. 91, 141918 (2007)

[15] R. E. Stahlbush, J. B. Fedison, S. D. Arthur, L. B. Rowland, J. W. Kretchmer and S. Wang, Mater. Sci. Forum 389-393, 427 (2002)

[16] T. Ohno, H. Yamaguchi, S. Kuroda, K. Kojimaa, T. Suzuki and K. Arai, J. Cryst. Growth 260, 209 (2004)

[17] H. Jacobson, J. Birch, R. Yakimova, M. Syvajarvi, J. P. Bergman, A. Ellison, T. Tuomi, and E. Janzen, J. Appl. Phys. 91, 6354 (2002)

[18] Y. Chen, M. Dudley, K. X. Liu, and R. E. Stahlbush, Appl. Phys. Lett. 90, 171930 (2007)

[19] J. D. Caldwell, R. E. Stahlbush, K. D. Hobart, O. J. Glembocki, and K. X. Liu, Appl. Phys. Lett. 90, 143519 (2007)

[20] P. Pirouz, M. Zhang, A. Galeckas, and J. Linnros, Mater. Sci. Forum 815, J6.1.1 (2004)

Mater. Res. Soc. Symp. Proc. Vol. 1069 © 2008 Materials Research Society   1069-D07-01

Propagation and Density Reduction of Threading Dislocations in SiC Crystals during Sublimation Growth

Ping Wu, Xueping Xu, Varatharajan Rengarajan, and Ilya Zwieback
Wide Bandgap Materials Group, II-VI Incorporated, 20 Chapin Rd., Suite 1005, Pine Brook, NJ, 07058

ABSTRACT

SiC single crystal wafers grown by sublimation exhibit relatively high dislocation densities. While it is generally known that the overall dislocation density tends to decrease throughout crystal growth, there has been a limited quantitative analysis of such trend. In this study, we measured the density of threading dislocations in the wafers sliced from several SiC boules. Although the dislocation density in the wafers sliced from different boules could differ by orders of magnitude, a consistent empirical relationship was found between the dislocation density (ρ) and the axial wafer position within the crystal (w): $\rho \propto w^{-0.5}$.

Monte Carlo simulations were performed based on two assumptions: (i) during growth the threading dislocations move randomly in the lateral directions, and (ii) two dislocations of opposite sign annihilate when they come within a critical distance between them. Good agreement was achieved between the model and experimental results. The critical distance determined from the simulations was in the range between a few hundred Å and a micron.

INTRODUCTION

Silicon carbide crystals grown by sublimation are used for the development and manufacturing of SiC and GaN semiconductor devices of new generation. In order to produce low-defect epilayers and high-quality devices, especially for SiC homoepitaxy, the substrate must have good crystal quality, including, low dislocation density.

Previously it has been reported [1,2] that threading screw dislocations as well as overall dislocation density decreases during growth. Generally, this was attributed to the generation of dislocations, especially screw dislocations, at the growth interface, followed by a gradual reduction. While the threading dislocations propagate towards the dome of the SiC crystal, those having opposite Burgers vectors can interact with each other and annihilate. However, there are limited data to quantitatively characterize this trend.

In this study, we studied changes in the threading dislocation densities through the growth of several 6H SiC crystals. Monte-Carlo simulations based on a simple annihilation model have been carried out to provide additional information on this trend.

EXPERIMENT

6H SiC single crystals were grown by the PVT technique and sliced into wafers normal to the c-axis using a multi-wire diamond saw. All wafers were nominally 400 micron thick, with the spacing between two adjacent wafers corresponds to 650 micron in the direction of crystal growth. The wafers were numbered in the direction from seed to the boule dome, that is, wafer # 1 was the one adjacent to the seed. The 6H crystals used in this study were either undoped or doped with vanadium to ~1×10^{17} cm^{-3}.

Characterization of the dislocation density was performed using etching in molten KOH. Sample wafers were polished on the silicon face and etched at 450°C for 2 - 3 minutes. Etching resulted in the removal of about 2 microns, mostly, from the carbon face, and formation of the dislocation-related etch pits on the silicon face. Etched samples were inspected under a microscope equipped with a PC-controlled X-Y table and CCD camera. An ImagePro software package was used for image processing, counting etch pits and creating wafer-scale dislocation density maps. Distinction between various dislocation types (threading edge, threading screw and basal plane dislocations) could be made on the basis of the etch pit shape. For the samples used in this study, threading type dislocations dominated, and the values reported herein include both threading screw and threading edge dislocations.

The measurement of dislocation density provides critical feedback for the improvement of crystal growth process. The effects of process modifications can be evaluated by analyzing a comprehensive set of dislocation density data.

To help understand the observed empirical relationship in reduction of dislocation density through the growth of a crystal boule, Monte-Carlo calculations were performed to simulate annihilation of dislocations during growth. In these calculations, a certain initial number of dislocations were distributed randomly in a two-dimensional box. Half of the dislocations were assumed to be of positive sign while the other half – of negative. At every incremental calculation step, each dislocation was allowed to move in a random direction, from 0 to 2π in angle, for a distance that was randomly chosen between 0 and a maximum distance D_m. Any dislocation that would move outside the boundary of the box would be redirected back into the box as if rebounded elastically. At the end of the calculation step, any two dislocations of opposite signs that moved within a critical distance of R_a from each other would be annihilated.

For each set of initial parameters, the simulation included between 25,000 and 5,000,000 steps. In order to reduce the variability in the simulation results, calculation was repeated 30 times for each set of starting parameters. The dislocation density values obtained at each step were used to calculate the median values, which are reported in this paper.

RESULTS AND DISCUSSION

Measured dislocation density

Figure 1 shows the measured dislocation density values on eight wafers from a 6H SiC crystal. The wafer number serves as a measure of the position in the boule, i.e., the distance from the original seed. These data can be best described by a power-law relationship, where the exponent is about –0.5. Similar measurements were performed on several other crystals, with three wafers from each boule. The results are shown in Figure 2. While the small sample size in each set increases the uncertainty in the derived parameters values, one can see that each of them can be reasonably fit by the power function with the exponent close to –0.5. It is interesting to note that a power-law relationship between the dislocation density and distance in the growth direction was reported for GaN [3] with an exponent of about –1.25. The more negative value corresponds to a more rapid reduction in the dislocation density with the growth thickness.

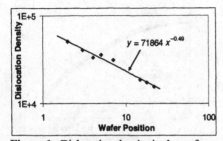

Figure 1. Dislocation density in the wafers sliced from different positions of a SiC crystal. The power-law relationship is shown by the linear trend line in the log-log scale.

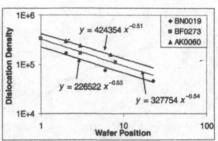

Figure 2. Dislocation density as a function of wafer position in crystal measured from three additional boules. The fitted trend lines are included.

Monte-Carlo simulations

We used Monte-Carlo simulations in order to gain additional insight on the process of dislocation density reduction. Mathis et al. [4] modeled dislocation annihilation reactions in GaN layers to explain the experimentally observed dislocation density as a function of epilayer thickness. They analyzed geometrically possible interactions between pairs of threading dislocations using an assumed annihilation radius. In our study, we also adopted the concept of annihilation radius, but employed a simpler model for numerical simulations.

As described in the Experiment section, the model requires only basic assumptions as to the behavior of dislocations. However, the choice of numeric parameters implies arbitrary units in all dimensions. When some of the typical experimental values are taken into consideration, it is possible to make some qualitative (order-of-magnitude) interpretation, as will be discussed later. The primary results of the simulation are the relationship between the calculated dislocation density and the number of steps, as shown in Figure 3. The simulation parameters included the dimensions of the square box (the length of the edge B), initial number of dislocations N_I, maximum movement of a dislocation in a step D_m, and the annihilation radius R_a. The values of these parameters are given in the captions to the figures. Since the selection of simulation parameters may seem arbitrary, we analyzed the effects of all of these parameters to be sure that the results are self-consistent.

First, the size of the steps was considered by setting D_m relative to R_a. Obviously, D_m should not be significantly larger than R_a, otherwise, the dislocation trajectories may cross without interaction. On the other hand, too small steps would slow down the progression in the simulation. Figure 4 compares results of four sets of simulations using different D_m settings. Logarithmic scales are used for both axes, and the number of steps was multiplied with a normalizing factor, $(D_m/R_a)^2$. One can see that the normalized results from all four sets agree well with each other. The need for a squared term in the normalizing factor can be easily understood. In the calculation, the movement of a dislocation at each step is randomized both in terms of distance and direction. The chance of annihilation with another dislocation is related to the area where the dislocation can be found compared to the area in which annihilation can happen. In the ensuing calculations, a value of D_m equal to that of R_a was used.

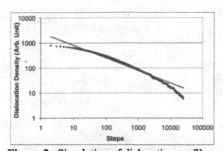

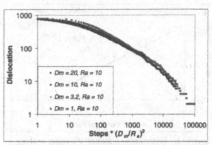

Figure 3. Simulation of dislocation profile, using the following parameters: $B = 1000$, $N_l = 1000$, $D_m = 10$, $R_a = 10$. For reference, the straight line corresponds to the following relationship: $D = A*Steps^{-0.5}$ (A=2500).

Figure 4. Normalized results of simulations with varying step size.

Figure 5 compares the results of two simulations in which the box dimensions differ by a factor of 5. The initial number of dislocations was different by a factor of 25, to keep the dislocation density the same. As can be expected, there is a very good agreement in the early portion of the simulated results. The difference becomes noticeable when the total number of dislocations in the smaller box decreases to about 20. Apparently, such a small number no longer supports a statistically meaningful result.

Next, the model assumes that there is a critical distance R_a within which two dislocations of opposite types will annihilate. While this is certainly an oversimplification, we think it is a reasonable starting point for analyzing the main effects. Numerically, this parameter works jointly with other dimensional parameters. Conceptually, the dislocation density may be expressed using the $1/R_a^2$ units. Therefore, all simulations were performed with a fixed R_a.

Several sets of simulations were performed to evaluate the dislocation profiles using different initial dislocation density values. The results are presented in Figure 6. Shown in the logarithmic scale, all profiles have a relatively flat initial section. This section is longer when the initial dislocation density is lower. The profiles gradually increase their slope, and in a fairly wide region they can be approximated by the function $y=Ax^{-0.5}$.

The insert in Figure 6 was plotted after offsetting each profile by dividing the dislocation density by the initial value and then multiplying the numer of steps by the same factor. Effectively, this moves the top profiles downward and to the right. One can see that the profiles overlap well. These results suggest that the relationship between the dislocation density and the axial position of the wafer within the boule is valid for the real characteristic dimensions of SiC crystal growth. The experimental measurement of dislocation density typically covers a range from a fraction of millimeter to a few centimeters, which corresponds to the region with steady slope in the simulation (10^2 to 10^6 steps).

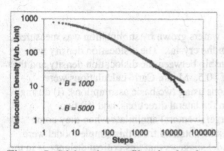

Figure 5. Dislocation profiles simulated using boxes of different sizes.

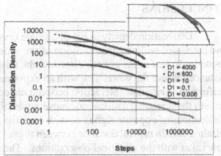

Figure 6. Simulated dislocation density profiles for various initial dislocation densities. The insert shows how the profiles overlap after offsetting.

Although the simulation was done using arbitrary units, we can learn more from this model by making some realistic estimation of the input parameters. Each step in the calculation corresponds to the increase in the crystal thickness that allows a threading dislocation to transverse a maximum distance of D_m. Previous studies by cross-sectional x-ray topography [5] and by step-wise etching and polishing removal [6] have shown that there can be significant bending of threading dislocations in sublimation grown SiC crystals. In other words, threading dislocations have a mobility in the lateral directions during crystal growth. At a scale of a few microns (as use in [] for polishing steps), the maximum lateral movement is on the same order of magnitude as the growth. Considering the profile shown in Figure 3 (the simulation used $D_m = R_a$), we can assume the thickness increase in each step is also comparable to R_a. Take a typical dislocation density (as shown in Figures 1-2) of 1×10^5, the model for Figure 3 (which is the same as the second series from top in Figure 6, D1=800) correspond to a box of 100μm×100μm, with $R_a = 1$μm. The range of thickness that would resemble the observed power-law relationship is approximate between 30 and 3×10^4 steps, or 30μm to 3cm. The typical range of wafer positions, from sub millimeter to a few centimeters, fits to this range on the higher side. On the other hand, the model shown as the last series in Figure 6 corresponds to $R_a = 0.05$μm. The section of the profile that resemble the observed power-law relationship is approximately between 5×10^3 and 5×10^6 steps, or 250μm to 25cm. The typical range of wafer positions fits to this range on the lower side. This analysis suggests that the critical distance for two dislocations of opposite types to annihilate is effectively between a few hundred Å to about a micron.

While this study employed a simple annihilation model for dislocations, there are other mechanisms for the reduction of dislocation density in real crystals. These may include the conversion of threading dislocations to basal plane ones with the same Bergers vector; the segregation of dislocations onto grain boundaries, and the reaction between two dislocations with Bergers vectors that are at a angle. It should also be noted that the discussion above implies the absence of any disturbances during growth. If there are polytype changes or other inclusions, we would expect corresponding changes in the dislocation density profile. Furthermore, the behavior of dislocations can be different for heavily doped crystals, since both the annihilation radius and mobility of the dislocations can be different. These differences may explain the discrepancies reported in the literature [7].

CONCLUSIONS

Dislocation density in 6H SiC single crystal wafers grown by sublimation was measured and correlated with the axial position of the wafer in the crystal. The dislocation density was found to decline in the growth direction; the relationship between the dislocation density and the distance follows the power-law with an exponent of –0.5. Monte-Carlo calculations were performed to simulate interaction between dislocations using two basic assumptions: (i) during growth the threading dislocations move randomly in the lateral directions, and (ii) two dislocations of opposite sign (i.e., with opposite Bergers vectors) annihilate when they come within a certain critical distance between them. Results obtained from this simple model were consistent with the empirical observations. The model suggests that this critical distance is in the range between a few hundred Å and a micron. Future investigation with more precise measurement of growth distance and focusing specifically on the first few hundred microns of growth will be able to expand the observation range and permit better comparison with the results of numeric modeling.

ACKNOWLEDGMENTS

The authors wish to thank I. Currier, C. Minto, S. Nahra, E. Semenas, and G. Wilson, for assistance in preparation of SiC samples.

REFERENCES

1. A. R. Powell, R. T. Leonard, M. F. Brady, St. G. Muller, V. F. Tsvetkov, R. Trussell, J. J. Sumakeris, H. McD. Mobgood, A, A. Burk, R. C. Glass and C. H. Carter, Jr., *Mater. Sci. For.*, **457-460**, 41 (2004).
2. P. Wu, M. Yoganathan and I. Zwieback, *J. Crystal Growth*, doi:10.1016/j.jcrysgro.2007.11.078. (2007).
3. X. Xu, R. P. Vaudo, A. Salant, J. Malcarne, J. S. Flynn, E. L. Hutchins, J. A. Dion and G. R. Brandes, Proc. High Temp. Electr. (2002).
4. S. K. Mathis, A. E. Romanov, L. F. Chen, G> E. Beltz, W. Pompe and J. S. Speck, *J. Crystal Growth*, **231**, 371 (2001).
5. X. Zhang, S. Ha, M. Benamara, and M. Skowronski, M. J. O'Loughlin and J. J. Sumakeris, *Appl. Phys. Let.*, **85**, 5209 (2004).
6. P. Wu, E. Emorhokpor, M. Yoganathan, T. Kerr, J. Zhang, E. Romano and I. Zwieback, *Mater. Sci. For.*, **556-557**, 247 (2007).
7. N. Ohtani, M. Katsuno, H. Tsuge, T. Fujimoto, M. Nakabayashi, H. Yashiro, M. Sawamura, T. Aigo and T. Hoshino, *J. Crystal Growth*, **286**, 55 (2006).
6. H. J. Queisser, *J. Crystal Growth*, **17**, 169 (1972).

Mater. Res. Soc. Symp. Proc. Vol. 1069 © 2008 Materials Research Society   1069-D07-04

Deep-Level Defects in Nitrogen-Doped 6H-SiC Grown by PVT Method

Pawel Kaminski[1], Michal Kozubal[1], Krzysztof Grasza[1,2], and Emil Tymicki[1]
[1]Institute of Electronic Materials Technology, ul. Wólczyñska 133, Warszawa, 01-919, Poland
[2]Institute of Physics Polish Academy of Sciences, Al. Lotników 32/46, Warszawa, 02-668, Poland

ABSTRACT

An effect of the nitrogen concentration on the concentrations of deep-level defects in bulk 6H-SiC single crystals is investigated. Six electron traps labeled as T1A, T1B, T2, T3, T4 and T5 with activation energies of 0.34, 0.40, 0.64, 0.67, 0.69, and 1.53 eV, respectively, were revealed. The traps T1A (0.34 eV) and T1B (0.40 eV), observed in the samples with the nitrogen concentration ranging from ~$2x10^{17}$ to $5x10^{17}$ cm^{-3}, are attributed to complexes formed by carbon vacancies located at various lattice sites and carbon antisites. The concentrations of traps T2 (0.64 eV) and T3 (0.67 eV) have been found to rise from ~$5x10^{15}$ to ~$1x10^{17}$ cm^{-3} with increasing the nitrogen concentration from ~$2x10^{17}$ to ~$2.0x10^{18}$ cm^{-3}. These traps are assigned to complexes involving silicon vacancies occupying hexagonal and quasi-cubic sites, respectively, and nitrogen atoms. The trap T4 (0.69 eV) concentration also substantially rises with increasing the nitrogen concentration and it is likely to be related to complexes formed by carbon antisites and nitrogen atoms. The midgap trap T5 (1.53 eV) is presumably associated with vanadium contamination. The presented results show that doping with nitrogen involves a significant change in the defect structure of 6H-SiC single crystals.

INTRODUCTION

The properties of silicon carbide (SiC), in particular the wide band gap, excellent thermal conductivity and high breakdown electric field, make this material very attractive for manufacturing electron devices operating with high-power and high-frequency signals and capable of working at high temperatures. However, the quality of SiC single crystals in terms of application in advanced electronics is strongly affected by point defects formed during the crystal growth. These defects introduce deep energy levels in the band gap and may reduce the device performance acting as traps or recombination centers. Thus, the knowledge on the electronic properties of defect centers and their microscopic structure is of great importance to control the material quality. Accordingly to technological conditions, various types of point defects are formed in bulk SiC crystals grown by Physical Vapor Transport (PVT) method [1]. These include non-stoichiometric native defects, such as carbon V_C and silicon V_{Si} vacancies, vacancy pairs $V_C V_{Si}$, as well as antisite defects Si_C and C_{Si} in carbon and silicon sublattice. There are also extrinsic point defects due to contamination of crystals with residual atoms of various chemical elements, mainly with boron, aluminum, titanium and vanadium [2]. The theoretical analysis has shown that the partial pressure of silicon over the growing crystal plays an important role in the formation of Schottky and Frenkel point defects in SiC single crystals [1]. With decrease in this pressure, the concentration of singly ionized Si vacancies increases linearly. In order to achieve a high electron concentration, the crystals are doped with nitrogen and the native defects can interact with nitrogen atoms producing various complexes.

So, the aim of this paper is to show how the nitrogen concentration affects the concentrations of deep-level point defects in SiC single crystals grown by PVT technique.

EXPERIMENT

The n-type 6H-SiC wafers used in our experiment were produced from 6H-SiC: N boules grown by the PVT method. The surface of the both sides of the wafers was mirror-polished. The wafers were (0001) oriented and their diameter was 2 inches and thickness ranged from 0.7 to 1.2 mm. The samples for Deep-Level Transient Spectroscopy (DLTS) measurements, with dimensions of approximately 10×10 mm^2, were cut from the central parts of the wafers prepared from the selected boules, labeled as A, B, and C. The samples were rinsed in boiling acetone and de-ionized water. To remove the oxide layer, they were then etched for 5 min in 5% hydrofluoric acid solution. The Schottky contacts were made by evaporating through a shadow mask a 20-nm layer of Cr and a 300-nm layer of Au on the front side (Si face) of the samples. The diameter of the Schottky contacts was 0.5 mm. Ohmic Al contacts of large area were deposited on the back side of the samples by using electron beam. Obtained from C-V measurements the net donor concentrations in the samples A, B, and C were equal to ~2.5×10^{16}, ~2.0×10^{17} and ~1.0×10^{18} cm^{-3}, respectively. The nitrogen and boron concentrations in the samples A, B, and C, determined by Secondary Ion Mass Spectroscopy (SIMS), were as follows: 2.1×10^{17} and 1.1×10^{17}cm^{-3}, 5×10^{17} and 1.2×10^{17}cm^{-3}, 1.8×10^{18} and 2.2×10^{17}cm^{-3}, respectively. The DLTS system used in the present work and the procedures used for determination of the trap activation energy, apparent capture cross-section and concentration have been described elsewhere [3]. The DLTS spectra were taken between 150 and 450 K at a reverse bias V_R ranging from -15 to -8V and a filling pulse amplitude $V_F = 5$ V.

RESULTS AND DISCUSSION

Figure 1 shows the results obtained by DLTS measurements for the samples A, B, and C with various nitrogen concentrations. It is seen that the broad DLTS peaks have been deconvoluted by fitting the spectra with the sum of Gaussian functions. The concentrations of detected traps are proportional to the heights of the Gaussian peaks obtained by the deconvolution. The spectrum for the sample A (figure 1(a)) reveals five deep electron traps, labeled as T1A (0.34 eV), T1B (0.40 eV), T2 (0.64 eV), T3 (0.67 eV) and T4 (0.69 eV), which were formed during the crystal growth. It should be noted, that the traps T1A (0.34 eV) and T2 (0.64 eV) are dominant and their concentrations in the sample are 6.0×10^{15} and 6.4×10^{15} cm^{-3}, respectively. In the sample B (figure 1(b)), with a higher nitrogen concentration ([N] = 5.0×10^{17} cm^{-3}), the same traps as in the sample A occur. However, the trap T3 (0.67 eV), whose concentration equals to 3.0×10^{16} cm^{-3}, is dominant in the sample B. For the sample C (figure 1(c)), with the highest nitrogen concentration ([N] = 1.8×10^{18} cm^{-3}), the traps T1A (0.34 eV) and T1B (0.40 eV) are not observed in the DLTS spectrum. On the other hand, a new midgap electron trap T5 with the activation energy of 1.53 eV appears. In this sample, the trap T3 (0.67 eV) is dominant and its concentration is 1.7×10^{17} cm^{-3}. It is worth noticing, that the trap T2 (0.64 eV) concentration is approximately two times lower and the concentration of the trap T4 (0.69 eV) is only slightly lower than that of the trap T3 (0.67 eV). The Arrhenius plots for all detected traps, illustrating the changes of log (T^2/e_n) as a function of the reciprocal of temperature T, where e_n is the thermal emission rate of electrons, are presented in figure 1(d).

Taking into account the thermal emission activation energies and electron-capture cross-sections (E_a, σ_n) for the traps T1A and T1B, that are equal to (0.34 eV, 2.5×10^{-16} cm^2) and (0.40 eV, 1.0×10^{-15} cm^2), respectively, we can identify these traps with defect levels E_c- 0.33 eV and E_c- 0.40 eV, related to defect centers known as E_1 and E_2 [4-6]. Although, the microstructure of these centers has not been established yet in detail, there is strong experimental evidence that they are native defects involving the negatively charged carbon vacancy V_C [5]. Moreover, the centers E_1 and E_2 are considered to be attributed to the same point defect but located at hexagonal (h) and cubic (k_1, k_2) lattice sites of 6H-SiC [6]. There are also suggestions

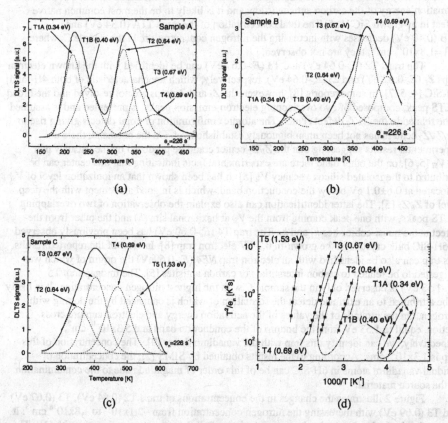

Figure 1. Results of DLTS measurements for the samples of 6H-SiC single crystals with various nitrogen concentrations. (a) DLTS spectrum of sample A, [N]=2.1×10^{17} cm^{-3}; (b) DLTS spectrum of sample B, [N]=5×10^{17} cm^{-3}; (c) DLTS spectrum of sample C, [N]=1.8×10^{18} cm^{-3}; (d) Arrhenius plots for electron emission from detected deep-level defects. Dashed peaks result from fitting the spectra with the sum of Gaussian functions.

that E_1/E_2 centers originate from a divacancy V_CV_{Si} or a V_CC_i complex [5, 6]. However, the most probable identity of the E_1/E_2 deep levels is based on the results of electron paramagnetic resonance measurements (EPR) indicating that these levels and the $P6/P7$ centers originate from the same defect [5]. This point defect has been proposed to be the complex composed of carbon vacancy and carbon antisite (V_CC_{Si}) [5]. It has been shown [5] that the V_CC_{Si} complexes are easily formed during annealing processes through the reaction $V_C + C_{Si} \rightarrow V_CC_{Si}$. In the bulk 6H-SiC crystal grown by the PVT technique, the carbon vacancies V_C arise due to an excess of silicon in the gas phase. The vacancies are mobile, even at temperatures below 500 °C, and can be captured by carbon antisites C_{Si} during the crystal cooling. It should be added, that the formation energy of the carbon antisite is low and it is likely to be the most common native defect in as-grown SiC [1, 5] The total concentration of the traps T1A (0.34 eV) and T1B (0.40 eV) decreases with increasing the nitrogen concentration. In the sample C, where [N] =1.8×10^{18} cm^{-3}, they are not observed.

The traps T2 (E_c-0.64 eV) and T3 (E_c-0.67 eV) can be identified with the known electron traps Z_1 (E_c-0.62 eV) and Z_2 (E_c-0.64 eV), respectively, which are characteristic of both 4H- and 6H-SiC [4, 5, 7]. In some reports [8], however, the Z_1 and Z_2 traps are not resolved and the broad DLTS peak, composed of two overlapping electron-emission peaks, is presented and is assumed to be related to the Z_1/Z_2 defect center. The atomic configuration of point defects giving rise to the Z_1/Z_2 centers has not been unambiguously established yet. On the one hand, there are experimental results suggesting that the Z_1/Z_2 center could be produced by the vacancy pair $V_C V_{Si}$ [5, 6], on the other hand there are experimental facts indicating that this center can be attributed to the isolated silicon vacancy V_{Si} [5]. It has been shown that an ionization level of V_{Si} is located at 0.6±0.1 eV below the conduction band, which is in good agreement with the deep level of Z_1/Z_2 [5]. The latter identification can also explain the observation of two overlapping DLTS peaks, with one peak coming from the V_{Si} at hexagonal site (h) and the other from the defect at the quasi-cubic sites k_1 and k_2. The trap T4 (E_c-0.69 eV) has been previously observed in 6H-SiC bulk crystals as the grown-in 0.7-eV electron trap [8]. In view of the reported results, this trap can also be identified with an electron trap $NE4$ (E_c-0.69eV) the origin of which was suggested to be related to carbon interstitials or carbon antisites [5]. The midgap trap T5 (E_c-1.53 eV) is observed only in the sample C with the highest nitrogen concentration. It is likely to be attributed to an extrinsic defect the formation of which is enhanced by the doping with nitrogen. On the grounds of the values of the activation energy and electron-capture cross-section, equal to 1.53 eV from the bottom of the conduction band and 5.3×10^{-10} cm^2, respectively, we can identify this trap with the vanadium donor [9]. The concentration of this trap is 1.3×10^{17} cm^{-3}. According to the results obtained by SIMS [2], the concentration of residual vanadium atoms in 6H-SiC can be of this order of magnitude due to the contamination of the source material.

Figure 2 illustrates the changes in the concentrations of traps T2 (0.64 eV), T3 (0.67 eV) and T3 (0.69 eV) with increasing the nitrogen concentration from ~2.1×10^{17} to 1.8×10^{18} cm^{-3}. It is seen that in the log-log scale, the concentrations of all these traps grow linearly with [N]. The presented dependences indicate that the traps T2 (0.64 eV), T3 (0.67 eV) and T3 (0.69 eV) are likely to be formed by complexes involving one nitrogen atom. Assuming that the former two traps are related to V_{Si} occupying hexagonal and cubic lattice sites, respectively, we can assign these traps to $V_{Si}(h)N_C$ and $V_{Si}(k_1,k_2)N_C$ complexes. Because of a difference in the electronic properties of the silicon vacancy located at the various lattice sites, the slop of the increase in the concentration of each trap is slightly different. In view of the dependence shown in figure 2, as well as the experimental results [5, 6] indicating involvement of the carbon antisite in the formation of the trap T3 (0.69 eV), we can finally identify this trap with the carbon antisite-nitrogen pair ($C_{Si}N_C$). It is worth adding, that both the reported experimental data and results of theoretical works give strong indication that native defect-nitrogen complexes play an important role in the defect structure formation of 6H-SiC single crystals [10]. The silicon vacancy nitrogen pair ($V_{Si}N_C$) in C_{1h} symmetry has been observed in 6H-SiC by EPR method as the so called $P12$ center [10].

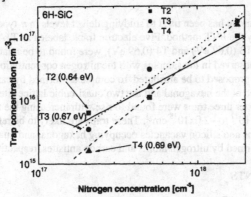

Figure 2. Effect of the nitrogen concentration on the concentrations of traps T2 (0.64 eV), T3 (0.67 eV) and T4 (0.69 eV) detected in 6H-SiC: N samples. Straight lines illustrate the changes in the traps concentrations determined by linear least-squares fitting.

The $C_{Si}N_C$ pairs are considered to be annealing products resulting from migration of carbon split interstitials. These pairs have been shown to be donors with a level in the upper part of the bandgap [10]. The properties of traps detected in the samples A, B and C of 6H-SiC: N single crystals grown by the PVT method are summarized in Table I.

Table I. Summary of activation energies and capture cross-sections for electron traps detected in the samples of 6H-SiC: N single crystals.

Trap label	Activation energy E_a (eV)	Capture cross-section σ_n (cm^2)	Sample	Tentative identification
T1A	0.34± 0.01	2.5x10^{-16}	A, B	$E1$, E_c - 0.33 eV, V_C (h)-C_{Si} [4-6]
T1B	0.40± 0.01	1.0x10^{-15}	A, B	$E2$, E_c - 0.40 eV, V_C (k_1, k_2)-C_{Si} [4-6]
T2	0.64± 0.02	1.4x10^{-15}	A, B, C	Z_1, E_c - 0.62 eV, V_{Si} (h)-N_C [4, 5]
T3	0.67± 0.03	3.0x10^{-16}	A, B, C	Z_2, E_c - 0.64 eV, V_{Si} (k_1,k_2) –N_C [4, 5]
T4	0.69± 0.03	4.0x10^{-17}	A, B, C	$NE4$, E_c- 0.68 eV, C_{Si}-N_C [5, 6]
T5	1.53± 0.03	5.5x10^{-10}	C	Residual vanadium atoms [2, 9]

The results shown in Table I indicate that non-stoichiometric native defects, dependent on the partial pressure of silicon vapor over the growing crystal, play an important role in defect reactions resulting in the formation of deep-level defects in 6H-SiC single crystals. It must be emphasized, however, that the proposed identification is not unambiguously proved and further investigations are needed for solving this challenging problem.

CONCLUSIONS

The DLTS technique has been used to studying defect levels in n-type bulk 6H-SiC single crystals grown by the PVT method. Five electron traps, labeled as T1A (0.34 eV), T1B (0.40 eV), T2 (64 eV), T3 (0.67 eV) and T4 (0.69 eV), were found to be related to native defects. The former two traps, observed in the samples with the nitrogen concentration not exceeding 5×10^{17} cm^{-3}, have been proposed to be attributed to complexes formed by carbon antisites and carbon vacancies located at the hexagonal and the two quasi-cubic lattice sites, respectively. The concentrations of the latter three traps were found to rise with increasing the nitrogen concentration from $\sim 2 \times 10^{17}$ to $\sim 2.0 \times 10^{18}$ cm^{-3}. These traps are likely to be related to complexes involving nitrogen atoms and silicon vacancies occupying hexagonal and quasi-cubic sites, as well as to complexes formed by nitrogen atoms and carbon antisites, respectively.

ACKNOWLEDGMENTS

These studies were supported by the Polish Ministry of Science and Higher Education within the framework of the Ordered Research Project No. PBZ-MEiN-6/2/2006.

REFERENCES

1. A. P. Garshin, E. N. Mokhov, and V. E. Shvaiko-Shvaikovskii, *Refractories and Industrials Ceramics* **44**, 199 (2003).
2. H. McD. Hobgood, R. C. Glass, G. Augustine, R. H Hopkins, J. Jenny, M. Skowronski, W. C. Mitchel, and M. Roth, *Appl. Phys. Lett.* **66**, 1364 (1995).
3. P. Kamiński, G. Gawlik, and R. Kozłowski, *Mat. Sci. Eng.* **B28**, 439 (1994).
4. M. O. Aboelfotoh, J. P. Doyle, *Phys. Rev. B* **59**, 10 823 (1999).
5. X. D. Chen, S. Fung, C. C.Ling, C. D.Beling, and M. Gong, *J. Appl. Phys.* **94**, 3004 (2003).
6. X. D. Chen, C. C.Ling, M. Gong, S. Fung, C. D.Beling, G. Brauer, W. Anwand, and W. Skorupa, *Appl. Phys. Lett.* **86**, 031903 (2003).
7. I. Pintilie, L. Pintilie, K. Irmscher, and B. Thomas, *Appl. Phys. Lett.* **81**, 4841 (2002).
8. A. Y. Polyakov, Q. Li, S. W. Huh, M. Skowronski, O. Lopatiuk, L. Chernyak, and E. K. Sanchez, *J. Appl. Phys.* **97**, 053703 (2005).
9. W. C. Mitchel, W. D. Mitchell, G. Landis, H. E. Smith, W. Lee and M. E. Zvanut, *J. Appl. Phys.* **101**, 013707 (2007).
10. U. Gerstmann, E. Rauls, Th. Frauenheim, and H. Overhof, *Phys. Rev. B* **67**, 205202 (2003).

Mater. Res. Soc. Symp. Proc. Vol. 1069 © 2008 Materials Research Society　　　　1069-D07-07

Stress Mapping of SiC Wafers by Synchrotron White Beam X-ray Reticulography

Ning Zhang[1], Yi Chen[1], Edward Sanchez[2], Michael F. MacMillan[2], and Michael Dudley[1]

[1]Department of Materials Science and Engineering, Stony Brook University, Stony Brook, NY, 11794-2275

[2]Dow Corning Compound Semiconductor Solutions, Midland, MI, 48611

ABSTRACT

Synchrotron white beam x-ray reticulography has been used to quantitatively map the residual stress/strain in SiC wafers. The basic principle of our study is that there exists a relationship between the stress state in a crystal and the local lattice plane orientation and that this relationship can be exploited in order to determine the full strain tensor as a function of position inside the crystal. The theoretical background involved in stress mapping using synchrotron white beam x-ray reticulography is introduced based on the change of the normal to a lattice plane due to the distortion associated with the residual strain. The stress in a region of a commercial 4*H* silicon carbide wafer has been studied using this technique and the results are discussed. This technique can in principle be used in any single crystal material.

INTRODUCTION

Silicon carbide (SiC) is well known as a potential semiconductor material for applications under extreme circumstances due to its outstanding properties such as high thermal conductivity, high breakdown voltage and high saturated electron drift velocity [1]. Accurate measurement of the residual stress/strain in as-grown SiC crystals is of great interest because measured values can allow validation of the thermal stress calculations used to design optimal crystal growth parameters. This is of particular importance in SiC since thermal stresses lead to the activation of basal plane slip systems resulting in significant densities of basal plane dislocations which are implicated in the forward voltage drop under forward bias in pin diodes. Other techniques that can be used to measure stress distributions in SiC crystals include high-resolution x-ray diffraction and Raman spectroscopy. However, mapping the full strain/stress tensor using these techniques can be extremely time-consuming. Here we identify a technique which can be used to map the complete strain/stress state of a large-area SiC wafer in a fast, effective and accurate way. This technique is based on the application of the ray-tracing principle to x-ray reticulography [2].

EXPERIMENT

A 3 in. 4H-SiC wafer that has been polished with diamond paste as small as 0.5 μm was studied. The measurement process and results on a 20 mm x 27 mm area are shown in this paper. This is the typical area that is choosed for the best efficiency and accuracy of the measuring process. It is easy and quick to duplicate the measurement on other areas and piece the results together in order to get the stress mapping of the whole wafer.

A fine scale x-ray absorbing mesh with periodicity of 1 mm placed on the x-ray exit surface of the sample splits the diffracted beams into an array of microbeams. The imaging was carried out at the Stony Brook Synchrotron Topography Station (Beamline X-19C) at

Brookhaven National Laboratory using Agfa Structurix D3-SC film at a specimen-film distance of 10 - 15 cm for transmission geometry.

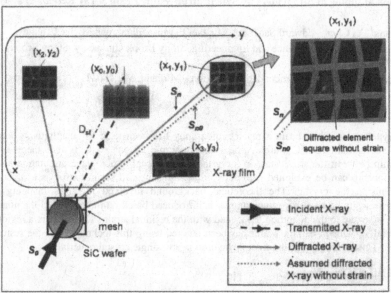

Figure 1. Schematic diagram showing the geometry used in the reticulography. (x_0, y_0), (x_1, y_1), (x_2, y_2) and (x_3, y_3) are the coordinates of the spots on the radiograph and three diffracted images. They are from the same location on the crystal surface.

Figure 1 is a schematic representation reticulography experiment. In a Cartesian coordinate system, the x axis is selected parallel to [11-20], and the y axis is oriented along [1-100]. $(x_0, y_0), (x_1, y_1), (x_2, y_2)$ and (x_3, y_3) are the coordinates of the image of the point of interest in the crystal on the radiograph and three diffraction images. Since the mesh is right on the crystal surface, these four points actually correspond to the same position of the crystal. S_0 is the unit vector of the incident beam. S_n is the vector of the diffracted beam and S_{n0} is the assumed vector of the diffracted beam if no strain exists in the SiC wafer.

DATA PROCESSING, RESULTS AND DISCUSSION

When the crystal is subjected to a displacement field, the original plane normal $\vec{n}^0$ for a given set of planes is modified to $\vec{n}$, according to the following equation

$$\vec{n}(x, y, z) = \vec{n}^0(x, y, z) - \nabla[\vec{n}^0(x, y, z) \cdot \vec{u}(x, y, z)] \tag{1}$$

where $\vec{n}^0(x, y, z)$ is the plane normal at (x, y, z) before distortion, $\vec{n}(x, y, z)$ is the plane normal at (x, y, z) after distortion and $\vec{u}(x, y, z)$ is the displacement vector at (x, y, z). Figure 2 gives the physical meaning of the equation (1). $\nabla[\vec{n}^0(x, y, z) \cdot \vec{u}(x, y, z)]$ is the vector point toward the direction of the greatest rate of increase of $\vec{n}^0(x,y,z) \cdot \vec{u}(x,y,z)$ and $\vec{n}^0(x,y,z) - \nabla[\vec{n}^0(x,y,z) \cdot \vec{u}(x,y,z)]$ is the vector (not unit vector) perpendicular to the plane after deformation.

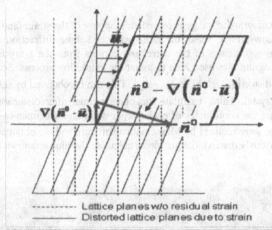

-------- Lattice planes w/o residual strain
———— Distorted lattice planes due to strain

Figure 2. Schematics showing the physical meaning of the equation (1). Dotted lines are lattice planes before distortion. Solid lines denotes lattice planes after distortion. $\vec{n}^0$ is the plane normal before distortion.

Note that the vector calculation in equation (1) must be carried out in the Cartesian coordinate system so that if the crystal structure is non-cubic, the crystal lattice vectors must first be expressed referred to such a Cartesian system. Calculate $\vec{n}(x,y,z)$ using equation (1), and separate it into three components.

$$\begin{cases} n_x = n_x^0 - \left(n_x^0 \dfrac{\partial u_x}{\partial x} + n_y^0 \dfrac{\partial u_y}{\partial x} + n_z^0 \dfrac{\partial u_z}{\partial x} \right) \\[2mm] n_y = n_y^0 - \left(n_x^0 \dfrac{\partial u_x}{\partial y} + n_y^0 \dfrac{\partial u_y}{\partial y} + n_z^0 \dfrac{\partial u_z}{\partial y} \right) \\[2mm] n_z = n_z^0 - \left(n_x^0 \dfrac{\partial u_x}{\partial z} + n_y^0 \dfrac{\partial u_y}{\partial z} + n_z^0 \dfrac{\partial u_z}{\partial z} \right) \end{cases} \qquad (2)$$

In equation (2), $\dfrac{\partial u_x}{\partial x}, \dfrac{\partial u_x}{\partial y}, \dfrac{\partial u_x}{\partial z}, \dfrac{\partial u_y}{\partial x}, \dfrac{\partial u_y}{\partial y}, \dfrac{\partial u_y}{\partial z}, \dfrac{\partial u_z}{\partial x}, \dfrac{\partial u_z}{\partial y}, \dfrac{\partial u_z}{\partial z}$ are nine unknown tensor components. Three groups of equations like equation (2), totally nine equations, can be generated if measurements are made on three topographic reflection spots thus determining the components of three modified plane normal: $n_{1x}, n_{1y}, n_{1z}, n_{2x}, n_{2y}, n_{2z}, n_{3x}, n_{3y}, n_{3z}$. The nine unknowns can be calculated by numerically determining the solutions of these nine equations. This process was computer-assisted by Mathematica v5.1. Then the strains can be calculated by its definition

$$\varepsilon_{ij} = \frac{1}{2}\left(\frac{\partial u_i}{\partial x_j} + \frac{\partial u_j}{\partial x_i}\right). \tag{3}$$

The corresponding stress component can be subsequently calculated, provided the shear modulus or Young's modulus.

After the reticulographic image is recorded, mapping of the strain (and thereby the stress) is carried out as follows. First, select the radiograph and three diffraction spots as shown in Figure 2. Extract the coordinates of the correlated corners from the x-ray film. An automatic image recognition program was used here to assist the extraction process. Second, calculate the plane normal before distortion, $\vec{n}^0$ $\left(\vec{n}^0 = \vec{S}_{n0} - \vec{S}_0\right)$. This can be obtained by selecting a minimum stress point in the crystal. Third, calculate the plane normal after distortion, $\vec{n}$. This can be obtained by measuring the coordinates of each corner point and specimen-to-film distance. The second and third steps were realized by Mathematica software and the obtained $\vec{n}^0$ and $\vec{n}$ will be plugged into the previously discussed equation to calculate the nine unknown tensor components.

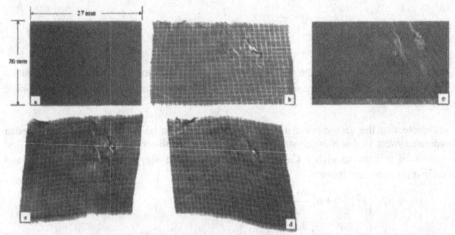

Figure 3. Transmitted radiograph (a), three diffracted images [(b), (c) and (d)] and SWBXT image (e) from the same crystal area.

Figure 3 shows a transmitted radiograph, three corresponding diffracted images and Synchrotron White Beam X-ray Topography (SWBXT) image. The transmitted radiograph in Figure 3(a) is undistorted while the diffracted images in Figure 3(b) – 3(d) are distorted. The diffracted images are distorted in different ways: Figure 3(b) is stretched horizontally while Figure 3(c) and 3(d) are skewed (also one is skewed in a different way from the other). These three diffracted images have been used in our calculation. In the diffracted images [(b) – (d)], a highly distorted region is visible near the upper right corner, which is due to low angle grain boundaries (see Fig. 3 (e)), indicating extremely high strain near this region. Nearly all the squares in the diffracted images are distorted. Such distortion contains not only the qualitative but also quantitative information regarding the residual strain.

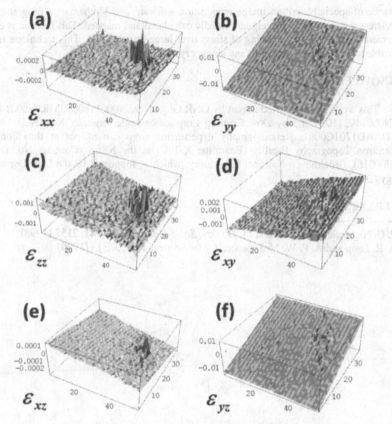

Figure 4. Strain mapping of a 20 mm x 27 mm region in a 3-inch SiC wafer. (a) – (f) are 3D mapping of the six strain components ε_{xx}, ε_{yy}, ε_{zz}, ε_{xy}, ε_{xz} and ε_{yz}, respectively.

Figure 4 gives the 3D mapping of all the six strain components ε_{xx}, ε_{yy}, ε_{zz}, ε_{xy}, ε_{xz} and ε_{yz}. Each strain component is larger near low angle grain boundaries than that away from them. The residual strain in the wafer is very complicate and it may change from compressive strain to tensile strain [e.g., in (a) – (c)], or it may change its sign from one region to the other. Further confirmation and data processing is under way to improve the accuracy and resolution of the results.

CONCLUSIONS

Synchrotron White Beam X-ray Reticulography (SWBXR) has been demonstrated to be a fast, effective and accurate method to map the strain/stress state in SiC wafer. With the assistance of specially written image recognition software and Mathematica programming, the strain/stress data can be accurately and rapidly calculated and mapped. This method is especially convenient to apply to the mapping of stress over large crystal areas. This technique is broadly applicable to stress measurement in any single crystal.

ACKNOWLEDGMENTS

This work was supported in part by ONR Grant Nos. N00011400100348, N000140110302, and N000140211014 and by Dow Corning Corporation under contract Nos. N0001405C0324 and DAAD1701C0081. Reticulography experiments were carried out at the Stony Brook Synchrotron Topography Facility (Beamline X-19C) at the NSLS (Contract No. DE-AC02-76CH00016), Brookhaven National Laboratory, which is supported by the U.S. Department of Energy (D.O.E).

REFERENCES

[1] P. G. Neudeck, W. Huang and M. Dudley, *Solid-State Electron.* **42**, 2157 (1998)
[2] A. R. Lang and A. P. W. Makepeace, *J. Synchrotron. Rad.* **3**, 313 (1996)

Mater. Res. Soc. Symp. Proc. Vol. 1069 © 2008 Materials Research Society

Effect of Annealing Temperature on SiC Wafer Bow and Warp

Xueping Xu, and Chris Martin
Wide Bandgap Materials Group, II-VI Incorporated, 20 Chapin Road, Suite 1005, Pine Brook, NJ, 07058

ABSTRACT

Single-side polished silicon carbide wafers could exhibit large bow and warp due to the presence of mechanical damage on the unpolished surface. In this study, we investigated the effect of thermal annealing on the wafer bow. Two commercial-grade, double-side polished 3" 6H SiC wafers with the bow less than 5 μm were lapped using 12 μm diamond grit. One wafer was lapped on the C-face and another on the Si-face. The lapped wafers were subjected to annealing in vacuum at temperatures between 500°C and 2040°C. The wafer bow and x-ray rocking curves were measured prior to annealing and after each annealing step in order to evaluate the extent of the surface damage and degree of healing. Thermal annealing led to a decrease in the wafer bow and sharpening of the x-ray rocking curves. However, the wafer bow generated by lapping was not completely removed until annealing temperature reached ~2000°C.

INTRODUCTION

Silicon carbide (SiC) is a wide bandgap semiconductor for applications in high frequency and high power devices [1]. Commercial-size SiC single crystals are grown by the technique of physical vapor transport (PVT). In this technique, the solid SiC source is vaporized at a high temperature and the Si- and C-containing volatile molecular species are transported to the seed driven by the temperature gradient in the growth system [2]. The grown SiC boule is subsequently oriented and sliced into wafers, which are polished. Depending on the customer requirements, SiC wafers can be double-side or single-side polished. Typically, single-side polished wafers have higher bow and warp due to the presence of a layer of surface damage on the unpolished wafer face. High wafer bow and warp are undesirable for epitaxial growth and device processing.

It is known that thermal annealing can lead to a stress reduction in SiC wafers. Wafer stress and associated wafer warp and bow are results of the internal stress from crystal growth and the stress from mechanic processes such as slicing, lapping and polishing. Okojie et al reported thermoplastic deformation of SiC wafers during annealing to 900°C [3-5]. Wafer bow was reduced during heating, but remained same during cool down. In this work, we studied the effects of annealing on bow and surface damage of single-side polished 6H SiC wafers. We extended the annealing temperature to 2040°C and used wafers with low internal stress in order to explicitly evaluate the effect surface damage.

EXPERIMENTAL

Two 3" commercial-grade, semi-insulating 6H SiC wafers were used in this study [6]. The 375 μm thick wafers were double-side polished with the bow and warp below 5 μm, indicative of low internal stress. The wafers were lapped using 12 μm diamond slurry: one wafer (wafer A) was lapped on the silicon face and the other (wafer B) on the carbon face, with the total material removal of 25 μm. The lapped wafers were annealed in vacuum in a resistively heated furnace. From the onset of annealing, the power to the furnace was controlled at a constant level and when the temperature reached the predetermined value the power was shut off. The heating rate was about 20°C/minute at the beginning of the heating cycle. After each annealing step, the wafer was characterized and then placed into the furnace again to reach a higher annealing temperature. The maximum annealing temperature was 2040°C.

Wafer bow and warp were measured using a Tropel interferometer. The subsurface damage was assessed using measurements of the x-ray rocking curves. The rocking curves were measured in the Ω mode (where Ω is the sample angle) using the symmetrical (0006) Bragg reflection. The measurements were carried out with an automated double-crystal Philips diffractometer: Cu Kα (λ=1.5406Å) and x-ray beam divergence of 10 arc-seconds. The beam incident upon the wafer surface was collimated to 1 x 1 mm^2, and the detector slit was completely open. The rocking curves were measured on both silicon and carbon faces. To better assess the degree of subsurface damage, both the full width at half maximum (FWHM) and the full width at 1% maximum (FW1%M) of the x-ray reflection were measured.

DISCUSSION

Effects of lapping

Single-side lapping introduced substantial damage to the wafer surface and caused the appearance of large wafer bow and warp. Measurements showed that wafer bow and warp increased from near zero for the double-side polished wafers to about 100 μm after lapping on one face. Both bow and warp had spherical symmetry with similar values, thus indicating uniform wafer stress of ~ 11 MPa [7]. Increase in the bow and warp caused by lapping of the C-face (wafer B) was similar to that caused by lapping of the Si-face (wafer A). This showed that lapping caused a similar degree of damage to both SiC faces.

Figures 1 and 2 compare the x-ray rocking curves for a double-side polished SiC wafer and for the same wafer after C-face lapping (wafer B). Figure 1 shows the rocking curves measured on the Si-face, and Figure 2 shows the curves measured on the C-face. As it follows from Figure 1, lapping of the C-face caused broadening of the x-ray reflection measured on the Si-face: the FWHM increased from 43 arc-secs prior to lapping to 73 arc-secs after lapping, i.e. by 30 arc-secs. Similarly when silicon-face was lapped (wafer A), the FWHM for the carbon face was broadened. The front faces in both cases were not damaged by lapping, and the measured increase in the reflection width was caused by the increase in the lattice curvature due to the wafer bow. The effect of lattice curvature on the x-ray reflection width is given by the following expression [8]:

$$B=b/R \qquad\qquad (1)$$

where B is the reflection broadening in radians, b is the beam width and R is the radius of lattice curvature. For a 3" wafer, a bow of 100 μm corresponds to the radius of curvature of 7 meters. For $R = 7m$ and $b = 1mm$, equation (1) predicts reflection broadening by 29 arc-sec. This calculated value agrees very well with the experimentally measured increase in the FWHM.

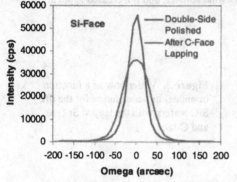

Figure 1. X-ray rocking curves measured on the Si-face of a 6H SiC wafer before and after lapping of the C-face.

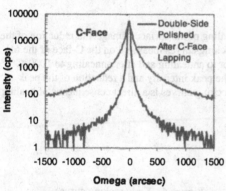

Figure 2. X-ray rocking curves measured on the C-face of a 6H SiC wafer before and after lapping of the C-face.

Figure 2 shows the rocking curves measured on the C-face prior to and after lapping of the C-face. The peaks demonstrate increase in the FWHM from 37 arc-secs to 94 arc-secs, i.e. by 57 arc-secs. In this case, both the lattice curvature due to the wafer bow and the subsurface damage contributed to the reflection broadening.

Increase in the reflection width measured on the lapped face was especially pronounced in the low intensity regions of the x-ray peaks. Measurements of the full width at 1% of the peak maximum (FW1%M) on the lapped face showed increase from 150 arc-secs to 1220 arc-secs for the lapped Si-face (wafer A) and from 260 arc-secs to 1170 arc-secs for the lapped C-face (wafer B). This broadening is a direct consequence of the lattice disorder and high concentration of defects in the damaged surface layer [9].

Effects of annealing

Annealing of the lapped wafers led to a marked reduction in bow and warp. Figure 3 shows wafer bow as a function of annealing temperature. Data for the wafers with lapped C- and

Si-faces are plotted separately using different symbols. One can see that increase in the annealing temperature leads to a monotonic reduction of bow. Based on the slope of the curves in Figure 3, one can distinguish three temperature regions of annealing. At temperatures below 1000°C, a relatively steep decrease in the wafer bow took place; at temperatures between 1000°C and 1600°C, the rate of bow annealing was somewhat smaller, and it increased again at temperatures above 1600°C.

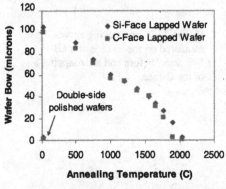

Figure 3. Wafer bow as a function of annealing temperature for the 6H SiC wafers with the lapped Si-face and C-face.

Annealing of bow can be attributed to healing of the surface damage and reduction of the associated wafer stress. Figure 4 shows x-ray rocking curves measured on the C-face of the 6H SiC wafer with the lapped C-face (wafer B), prior to annealing and after annealing to 1000°C and 1750°C. Annealing leads to an increase in the peak intensity and a reduction of the peak width and background. This sharpening of the rocking curves is a direct consequence of healing of the surface damage.

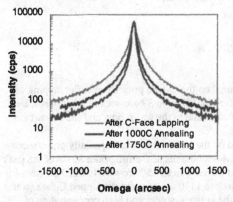

Figure 4. X-ray rocking curves measured on the C-face of the 6H SiC wafer with the lapped C-face prior to annealing and after annealing to 1000°C and 1750°C.

Figure 5 shows the FW1%M versus annealing temperature for the wafers with the lapped Si- and C-faces. In both cases, the FW1%M reduced monotonically with annealing temperature. Similarly to the annealing of bow shown in Figure 3, the dependence of FW1%M on the annealing temperature has also three temperature regions. At T < 1200°C, the FW1%M decreased rapidly with annealing temperature; in the region between 1200°C and 1750°C, the

FW1%M reduced more slowly, and at temperatures above 1750°C, the rate of the FW1%M reduction was higher again.

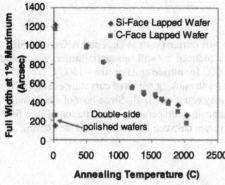

Figure 5. Dependence of FW1%M on annealing temperature for the wafers with the lapped C- and Si- faces (measured on the lapped face).

Various healing mechanisms can be at work during annealing. At relatively low temperatures (*e.g.*, below 1200°C), annealing may lead to annihilation of dislocations and other defects generated by lapping. At higher annealing temperatures (*e.g.*, above ~1700°C), the degree of surface damage may be further reduced by decomposition and evaporation of the SiC surface. In order to evaluate the evaporative losses from the surface, the wafer weight was measured after each annealing step using a high-precision balance. Figure 6 shows relative wafer weight plotted against annealing temperature. At temperatures of 1750°C and higher, a noticeable weight loss was observed. The wafers annealed at T > 1750°C also appeared darker due to the losses of Si from the surface and formation of a C-rich surface layer.

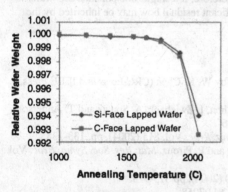

Figure 6. Evaporative weight losses versus annealing temperature for the 6H SiC wafers with the lapped C-face and Si-face.

Differences between C-face lapping and Si-face lapping

At the annealing temperatures below 1600°C, there was no substantial difference in the effect of annealing on Si-face lapped wafer (wafer A) and C-face lapped wafer (wafer B). Both wafers had negligible weight loss and similar reductions of wafer bow and FW1%M. However, at annealing temperatures above 1750°C, the evaporative losses were higher for wafer B with the lapped C-face. In addition, wafer B had slightly lower wafer warp and sharper x-ray peaks with the annealing temperature above 1750°C, indicating better elimination of surface damage from the lapped carbon face (wafer B) than from the lapped silicon face (wafer A).

Implication of wafer bow on epitaxial growth

SiC wafers are used as substrates for group-III nitride (such as GaN and AlGaN) or SiC epitaxy. If the starting substrate has a large bow, a reduced but still substantial wafer bow would be present at typical growth temperatures of ~1100°C for nitride epitaxy and ~ 1600°C for silicon carbide epitaxy. It has been reported in literatures using radius of curvature change to calculate the stress in epitaxial films based on Stoney equation [3, 4]. Since bow of single-side polished SiC wafer reduces monotonically with annealing temperature, a calculation of thin film stress based on the difference of the radius of curvature between the epilayer and the starting substrate could lead erroneous results.

CONCLUSION

Single-side lapping of SiC wafers using 12 μm diamond grit caused substantial damage to the lapped face, which led to increase in the wafer stress to about 11 MPa with wafer bow and warp reaching 100 μm. For the wafers with the lapped C- and Si-faces the values of bow and warp were similar.

Wafer annealing led to decrease in the wafer bow, as well as sharpening of the x-ray reflection. The latter was manifested in the increased peak intensity and reduced values of both FWHM and FW1%M.

Wafer stress induced by single-side lapping was not completely eliminated until the annealing temperature reached about 2000°C. Therefore, if a single-side polished SiC wafer is used as a substrate in GaN or SiC epitaxy, a significant residual bow may be inherited by the epitaxial structure.

REFERENCES

1. P. G. Neudeck, in *The VLSI Handbook*, edited by W. K. Chen (CRC Press and IEEE Press, Boca Raton, Florida, 2000), p. 6.1-6.24
2. A. Gupta M, E. Semenas, E. Emorhokpor, J. Chen, I. Zwieback, A. Souzis and T. Anderson, *Mater. Sci. Forum* Vols. 527-529 (2006), pp 43-46.
3. R. S. Okojie and M. Zhang, *Mat. Res. Soc. Symp. Proc.* Vol. 815 (2004), pp. 133-138.
4. R. S. Okojie, X. Huang, M. Dudley, M. Zhang and P. Pirouz, *Mat. Res. Soc. Symp. Proc.* Vol. 911 (2006), pp 145-150.
5. R. S. Okojie, *Mater. Sci. Forum* Vols. 457-460 (2004), pp 375-378.
6. II-VI., 20 Chapin Rd, Suite 1005. Pine Brook. NJ 07058
7. Calculated from formula $\sigma=0.5EdR^{-1}$ where σ is the wafer stress, E is Young's module, d is wafer thickness and R is radius of curvature of the lapped wafer.
8. M. Yoganathan, P. Wu and I. Zwieback, *Mater. Sci. Forum*, in press.
9. See, for example, F. M. Anthony, A. M. Khounsary, D. R. McCarter, F. Krasnicki and M. Tangedahl, *Proc. SPIE*, Vol. 4115 (2001), pp 37-44.

Epitaxial Material and Characterization

Mater. Res. Soc. Symp. Proc. Vol. 1069 © 2008 Materials Research Society 1069-D03-05

Residual Stress in CVD-grown 3C-SiC Films on Si Substrates

Alex A. Volinsky[1], Grygoriy Kravchenko[1], Patrick Waters[1], Jayadeep Deva Reddy[1], Chris Locke[2], Christopher Frewin[2], and Stephen E. Saddow[2]

[1]Department of Mechanical Engineering, University of South Florida, 4202 E. Fowler Ave. ENB118, Tampa, FL, 33620

[2]Department of Electrical Engineering, University of South Florida, 4202 E. Fowler Ave. ENB118, Tampa, FL, 33620

ABSTRACT

Having superior mechanical properties, 3C-SiC is one of the target materials for power MEMS applications. Growing 3C-SiC films on Si is challenging, as there is a large mismatch in lattice parameter and thermal expansion between the SiC film and the Si substrate that needs to be accommodated, and results in high residual stress. Residual stress control is critical in MEMS devices as upon feature release it results in substantial deformation.

3C-SiC single crystalline films were deposited on 50 mm (100) and (111) Si substrates in a hot-wall CVD reactor. The film tensile residual stress was so high that it fractured on the (111) Si wafer. The resulting film thickness on the (100) Si wafer was non-uniform, having a linear profile along the growth direction. This presented a challenge of using the substrate curvature method for calculating residual stress. Finite Element Method correction was applied to the Stoney's formula for calculating the residual stress along the wafer radius. Suggestions for reducing the amount of residual stress are made.

INTRODUCTION

SiC is the material of choice for power MEMS applications, since it has much better mechanical performance at high temperatures, compared to Si. Deposition and processing techniques must be mastered before a functioning device can actually be built. It is advantageous to deposit SiC on Si wafers due to their high quality and low cost, in comparison to SiC substrates. However, one has to overcome the 22% lattice mismatch between the 3C-SiC film and the (100) Si substrate, and additional 8% mismatch in thermal expansion. Any temperature variation will cause stress. 3C-SiC films on (111) Si wafers have larger residual stress, although there is a slightly smaller lattice mismatch of about 19.7%. In both instances 3C-SiC can be hetero-epitaxially grown on a highly defective SiC buffer layer formed during the so-called carbonization step, which could accommodate some of the mismatch strain.

EXPERIMENT

Single crystal 3C-SiC films were grown hetero-epitaxially in a hot-wall chemical vapor deposition (CVD) reactor on (100) and (111) 50 mm Si substrates [1]. The 3C-SiC on Si deposition process was developed using the two step carbonization and growth method. C_3H_8 and SiH_4 were used as the precursors to provide the carbon and silicon sources, respectively. The carrier gas was ultra-high purity hydrogen, purified in a palladium diffusion cell. Prior to growth, the substrates were prepared using the RCA cleaning method [2], followed by a 30 second immersion in diluted hydrofluoric acid to remove the surface oxide and possible contaminants.

The samples were then thoroughly dried in N_2 and loaded into the CVD reactor, which was then evacuated of all residual gases prior to processing.

The carbonization process step involved heating either the (100) Si or the (111) Si substrate from room temperature to 1135 °C at 400 Torr pressure in a flow of 16 standard cubic centimeters per minute (sccm) of C_3H_8 and 10 standard liters per minute (slm) of H_2. This temperature was maintained for two minutes to carbonize the substrate surface. After carbonization, the temperature was gradually ramped to 1380 °C, which began the growth stage. During the ramp, the propane flow was linearly decreased to 8.0 sccm for a (100) Si substrate or to 5.6 sccm for a (111) Si substrate. A 10% silane/hydrogen mixture (SiH_4), which is the silicon precursor, was slowly added until a final flow rate of 270 sccm was achieved at the end of the second thermal ramp when using a (100) Si substrate. When growing on (111) Si, the 10% silane/hydrogen mixture flow rate was linearly increased during the second ramp to a final value of 218 sccm. The silicon to carbon ratio, Si/C, for the growth stage was 1.13 and 1.30 for (100) Si and (111) Si, respectively. For growth on a (100) Si wafer, the pressure was reduced from 400 to 100 Torr and the hydrogen flow was linearly increased to 30 slm at 65 °C before the ramp was completed in order to transport the cracked growth species more effectively to the growth zone located 60 mm upstream from the hot-zone outlet. However, for (111) Si, the pressure was maintained at 400 Torr and the hydrogen flow was increased to 40 slm at 1320 °C. The temperature and gases flow were then held constant at the end of the second ramp, allowing the continuous epitaxial growth of 3C-SiC on the carbonized Si buffer layer [3]. After the growth process was completed, the wafer was cooled to room temperature in an Ar atmosphere [4]. The process flow is shown schematically in Figure 1, and the details of both the reactor construction and the growth process can be found in reference [5].

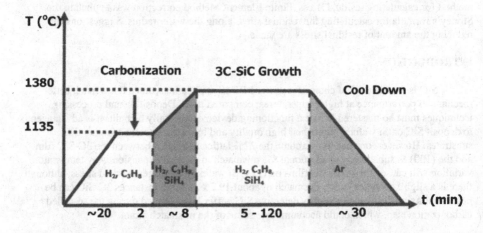

Figure 1. Single crystal 3C-SiC films growth process [5]. The resulting growth rate was 10 μm/h.

The crystal orientation of the grown films was determined by X-ray diffraction (XRD) using a Philips X-Pert X-ray diffractometer. Figure 2 shows powder (θ-2θ) scans obtained from

SiC films grown on (100) and (111) Si substrates. The orientation of the single crystal SiC film is (100) on (100) Si wafers and (111) on (111) Si wafers. The (100) 3C-SiC film is of higher quality, as it exhibits a sharp (200) peak, while the (111) 3C-SiC film peak on the (111) Si wafer was split into three peaks due to the high residual stress and its relief. This stress caused visible (111) Si wafer curvature and was partially relieved by film fracture and delamination, causing a triple peak. XRD rocking curve data on the (100) 3C-SiC film showed that the films were single crystal with an X-ray rocking curve peak full width half maximum (FWHM) of approximately 300 arcseconds [5, 6].

Mechanical properties of 3C-SiC on (100) Si were measured, and it was determined that single crystalline films have an elastic modulus of 430 GPa and a hardness of 30 GPa [6]. The elastic properties data was used for the Finite Element Modeling. The mechanical properties of the polycrystalline films were slightly higher, making them more attractive for power MEMS applications.

In the reactor used for the growth no wafer rotation is available, therefore the film thickness varied substantially across the wafer. Figure 3a shows interference fringes originating from the film thickness non-uniformity and the (100) Si wafer curvature due to residual stress observed when using 1/10 wave optical flat on the 50 mm wafer. Vertical film thickness profile is presented in Figure 3b obtained from an Accent QS-1200 Fourier transform infrared spectrometer (FTIR). The thickness measurements were taken along the wafer diameter starting from the primary flat and ending on the opposite wafer side. Using the FTIR graduated mounting stage, each thickness measurement was made in 5 mm intervals.

The gases were introduced from the top of the wafer, which was specified in these experiments as the side of the wafer opposite the primary wafer flat, resulting in a linear film thickness profile along the y direction (Figure 3b).

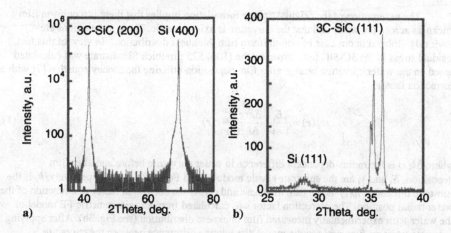

Figure 2. XRD powder (θ-2θ) scan for 3C-SiC films grown on a) (100) and b) (111) Si substrates. In both instances the films are single crystalline, with the (111) film having triple peak due to high residual stress and its relief by fracture.

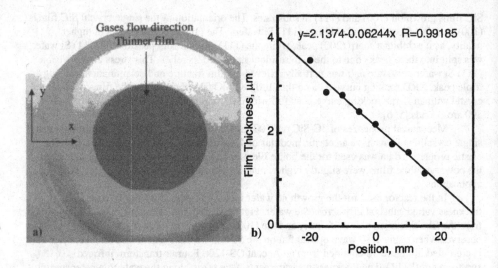

Figure 3. a) Photograph of the 50 mm (100) Si wafer with 3C-SiC film showing interference fringes due to film thickness non-uniformity and wafer curvature; b) 3C-SiC film thickness vertical profile measured using FTIR spectrometry.

DISCUSSION

The assumption of the original Stoney formulation implies that there is a uniform film thickness across the wafer and that the curvature is axisymmetric. These assumptions are obviously violated in our case of non-uniform film thickness distribution. In view of this fact, residual stress in the 3C-SiC film grown on the (100) 525 μm thick Si substrate was calculated based on the wafer curvature change after film deposition utilizing the Stoney equation [7] with a correction factor:

$$\sigma_R(r) = \frac{E_s}{1-v_s} \frac{\Delta k(r) h_s^2}{6 h_f(r)} c(r)$$ (1),

where $\Delta k(r)$ is the radius-dependent difference in wafer curvature before and after film deposition, E_s and v_s are the substrate elastic modulus and Poisson's ratio, respectively, h_s is the substrate thickness, $h_f(r)$ is the film thickness and $c(r)$ is the correction factor as a function of the wafer radial position. The correction factor was calculated from an axisymmetric FE model of the wafer with experimentally measured film thickness distribution (see Fig. 3b). After applying a dummy uniform film stress to the model, the relative difference between the curvature calculated by the Stoney's formula and the FEM can be obtained, giving the correction factor. It turns out that the Stoney's formula overestimates the stress by over 7-10% for the 50% film thickness variation across the wafer (Figure 4a). Figure 4b shows the residual stress profile of

3C-SiC film in the vertical direction calculated based on the wafer curvature change and the FEM correction to the Stoney's formula. The stress ranges from 215 to 450 MPa across the 50 mm wafer diameter.

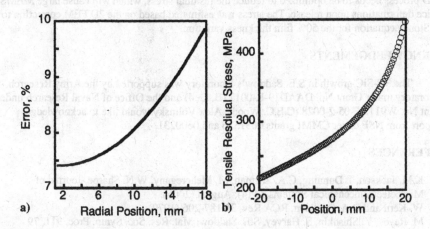

Figure 4. a) FEM correction to the Stoney's formula for the measured film thickness variation; b) Residual stress in the 3C-SiC film grown on (100) Si substrate based on the curvature method with the FEM correction.

The 2D finite element correction assumed that the curvature change is axisymmetric, which is also not the case. Further model development is needed to provide a 3D solution. In the middle of the wafer the tensile equibiaxial stress is 265 MPa. The (111) Si wafer curvature after 3C-SiC film growth was not measured because the film fractured due to the high tensile residual stress relief. Wafer curvature was noticeable to the naked eye. The wafer warped upwards indicating tensile residual stress in the film.

Numerous approaches have been investigated to reduce the lattice mismatch between 3C-SiC and Si. One of the problems is that SiC does not alloy, as is the case of the SiGe system for example. While buffer layers are used in all known SiC growth processes, the so-called carbonization step produces a highly defective buffer layer, which helps accommodate the misfit. Unfortunately it can not help in reducing the thermal stresses. Complaint substrates have been investigated using porous Si buffer layers [8], nano-grooved surfaces [9], and recent work has focused on growing 3C-SiC on oxide layers to provide strain relief [10]. One method used in our reactor to achieve a more uniform film thickness is to mechanically rotate the wafer by 180° and continue the growth. This has resulted in a highly uniform film thickness but the stress is unknown at this point. These films are being studied as per this paper and the results will be provided in the near future.

CONCLUSIONS

3C-SiC thin films deposited on Si wafers for MEMS applications have large residual tensile stress that varies substantially across the 50 mm wafer diameter, from 215 to 450 MPa. CVD process needs to be optimized to reduce the residual stress, which will cause large MEMS device deformations upon release. The stress was estimated based on the 2D FEM correction to the Stoney equation for the 50% film thickness variation.

ACKNOWLEDGEMENTS

The 3C-SiC growth in S.E. Saddow's laboratory was supported by the Army Research Laboratory under Grant No. DAAD19-R-0017 (B. Geil) and the Office of Naval Research under Grant No. W911NF-05-2-0028 (C.E.C. Wood). Alex Volinsky would like to acknowledge support from NSF under CMMI grants 0631526 and 0600231.

REFERENCES

1. K.M. Jackson, J. Dunning, C.A. Zorman, M. Meheregany, W.N. Sharpe, Journal of Microelectromechanical Systems **14** (4), August (2005).
2. W. Kern and D.A. Puotinen, RCA Rev. **31** 187-206 (1970).
3. M. Reyes, Y. Shishkin, S. Harvey, S.E. Saddow, Mat. Res. Soc. Symp. Proc. **911**, 79 (2006).
4. R.F. Davis, G. Kelner, M. Shur, J.W. Palmour and J. Edmond, Proc. IEEE **79** 677 (1991).
5. R.L. Myers, Y. Shishkin, O. Kordina, and S.E. Saddow, Journal of Crystal Growth **285** (4) 483-486 (2005).
6. J. Deva Reddy, A.A. Volinsky, C. Frewin, C. Locke, S.E. Saddow, Mat. Res. Soc. Symp. Proc. **1049** AA3.6 (2008).
7. G.G. Stoney, Proc. Roy. Soc. Lond. **A82** 72 (1909).
8. S.E. Saddow, Y. Shishkin and R.L. Myers-Ward, Chapter 3 in Porous Silicon Carbide and Gallium Nitride: Epitaxy, Catalysis, and Biotechnology Applications, edited by R. M. Feenstra and C. E. C. Wood, Wiley, New York (2008).
9. H. Nagasawa, K. Yagi, T. Kawahara, N. Hatta, Mater. Sci. Forum **433–436** 3–8 (2003).
10. C.L. Frewin, C. Locke, J. Wang, P. Spagnol, and S.E. Saddow, submitted to Appl. Phys. Lett., April (2008).

Mater. Res. Soc. Symp. Proc. Vol. 1069 © 2008 Materials Research Society   1069-D04-01

Silicon Carbide Hot-Wall Epitaxy for Large-Area, High-Voltage Devices

M. J. O'Loughlin, K. G. Irvine, J. J. Sumakeris, M. H. Armentrout, B. A. Hull, C. Hallin, and A. A. Burk, Jr.
Cree, Inc., 4600 Silicon Dr., Durham, NC, 27703

ABSTRACT

The growth of thick silicon carbide (SiC) epitaxial layers for large-area, high-power devices is described. Horizontal hot-wall epitaxial reactors with a capacity of three, 3-inch wafers have been employed to grow over 350 epitaxial layers greater than 100 μm thick. Using this style reactor, very good doping and thickness uniformity and run-to-run reproducibility have been demonstrated. Through a combination of reactor design and process optimization we have been able to achieve the routine production of thick epitaxial layers with morphological defect densities of around 1 cm^{-2}. The low defect density epitaxial layers in synergy with improved substrates and SiC device processing have resulted in the production of 10 A, 10 kV junction barrier Schottky (JBS) diodes with good yield (61.3%).

INTRODUCTION

In recent years, there has been significant progress in the development of silicon carbide (SiC) power devices. SiC Schottky Diodes with blocking voltages between 300 and 1200 V are commercially available, and higher voltage devices are being introduced or developed. There have been some demonstrations of high-voltage diodes using SiC, but in most instances these have been small area devices [1]. The ultimate benefits of using SiC for high-voltage diodes can only be achieved with large-area, high-current devices. For example, Cree's 1200 V-50 A Schottky diodes have an active device area of 0.33 cm^2 [2], and we have demonstrated 10 kV, 20 A junction barrier Schottky (JBS) diodes with an active area of 1.5 cm^2 [3]. To achieve significant yields of these large area devices, defect densities of less than 1 cm^{-2} are necessary. These low defect densities have been achieved only through significant advances in substrate and epitaxial layer quality. The subject of this paper will be the development of hot-wall epitaxy reactors and process capability to grow large quantities of low-defect, uniform SiC epitaxial layers greater than 100 μm thick.

The hot-wall epitaxy reactors that we employ have a capacity of three, 3-inch diameter wafers. For epilayers thicker than about 25 μm, we typically grow on 8° off-axis substrates although we are not constrained by that angle. For both p and n-type intentional doping, layers can be reproducibly grown from 2×10^{14} to over 1×10^{19} cm^{-3}, permitting complex structures, like GTOs, to be grown in a continuous run. Thickness and doping uniformity are generally on the order of 1% or better and 5-8% (std. dev./mean), respectively. Most pertinent to the theme of this presentation, we utilize an optimized buffer process to dramatically reduce the density of morphological defects, particularly "carrot" defects, in subsequent layers [4]. Using this process, we have grown hundreds of thick epilayers with a median defect density of 1.38 cm^{-2} (including substrate related defects). Some of the better wafers have yielded greater than 80% for 10 kV, 10 A JBS diodes (0.88 cm^2), consistent with a device killing defect density of 0.25 cm^{-2} [3].

EXPERIMENT

Custom-built, horizontal hot-wall reactors were utilized for all growths described in this paper. Each reactor can be interchangeably used for 5x2", 3x3", 1x100 mm, and eventually 1x150 mm epitaxial growth of silicon carbide. Each wafer configuration includes a main axis of rotation, and the multi-wafer configurations also include individual wafer rotation. A plan view schematic of the reactor in the 3x3" configuration is shown in Figure 1.

Figure 1 Plan-view schematic of 3x3" reactor including individual wafer rotation.

We have developed a robust susceptor design that can be used without significant degradation of epilayer quality for about 2.5 mm of cumulative growth before refurbishing (twenty 125 μm thick growth runs). After refurbishing, we can achieve background doping levels of less than $1x10^{14}$ cm^{-3} after about two hours of growth, making the total maintenance cycle less than one day. During an epitaxial growth campaign, we have done as many as nine consecutive 125 μm growth runs with good result.

Depending on layer specifications, epitaxial layers are typically grown at temperatures between 1500 and 1650°C and pressures between 30 and 90 torr. Silane and propane are used as precursors, and the silicon-to-carbon ratio is normally kept in a fairly narrow range about unity. The thick epitaxial layers described in this study were grown at about 14 μm/hr. Growth rate is a linear function of precursor flow up to at least 30 μm/hr. At this point, the growth rate is limited only by mass flow controller capacity. A separate gas injector is used to flow additional carrier gas along the ceiling of the growth cell. This ceiling purge reduces parasitic deposits on the surfaces above the wafers, and thus reduces particulate related defects in the epilayers and extends the maintenance interval of the reactor [5].

N-type, Ultra-Low Micropipe 4H SiC substrates were used for epitaxial layer growths. The vast majority of the 3-inch wafers were oriented 8° off axis, and all of the 100 mm wafers were oriented 4° off axis. After growth, the epilayer thickness was measured by FTIR and/or mass gain, and doping was measured using an automated mercury probe C-V profiler with a mercury contact area of 0.022 cm^2. An optical surface analyzer (OSA) was used to locate and count surface imperfections [6]. Large area, high-voltage devices were then fabricated and tested.

DISCUSSION

Data was collected for over 356 thick epi wafers (10kV class) grown in over 123 runs. Thickness and doping uniformity and reproducibility will be presented, followed by a discussion of surface morphology, defect density, and large area device yield.

Doping

In order to reproducibly grow thick layers with intentional doping in the range $2x10^{14}$ to $6x10^{14}$ cm^{-3}, it is desirable for the unintentional (background) doping to be well under $1x10^{14}$ cm^{-3}. Furthermore, for high-throughput, it is desirable to achieve those low backgrounds quickly after susceptor maintenance. In Figure 2, the background, as measured by Hg probe C-V, is

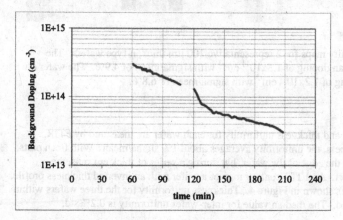

Figure 2 Time decay of background doping after susceptor rebuild. The background dopant concentration was measured by C-V for two consecutive runs.

plotted as a function of growth time after susceptor replacement. For this figure, data from two consecutive two-hour growths have been combined. After the first two-hour growth, the background is already approaching $1x10^{14}$ cm^{-3} (n-type) and we would normally commence product growth.

Reported values of doping uniformity depend on numerous factors. For example, the number and location of measurements, edge-exclusion, and the mean dopant concentration can all affect the uniformity value (usually std. dev./mean). For most production samples, we measure Hg-probe C-V at a relatively small number of locations. For thick epi, we typically use a nine-point cross pattern with 1 point in the center, 4 points at 1/3 R, and 4 points at 2/3 R. For that pattern, the mean uniformity of over 250 thick epi samples is 4.7%. On select wafers, a more extensive C-V measurement was performed. On these wafers the C-V was measured in the center point of each die in a hypothetical 8x8 mm die pattern. Results for two, greater than 100 μm thick, n-type epilayers is presented in Figure 3. The wafers have mean dopant concentrations of 1.8×10^{14} cm^{-3} and 7.2×10^{14} cm^{-3}, and doping uniformities of 4.9% and 6.8% respectively. The median inter-wafer uniformity, for the three wafers within a run, is 1.05% std. dev./mean.

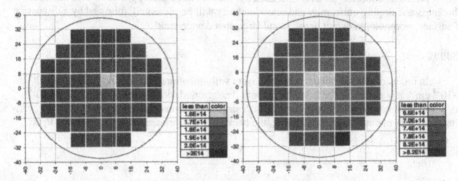

Figure 3 Doping uniformity maps for 2 representative 100 μm thick, n-type wafers. The wafer on the left has a mean doping of 1.8×10^{14} cm^{-3} with sigma/mean of 4.9%. The wafer on the right has a mean doping of 7.2×10^{14} cm^{-3} with sigma/mean of 6.8%.

Thickness

Epilayer thickness and thickness uniformity for each wafer are measured by FTIR. Irrespective of total thickness, the uniformity averages about 1% sigma/mean. With fresh parts, the layers are thicker near the edge of the wafer, but during a series of thick epi runs, the thickness profile will invert. An FTIR contour map of a wafer with an inverted thickness profile, *i.e.* thickest in the center, is shown in Figure 4. Thickness uniformity for the three wafers within a run is typically very good. The median value for inter-wafer uniformity is 0.2% std. dev./mean.

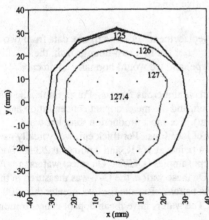

Figure 4 FTIR thickness map for a thick epilayer. For this wafer, the target thickness was 128 μm.

Reproducibility

Thickness and doping data have been collected from a large number of runs and wafers. For comparison purposes, the data from all runs has been transformed to deviation from target value expressed as a fraction, *e.g.* the target would have a value of 0, and 0.10 is equivalent to a 10% deviation from target. Histograms of the transformed data are presented in Figure 5. For thickness, the upper and lower specification limits, USL and LSL, are ±10%. For doping, USL and LSL are ±25%, and in this case a small fraction, 3.5%, of the wafers do not meet specification.

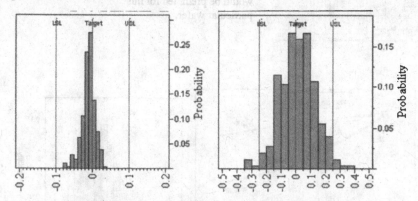

Figure 5 Histogram of thickness, expressed as fractional (%/100) deviation from target, on the left, and doping on the right.

Morphological Defect Density

Most of the epitaxial layers grown in this study were intended for 8x8 mm 10 A, 10 kV JBS or DMOS devices. To achieve greater than 50% yield on one of these wafers, it is necessary to keep the density of detrimental morphological material defects below 1 cm^{-2}. All wafers were grown using our carrot reduction process (CRP) buffer. Each wafer is scanned with an optical surface analyzer (OSA) to locate, count, and to some extent classify surface imperfections. Since defects are often clustered, we reanalyze the OSA data to determine a hypothetical die yield. An example of a die yield map is shown in Figure 6. In that figure, the point defect map is combined with an 8x8 mm grid. Although there are 15 total defects on this particular wafer, only 5 die (shaded) are affected. Similar data has been collected from a large number of thick epitaxial layers. The data has been summarized in Figure 7 where the calculated yield from the defect map has been plotted as a function of device area. Median yields for ~10 A (0.64 cm^2) and 20 A (1.5 cm^2) die sizes are 61.5% and 34.8% respectively. Relatively high defect densities on the wafers representing the minimum yield line in Figure 7 are almost always due to poor wafer preparation, handling, or reactor condition. Most of the defects on those wafers are particulate related, *e.g.* cubic inclusions and arrow defects.

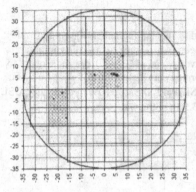

Figure 6 Hypothetical yield map for a 8x8 mm, 10 kV JBS diode epiwafer. Defects are detected and analyzed using an optical surface analyzer. Actual device yields correlate well with defect counts on wafers. A 90% maximum yield would be predicted for this particular wafer.

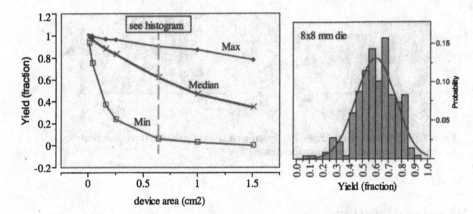

Figure 7 Hypothetical yield as a function of device area is shown on the left. Data is generated from defect maps on a large number of thick epitaxial layers. The complete distribution for 8x8 mm die, 0.64 cm^2, is shown at right.

Large-Area Device Yield

Results for large-area, 10 kV JBS diodes have been described in detail elsewhere [3]. In particular, results were described for several lots of thick epitaxial layers grown on substrates with significantly lower density of 1-C screw dislocations [7]. For those lots of 10 A, 10 kV JBS diodes, the average die yield was 61.3% with the best wafer yielding 80.6%. The median reverse bias leakage current for all diodes was 3.1 μA at 10 kV.

CONCLUSIONS

Production quantities of greater than 100 μm thick SiC epitaxial layers were grown in horizontal hot-wall reactors. The epitaxial layers exhibit good doping and thickness uniformities, and we have demonstrated estimable process control. A median morphological defect density of 1.38 cm^{-2} has been demonstrated, and defect densities of less than 1 cm^{-2} can be routinely achieved. These very low defect densities have made possible the fabrication of large-area JBS and DMOS devices with exemplary yield (>60%).

ACKNOWLEDGMENTS

This work was supported in part by DARPA Contract N00014-05-C-0202, monitored by Dr. Sharon Beermann-Curtin and Dr. Harry Dietrich, and by Pennsylvania State University Contract 0275-5-0141, monitored by Dr. Joseph Flemish.

REFERENCES

1. C. Hecht, B. Thomas, and W. Bartsch, Mat. Sci. Forum Vols. 527-529 (2006) pp 239-242.
2. Product Specifications, Cree, Inc., 4600 Silicon Dr., Durham, NC, USA.
3. B. Hull, J. Sumakeris, M. O'Loughlin, Q. Zhang, J. Richmond, A. Powell, M. Paisley, V. Tsvetkov, A. Hefner, and A. Rivera, presented at the 2007 International Conference on Silicon Carbide and Related Materials, Otsu, Japan, 2007 (to be published).
4. M. O'Loughlin and J. Sumakeris, U.S. Patent No. 7 230 274 (12 June, 2007)
5. J. Sumakeris, M. Paisley, and M. O'Loughlin, U.S. Patent No. 7 118 781 (10 October, 2006)
6. Candela CS2, KLA-Tencor, Inc., Freemont, CA, USA.
7. R. Leonard, Y. Khlebnikov, A. Powell, C. Basceri, M. Brady, I. Khlebnikov, J. Jenny, D. Malta, M. Paisley, V. Tsvetkov, R. Zilli, E. Deyneka, H. Hobgood, and C. Carter, presented at the 2007 International Conference on Silicon Carbide and Related Materials, Otsu, Japan, 2007 (to be published).

Mater. Res. Soc. Symp. Proc. Vol. 1069 © 2008 Materials Research Society

Influence of Growth Conditions and Substrate Properties on Formation of Interfacial Dislocations and Dislocation Half-loop Arrays in 4H-SiC(0001) and (000-1) Epitaxy

Hidekazu Tsuchida[1], Isaho Kamata[1], Kazutoshi Kojima[2], Kenji Momose[3], Michiya Odawara[3], Tetsuo Takahashi[2], Yuuki Ishida[2], and Keiichi Matsuzawa[3]

[1]Central Research Institute of Electric Power Industry (CRIEPI), 2-6-1 Nagasaka, Yokosuka, Kanagawa, 240-0196, Japan

[2]National Institute of Advanced Industrial Science and Technology (AIST), 1-1-1 Umezono, Tsukuba, Ibaraki, 305-8568, Japan

[3]SHOWA DENKO K.K. (SDK), 1505 Shimokagemori, Chichibu, Saitama, 369-1871, Japan

ABSTRACT

Formation of interfacial dislocations (IDs) and dislocation half-loop arrays (HLAs) and their appearance in 4H-SiC epi-wafers are investigated by X-ray topography and KOH etching analysis. Synchrotron reflection X-ray topography demonstrates the ability to image IDs and HLAs simultaneously and reveal their densities as well as spatial distributions in the epi-wafers. The vertical location of IDs in the epi-wafer is also examined by this technique. The influence of wafer warp, in-situ H_2 etching prior to epitaxial growth, substrate off-angle as well as the growth face (Si-face and C-face) on the densities and spatial distributions of IDs and HLAs are discussed.

INTRODUCTION

Silicon carbide (SiC) power devices are expected to be used in low-loss power conversion equipment in wide-ranging industrial areas. Although 4H-SiC Shottky diodes have been launched on the commercial market and many results obtaining outstanding performance of 4H-SiC switching devices have been reported, a reduction of extending defects in the material is expected to fabricate large devices and ensure the reliability of devices.

The formation of interfacial dislocations (IDs) with a Burgers vector of [11-20] type and a pure edge character near the substrate and epilayer interface, in which the difference in the nitrogen doping concentration between the substrate and epilayer exceeds a certain level, is reported by Jacobson et al. [1, 2]. Zhang et al. suggested the majority of IDs are formed by the sideway glide of preexisting basal plane dislocations (BPDs) in the epilayers [3]. More recently, Zhang et al. reported that the dislocation half-loop arrays (HLAs) [4, 5] are generated during epitaxial growth in the course of the sideway glide of BPDs [6]. In this paper, we report on the imaging of IDs and HLAs by synchrotron reflection X-ray topography and the spatial distribution of IDs and HLAs in epi-wafers in which the epitaxial growth is performed on substrates with different properties and growth conditions.

EXPERIMENT

Epitaxial growth was performed in a multi-wafer planetary reactor using a $SiH_4+C_3H_8+H_2$ system. For n-type doping, N_2 was introduced. Commercial n-type 4H-SiC (0001)Si-face and (000-1)C-face substrates with an off-cut angle of $4°$ or $8°$ towards [11-20] were used in the experiments. The doping concentration of the substrates is ~3×10^{18} cm^{-3}. Reflection X-ray topography was performed using a synchrotron beam source (SPring-8 BL-16) and a X-ray rotation target to investigate IDs and HLAs in the epilayer. Molten KOH was employed for defect selective etching to reveal HLAs. Wafer warp was monitored by the Corning Tropel Ultra Sort. The wafer warp measurement was made with an exception of 3 mm from the edges.

RESULT and DISCUSSION

A synchrotron X-ray topography image with a g-vector of 11-28 for a n⁻ 4H-SiC 8° off (0001)Si-face epilayer is shown in Fig. 1. The thickness of the epilayer is ~10 μm and the X-ray penetration depth is comparable to the epilayer thickness. In the imaged area, a high density of dark linear contrast perpendicular to the step flow [11-20] direction (parallel to [1-100]) is found. Some of the dark contrast lines curve towards the step flow direction at one end. This appearance of dislocations in X-ray topography matches with BPDs forming IDs reported by Jacobson et al. and Zhang et al. [2, 3]. Arrays of bright spots, which incline with a small angle from [1-100], are also confirmed in Fig. 1. One end of the bright spots arrays is connected to one end of a BPD forming an ID. We have confirmed that each of the bright spots corresponds to a pair of TEDs forming a TED pair array (HLA) by KOH etching analysis. These observations are perfectly consistent with the model of HLA formation reported by Zhang et al. in which a HLA is generated during epitaxial growth while a BPD glides sideway on the basal plane forming an ID [6]. Our results demonstrate that synchrotron reflection X-ray topography is able to non-destructively image IDs and HLAs at the same time without sample treatment requirements.

Figure 1. Synchrotron reflection X-ray topography image (g=11-28) of n⁻ 4H-SiC 8° off (0001)Si-face epi-wafer. The epilayer thickness is ~10 μm.

The vertical location of IDs is verified for an epi-wafer consisting a three-layer system of n⁺ substrate, n⁺ buffer epilayer (11 μm-thick) and n⁻ epilayer (11 μm-thick). The doping concentration of the n⁺ buffer epilayer (3×10^{18} cm^{-3}) is set to be the same as that of the n⁺ substrate (~3×10^{18} cm^{-3}). The doping concentration of the n⁻ epilayer is 1.5×10^{16} cm^{-3}, significantly lower than that of the n⁺ buffer epilayer. Figure 2 shows a synchrotron reflection X-ray topography image taken from the epi-wafer with the three-layer system. Dark horizontal topography contrast lines correspond to IDs. One end of the IDs curves towards the step flow direction as marked BPD1-BPD6 in Fig. 2. After KOH etching, oval BPD pits are confirmed at the surface end position of BPD1-BPD4, as shown in the inset of Fig. 2, while hexagonal TED pits are confirmed at the surface end position of BPD5 and BPD6. This confirms that BPD1-BPD4 reaches the n⁻ epilayer surface as a BPD, although BPD5 and BPD6 may have transformed into TEDs in the n⁻ epilayer. For BPD1-BPD4, the depth of IDs from the n⁻ epilayer surface is determined by the distance (marked d in Fig. 2) from the surface end of the BPDs to the corresponding IDs, taking account of the off-cut angle. All the depths of the four IDs of BPD1-BPD4 from the surface are identical as 10.7 μm, corresponding to the n⁻ epilayer thickness of 11 μm. This confirms that the IDs in Fig. 2 are lying at or near the interface of two different doping levels and not at the interface between the substrate and epilayer (buffer).

Density and spatial distributions of IDs and HLAs vary largely in the different wafers. Figure 3 shows reflection X-ray topography images (g=11-28) for 3-inch epi-wafers. The epilayers were grown on 4H-SiC 8° off (0001)Si-face substrate and (000-1)C-face substrate, respectively. The two substrates are sliced from the same bulk. The dotted ellipses are a guide for eyes and indicate areas exhibiting IDs in the topography images. The (0001)Si-face epi-wafer exhibits IDs at the central region of the wafer, although the (000-1)C-face epi-wafer exhibits IDs in the regions near the primary orientation flat and the opposite edge. This trend is typical when using substrates purchased from a source. We note that the (0001)Si-face substrates are systematically concave and (000-1)C-face substrates are convex towards the respective growth faces.

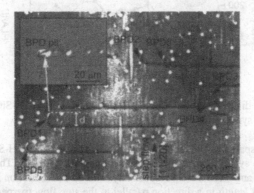

Figure 2. Synchrotron reflection X-ray topography image (g=11-28) of a 4H-SiC 8° off (0001)Si-face epi-wafer. The epi-wafer consists of a three-layer system of n⁺ substrate (~3x10¹⁸ cm⁻³ doping), n⁺ buffer epilayer (~3x10¹⁸ cm⁻³ doping, 11 μm) and n⁻ epilayer (~1.5x10¹⁶ cm⁻³ doping, 11 μm). The inset shows the KOH pit appearance at the surface end of BPD1, while the distance d corresponds to the thickness of the n⁻ epilayer.

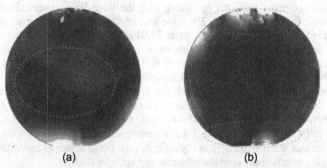

(a)　　　　　　　　(b)

Figure 3. Reflection X-ray topography images (g=11-28) of (a) 4H-SiC (0001)Si-face epi-wafer and (b) 4H-SiC (000-1)C-face epi-wafer. The dotted ellipses represent a guide for the eyes surrounding areas exhibiting IDs in the topography images. The (0001)Si-face wafer is concave (warp=57 μm) and the (000-1) wafer is convex (warp=62 μm) towards respective growth faces at room temperature before epitaxial growth.

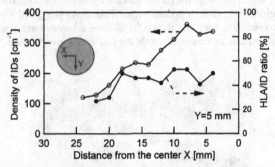

Figure 4. The spatial distributions of IDs and HLAs in a 3-inch 4H-SiC 8° (0001)Si-face epilayer obtained from synchrotron reflection X-ray topography images.

Figure 4 shows the spatial distributions of IDs and HLAs in a 3-inch 4H-SiC 8° (0001)Si-face epilayer obtained from synchrotron reflection X-ray topography images. The epilayer thickness and doping are 10 μm and 7×10^{15} cm^{-3}, respectively. The ID concentration is determined as the number of IDs per unit length in a direction parallel to the step flow (perpendicular to the IDs). The X direction in the inset of Fig. 2 is parallel to the step flow direction. The concentration of IDs peaks near the center of the wafer and gradually decreases with the distance from the center. The particular epilayer forms a high density of IDs exceeding 300 cm^{-1} near the center, although the maximum density of IDs is ~30 cm^{-1} for another investigated sample. The length of IDs extends as their density increases. Meanwhile, the spatial distribution of IDs indicates that the central wafer region was more strained than the outer region, since the density of IDs could relate to the magnitude of relaxed strain caused by the insertion of the dislocations. The density of IDs of 300 cm^{-1} and 30 cm^{-1} correspond to relative strains of $~1 \times 10^{-5}$ to $~1 \times 10^{-6}$ respectively, taking account of the <11-20> type Burgers vector of the IDs. We defined the HLA/ID ratio as a percentage of IDs forming a HLA. The HLA/ID ratio along the X direction is shown in Fig. 4.

The maximum HLA/ID ratio exceeds 50% in the central region of the wafer. The HLA/ID ratio is found to increase with the ID density and shows a larger value near the center of the wafer. This implies that HLA generation is preferentially promoted in the highly strained area.

In our experiment, wafer warp is found to influence the density and spatial distribution of IDs. Figure 5(a) compares the wafer warp and HLA density for the 4H-SiC (0001)Si-face epi-wafers. The HLA densities were obtained by counting the KOH etch pit array in a direction parallel to the step flow (perpendicular to the HLAs). The investigated wafers are concave towards the (0001)Si-face, while substrates having a larger warp value tend to result in a higher density of HLAs. If the temperature of the substrate backside is higher than the front side due to heating from the susceptor during epitaxy, concaved substrates may become more concave at a high temperature while convex substrates may become less convex or eventually concave. Such wafer warp can affect crystal bending of the {0001} growth faces.

The densities of IDs for 4H-SiC 8° (0001)Si-face epi-wafers grown with and without intentional in-situ H_2 etching are compared in Fig. 5(b). In-situ H_2 etching was performed at 1515°C for ~30 min. While epitaxial growth without intentional H_2 etching results in a high density of IDs of 19 cm^{-1}, few IDs (less than 1 cm^{-1}) are found when applying the intentional H_2 etching. Comparison of ID densities for 4H-SiC (0001)Si-face epi-wafers grown on 8° off substrate and 4° off substrate is shown in Fig. 5(c). Epitaxial growth on 4° off substrates results in a low density of IDs compared with growth on 8° off substrate. It has been reported that propagation of BPDs from the substrate into the epilayer can be suppressed by applying in-situ H_2 etching prior to epitaxial growth as well as using 4° substrates [7, 8], and the results in Fig. 5(b) and (c) indicates that the density of IDs is influenced by the density of BPDs in the epilayers. A difference in BPD propagation ratio between the (0001)Si-face and (000-1)C-face [7] can also influence the difference in the appearance of IDs in the epi-wafers shown in Fig. 3. In the synchrotron reflection X-ray topography images shown in Figs. 1 and 2, all the IDs are formed by the sideway glide of a BPD propagated from the substrate and no generation of grown-in new BPDs forming a ID is confirmed. A similar situation in which the majority of IDs are formed by BPDs propagated from the substrate has been reported by Zhang et al. [6]. Therefore, a reduction in the density of BPDs propagated from the substrate into the epilayer can be an effective way to reduce the density of IDs as well as HLAs.

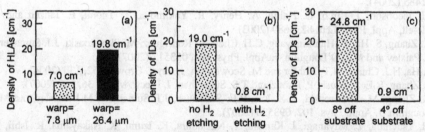

Figure 5. Comparison of (a) HLA densities for 4H-SiC 8° (0001)Si-face epilayers grown on substrates with different warp levels, (b) ID densities for 4H-SiC 8° (0001)Si-face epilayers grown with and without the intentional H_2 etching and (c) ID densities for 4H-SiC (0001)Si-face epilayers grown on 8° off and 4° off substrates.

According to the location of IDs identified in the three-layer system, the difference in nitrogen doping can be claimed as a source of stress arranging the IDs in a straight line at the interface. As discussed in the earlier report, incorporation of nitrogen shrinks the lattice constant of SiC and result in compressive stress for the n⁻ epilayer grown on n⁺ substrates [1]. In our experiment, wafer warp as well as growth faces (Si-face and C-face) are found to influence the density and spatial distribution of ID and HLA formation. The observed spatial distribution and the face difference are difficult to explain solely by the nitrogen induced lattice mismatch. A concaved warp (concaved crystal bending) can result in compression of the substrate lattice constant on the growth face and enhances compressive stress for the epilayer. Moreover, the temperature of the central region of the wafer can become higher than that of the outer region if the wafer is concave. A difference in BPD propagation from the substrate to epilayer can also influence the density of IDs. More systematic analysis of the sources of stress and BPD propagation are needed to explain the appearance of IDs and HLAs.

CONCLUSIONS

We have demonstrated that synchrotron reflection X-ray topography is able to image IDs and HLAs simultaneously. The vertical location of IDs in a three-layer system of n⁺ substrate (~3×10^{18} cm⁻³ doping), n⁺ buffer epilayer (3×10^{18} cm⁻³ doping) and n⁻ epilayer (1.5×10^{16} cm⁻³ doping) is found to be at or near the interface between n⁺ buffer epilayer and n⁻ epilayer. Spatial distributions and densities of IDs and HLAs are also investigated and found to be influenced by wafer warp, in-situ H_2 etching prior to epitaxial growth, substrate off-angle as well as growth face (Si-face and C-face).

ACKNOWLEDGMENTS

The authors wish to thank Drs. H. Okumura of AIST, Y. Saito and T. Sato of SDK for their support and discussion throughout this work.

REFERENCES
1. H. Jacobson, J.P. Bergman, C. Hallin, E. Janzèn, T. Tuomi and H. Lendenmann, J. Appl. Phys. 95, 1485 (2004).
2. H. Jacobson, J. Birch, C. Hallin, A. Henry, R. Yakimova, T. Tuomi, E. Janzèn and U. Lindefelt, Appl. Phys. Lett. 82, 3689 (2003).
3. X. Zhang, S. Ha, Y. Hanlumnyang, C.H. Chou, V. Rodriguez, M. Skowronski, J.J. Sumakeris, M.J. Paisley and M.J. O'Loughlin, J. Appl. Phys. 101, 053517 (2007).
4. S. Ha, H.J. Chung, N.T. Nuhfer and M. Scowronski, J. Cryst. Growth 262, 130 (2004).
5. Z. Zhang, R.E. Stalbush, P. Pirouz and T.S. Sudarshan, J. Electr. Mater. 36, 539 (2007).
6. X. Zhang, M. Skowronski, K.X. Liu, R.E. Stahlbush, J.J. Sumakeris, M.J. Paisley and M.J. O'Loughlin, J. Appl. Phys. 102, 093520 (2007).
7. H. Tsuchida, T. Miyanagi, I. Kamata, T. Nakamura, K. Izumi, K. Nakayama, R. Ishii, K. Asano and Y. Sugawara, Mater. Sci. Forum 483-485, 97 (2005).
8. K. Kojima, T. Kato, S. Kuroda, H. Okumura and K. Arai, Mater. Sci. Forum 527-529, 147 (2006).

Mater. Res. Soc. Symp. Proc. Vol. 1069 © 2008 Materials Research Society   1069-D04-04

Suitability of 4H-SiC Homoepitaxy for the Production and Development of Power Devices

Christian Hecht, Bernd Thomas, Rene Stein, and Peter Friedrichs
SiCED Electronics Development GmbH & Co. KG, Guenther-Scharowsky-Strasse 1, D-91058
Erlangen, Germany

ABSTRACT

In this paper, we present results of epitaxial layer deposition for production needs using our hot-wall CVD multi-wafer system VP2000HW from Epigress with a capability of processing 6×100mm wafers per run. Intra-wafer and wafer-to-wafer homogeneities of doping and thickness for full-loaded 6×100mm runs will be shown and compared to results of the former 7×3" setup. The characteristic of the run-to-run reproducibility for the 6×100mm setup will be discussed. To demonstrate the suitability of the reactor for device production results on Schottky Barrier Diodes (SBD) processed in the multi-wafer system will be given. Furthermore, we show results for n- and p-type SiC homoepitaxial growth on 3", 4° off-oriented substrates using a single-wafer hot-wall reactor VP508GFR from Epigress for the development of PiN-diodes with blocking voltages above 6.5 kV. Characteristics of n- and p-type epilayers and doping memory effects are discussed. 6.5 kV PiN-diodes were fabricated and electrically characterized. Results on reverse blocking behaviour, forward characteristics and drift stability will be presented.

INTRODUCTION

Within the last 10 years the market for SiC-based power devices has been developed very rapidly. Schottky Barrier Diodes (SBD) with blocking voltages up to 1200 V are on the market, SBD for higher reverse voltages, high-voltage PiN-Diodes, JFETs and MOSFETs are evaluated and qualified for potential applications. The cost/performance ratio was improved due to progress in the quality and size of the wafer material as well as by significant advances in the epitaxial growth of active layers.

The requirement of a cost-effective, reproducible and reliable epitaxial process led to the introduction of multi-wafer systems in 1998. Starting with multiple 35 mm and 2" wafer capability, multi-wafer systems for processing large-area wafers with diameters up to 100 mm are used today [1,2]. Important key parameters like homogeneities of doping and thickness, run-to-run reproducibility and wafer throughput must be considered when evaluating epitaxial reactors and processes for industrial applications. For the production of SBD and other power devices in the blocking voltage range up to 3 kV, multi-wafer systems can fulfill the criteria given above, whereas single-wafer hot-wall reactors are mostly used for the development of high-voltage PiN-diodes with their significantly thicker epitaxial layer.

The reduction of the off-orientation from 8° to 4° for 3" and 100 mm substrates led to new challenges for the epitaxial process. The commercially available SiC devices with their relatively thin epitaxial layer are fabricated on the cheaper 4°-off substrates. For thicker epitaxial layers, especially for the voltage range above 3 kV blocking voltage, the more expensive 8° off-oriented wafers are still most commonly used in order to maintain a good surface morphology by avoiding surface degeneration and step-bunching. In 2005, SiCED demonstrated the successful processing of PiN-diodes blocking more than 6.5 kV on 4° off-substrates [3].

In the first part of the paper, we present results of epitaxial layer deposition for the fabrication of SBD using our hot-wall multi-wafer reactor capable of processing 7×3" or 6×100mm wafers. New results using the 6×100mm configuration like intra-wafer and wafer-to-wafer homogeneities of doping and thickness will be shown and compared to our previously reported 7×3" results [4]. Results on SBD processed in the multi-wafer system will be given to demonstrate the suitability for device production.

In the second part, we show some results for n- and p-type SiC homoepitaxial growth on 3", 4° off-oriented substrates using a single-wafer hot-wall reactor VP508GFR from Epigress for the development of PiN-diodes with blocking voltages above 6.5 kV. Characteristics of n- and p-type epilayers and doping memory effects are discussed. For the characterization of epitaxially grown pn-junctions 6.5 kV PiN-diodes were fabricated and electrically characterized. Results on reverse blocking behavior, forward characteristics and drift stability will be presented.

EXPERIMENTAL DETAILS

A multi-wafer VP2000HW Planetary Reactor® with an actively heated ceiling built by Epigress was used for epitaxial layer deposition for the production of SBD in the voltage range up to 1700 V. The experimental details of this system are published elsewhere [5]. Recently the system was upgraded together with Epigress to a 100mm setup suited for a new 7×3" and 6×100mm susceptor configuration. For the development of high-voltage PiN-diodes a VP508GFR hot-wall reactor from Epigress with a capacity of 3×2", 1×3" or 1×100mm was used for the deposition of n- and p-type layers. The details of this reactor are described in [6]. Epitaxial layers were grown at temperatures between 1550 °C and 1650 °C using silane and propane as precursors, hydrogen as carrier gas, and argon for the Gas Foil Rotation of the wafer carrier. Nitrogen served as n-dopant. In the VP508GFR reactor tri-methyl-aluminum (TMA) was used as precursor for p-doping.

Commercially available 3-inch and 100 mm n-type 4H-SiC wafers with (0001) orientation (Si-face) and an off-orientation of 4° towards the <11-20> - direction were used as substrates.

RESULTS

A. MULTI-WAFER CVD

Overview of reactor setups

Starting with a 7×2" reactor setup in 2001, the wafer capacity of our multi-wafer reactor VP2000HW was increased to 5×3" followed by 7×3" up to 6×100mm today. Table I summarizes the properties of the different setups. With the increase in wafer diameter from 2" to 100 mm the total wafer area that can be processed was enlarged by more than a factor of 3 while maintaining the layer properties necessary for the cost-effective production of SBD in the voltage range up to 1700 V. A further increase of wafer throughput is planned by installing a new epitaxial reactor at our facilities built by Epigress until the end of 2008 that is capable of processing 10×100mm or even 6×150mm wafers.

Table I. Properties of epitaxial layers processed in different reactor setups since 2001.

Reactor setup	7×2"	5×3"	7×3"	7×3" new	6×100mm
Year of setup change	2001	2003	2005	2007	2007
Total wafer area [cm^2]	142	228	319	319	471
Intra-wafer doping homogeneity	6.0 %	10.0 %	4.1 %	6.5 %	4.4 %
Intra-wafer thickness homogeneity	0.4 %	2.0 %	1.5 %	2.2 %	2.4 %
Wafer-to-wafer doping homogeneity	2.7 %	3.3 %	1.4 %	4.8 %	2.3 %
Wafer-to-wafer thickness homogeneity	0.4 %	1.6 %	2.2 %	1.0 %	0.9 %
Results presented in reference	[5]	[2,5]	[4]	[7]	this work

Intra-wafer doping and thickness homogeneity

Figure 1 shows a typical doping uniformity profile of a 14 µm thick layer on a 100 mm wafer at a depth of 1 µm from the surface. From the center of the wafer towards the wafer edge the doping concentration slightly increases until approximately 1/3 r (r: radius of the wafer) and then decreases reaching its minimum near the wafer edge. Clearly, the axial symmetric distribution, caused by the wafer rotation, can be seen. Considering an edge exclusion of 3 mm the mean doping value N_D-N_A and sigma/mean are $3.94×10^{15}$ cm^{-3} and 4.4 %, respectively.

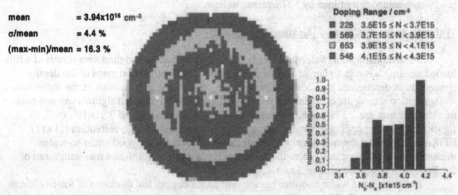

Figure 1. Intra-Wafer doping uniformity of a 14 µm thick epitaxial layer on a 100 mm wafer (6×100mm setup) determined by CV measurements.

A typical thickness mapping of a 14 µm epitaxial layer on a 100 mm wafer is shown in figure 2. Again, the axially symmetric distribution, caused by the wafer rotation, can be seen. Like the doping distribution, the maximum thickness is located at approximately 1/3 r, whereas the minimum thickness can be seen at the wafer edge. Typical values of sigma/mean and max-min/mean with an edge exclusion of 3 mm are 2.4 % and 7.0 %, respectively.

Compared to the data we reported in [7] a significant improvement of the intra-wafer doping homogeneity while maintaining the very good thickness homogeneity can be noticed for the 6×100mm setup, mainly caused by ongoing process optimization and increased wafer throughput after switching a part of the production to 100mm wafers. Now the results are comparable to the excellent data we reported for the old 7×3" setup with the smaller susceptor [4].

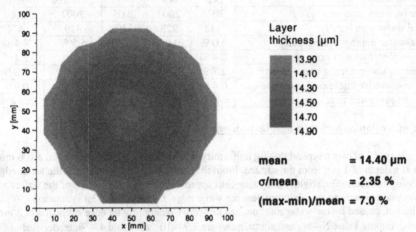

Figure 2. Intra-Wafer thickness uniformity of a 14 μm thick epitaxial layer on a 100 mm wafer (6×100mm setup) determined by FTIR measurements.

Wafer-to-wafer doping and thickness homogeneity

To demonstrate the wafer-to-wafer homogeneity of doping and thickness results of a full-loaded run are shown in figure 3 to figure 5. In figure 3 high-resolution maps of the doping concentration determined by CV measurements are presented with indication of the individual wafer position during the epitaxial run. With an edge exclusion of 3 mm sigma/mean and max-min/mean values are 2.3 % and 6.6 %, respectively, at a mean doping of 3.91×10^{15} cm^{-3}. In figure 4, the corresponding thickness homogeneity is presented by high-resolution (11×11) FTIR-maps showing the individual intra-wafer uniformities. A very good wafer-to-wafer thickness homogeneity can be seen, indicated by sigma/mean of 0.9 % and max-min/mean of 2.3 % with an edge exclusion of 3 mm.

In figure 5, the wafer-to-wafer homogeneities for doping and thickness of the 6×100mm setup (a) and the new 7×3" setup (b) are shown. The normalized average thickness and doping results for each wafer are depicted, labeled wih the wafer position in the susceptor. It can be seen that the wafer-to-wafer doping homogeneity for the 6×100mm setup is significantly better than the one for the 7×3" setup. This can be contributed to further hardware and process improvements performed after the last change of the susceptor to 6×100mm .

132

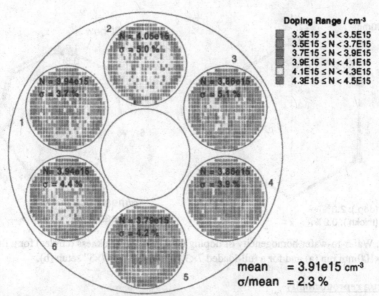

Figure 3. Schematic of wafer position and doping homogeneity of a full-loaded 6×100mm run.

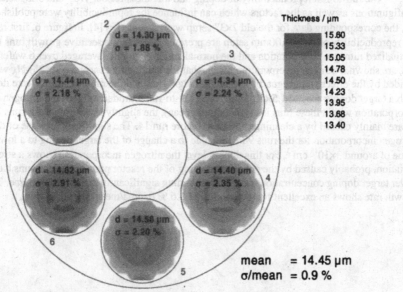

Figure 4. Schematic of wafer position and thickness homogeneity of a full-loaded 6x100mm run.

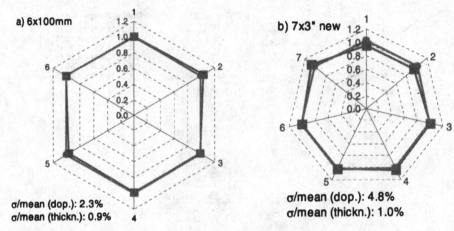

Figure 5. Wafer-to-wafer homogeneity of doping (squares) and thickness (circles) for a full-loaded 6×100mm run (a) and for a full-loaded 7×3" run in the new 7×3" setup (b).

Run-to-run reproducibility

The run-to-run reproducibility of doping and thickness for our 7×2" and 5×3" reactor configurations as well as the factors which can influence the reproducibility were published in [2], the corresponding data for the old 7×3" setup was presented in [4]. In figure 6, first results of the reproducibility of the 6×100mm setup are presented. For 15 consecutive growth runs the normalized nitrogen incorporation and the normalized growth rate, averaged over 6 wafers per run, are shown. Nitrogen incorporation means that the net doping concentration N_D-N_A was divided by the effective nitrogen flow. The first eight runs and the runs 14 and 15 were done with a target doping around $1.5×10^{16}$ cm^{-3}. The run-to-run reproducibility of the nitrogen incorporation within these runs is 4.4 % (sigma/mean), the slightly lower values for runs 14 and 15 are mainly caused by a cleaning procedure before run 14. The significant increase of the nitrogen incorporation for the runs 9 to 13 is due to a change of the target doping to a lower value of around $3×10^{15}$ cm^{-3}. For this doping level the nitrogen incorporation shows a stronger variation, probably caused by the enhanced coating of the reactor parts during the runs having a lower target doping concentration and a corresponding significantly longer growth time. The growth rate shows an excellent reproducibility of 1.6 % (sigma/mean) for all 15 runs.

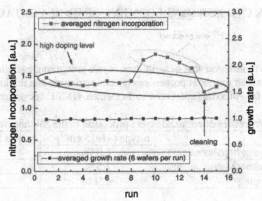

Figure 6. Nitrogen incorporation and growth rate reproducibility of the 6×100mm setup.

Results for Schottky barrier diodes

With the 7×3" setup used since 2005 several thousands of wafers were processed for the production of SBD with blocking voltages between 300V and 1700V. More than 99 % of the wafers did meet the epi layer specifications given by the device designers demonstrating an excellent reproducibility for growth rate and nitrogen incorporation.

As a measure of the overall "killer-defect" , density the reverse blocking yield for a 3" wafer with 1200 V SBD is shown in figure 7a. In spite of the large active area of 23 mm^2 more than 70 % of the devices passed the yield criteria (filled squares) representing a "killer-defect" density of less than 1.6 cm^{-2}. In figure 7b the distribution of the epi yield (low-voltage blocking capability measured directly after the epitaxial process with a chip area of 2.08 mm^2) for 100 wafers with epitaxial layers for SBD with blocking voltages of 1700V is given. The mean value of 92.6 % reflects an average defect density of 3.7 cm^{-2} demonstrating that the multi-wafer epitaxial process is well-suited for the volume production of high-power SBD.

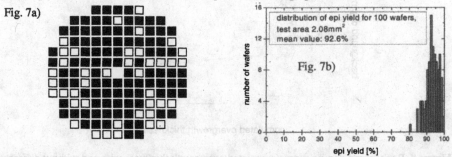

Figure 7. a) Low-voltage blocking yield of a 3" wafer with 1200 V SBDs and an active area of 23 mm^2 (open squares: failed chips, filled squares: good chips), and b) Distribution of low-voltage blocking yield after epitaxial processing of 3" wafers for 1700 V SBD with an active area of 2.08 mm^2.

B. SINGLE-WAFER CVD FOR DEVELOPMENT OF HIGH-VOLTAGE PIN-DIODES

Epitaxial results.

We already demonstrated that n- and p-type layers grown in the single-wafer hot-wall reactor VP508GFR exhibit excellent properties concerning doping and thickness homogeneity [3,4]. Typical values for epitaxial layers on 4° off-oriented 4H-SiC substrates are summarized in table II.

Table II. Properties of epitaxial layers processed in the SW hot-wall reactor.

Dopant	n-type (~1e15 cm⁻³)	p-type (~1e18 cm⁻³)
Intra-wafer doping homogeneity	8.0 %	2.8 %
Intra-wafer thickness homogeneity	2.5 %	1.5 %
Background doping	< 5e14	< 5e14

Doping memory effect

For the growth of low-doped n-type and highly-doped p-type layers in the same reactor it is essential to minimize the doping "memory effect" leading to an increase of the background doping after p-doped runs in the high doping range [4]. Special procedures were applied to maintain the low background doping given in table II. In figure 8 the residual background doping after a highly p-doped run and consecutive overgrowth runs is shown. With special cleaning procedures the background could be decreased within a single-run below $5e14 \text{ cm}^{-3}$ (s. triangles & dotted line in figure 8).

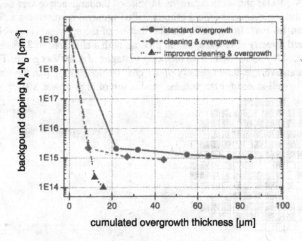

Figure 8. Background doping versus cumulated overgrowth thickness with and without different cleaning procedures.

Device fabrication and electrical measurements

6.5 kV PiN-diodes with an active device area of 5.7 mm^2 were fabricated on n-type drift layers with a continuously grown p-type emitter layer. In figure 9a, the averaged reverse characteristic of 184 diodes is shown. Up to the onset of avalanche, the devices exhibit extremely low leakage currents. About 50 % of the devices fulfill our strict yield criteria at 6.5 kV. The forward characteristic of a representative mounted diode at 25 °C and 150 °C before and after current stressing at 100 A/cm^2 for 1h is shown in figure 9b. The diode exhibits a low forward voltage drop of 3.50 V @ 25 °C and 100 A/cm^2. After stressing the device by applying a DC forward current only a slight increase of the forward voltage drop V_F (Δ=0.04 V) at 25 °C and no change of V_F at 150 °C could be observed demonstrating an excellent drift stability.

Generally, the overall basal plane dislocation (BPD) density has been made responsible for the degradation of bipolar SiC devices [8,9]. Sumakeris et al. reported that several measures like e.g. a selective KOH-etch procedure can improve the V_F stability [10]. For our diodes, no special pretreatments were performed apart from the continuous growth process that also leads to a decrease of the BPD density [10]. Further experiments in order to reduce the BPD density by wafer pretreatments and variations of the epitaxial process are currently under investigations. Parts of the results were presented by Kallinger et al. in [11], showing that a special combination of growth parameters can lead to a very low BPD density which is comparable to the results reported in [10] for the KOH-etch procedure. Apart from the BPD density, the design of the PiN-diode like e.g. the thickness of the p-emitter may also significantly influence the drift stability. Details concerning the influence of these parameters can be found in [12].

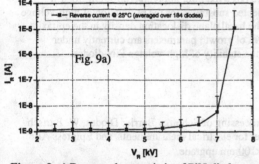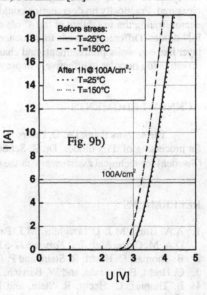

Figure 9. a) Reverse characteristic of PiN-diodes at 25 °C, averaged over 184 diodes, and b) Forward characteristics of a mounted PiN-diode at 25 °C and 150 °C before and after current stressing with 100 A/cm^2 for 1 h.

SUMMARY

Epitaxial SiC layers on 100 mm wafers were grown by CVD using a VP2000HW multi-wafer reactor with an upgraded 6×100mm setup. The intra-wafer thickness and doping uniformities of 2.4 % and 4.4 %, respectively, fulfill the demands on layer properties for high-voltage power devices. Excellent values for the wafer-to-wafer uniformity of thickness and doping (0.9% and 2.3%, respectively) could be presented. First data on the reproducibility of the 6×100mm setup demonstrated a stable performance of the reactor especially for the growth rate and the nitrogen incorporation at higher doping levels. For the former 7×3" setup the reproducibility of the epitaxial growth meets the requirements for production process management, demonstrated by a failure rate of less than 1 % averaged over more than thousand wafers and an average epi yield of 92.6 % for one hundred wafers with epitaxial layers for 1700 V SBDs. An overall defect density of less than 1.6 cm^{-2} could be demonstrated with a reverse blocking yield of more than 70 % for a wafer with 1200 V SBD having an active area of 23 mm^2.

N- and p-type epitaxial layers were grown on 3-inch 4° off-oriented 4H-SiC substrates in a VP508 GFR single-wafer reactor. Excellent doping and thickness homogeneities for both types of layers could be achieved. To enable the continuous growth of pn-layers for PiN-diodes a special cleaning procedure was applied preventing long-time overgrowth runs caused by the Al-doping memory effect. PiN-diodes with blocking voltages of more 6.5 kV were fabricated on continuously grown pn-layers showing excellent reverse characteristics with a blocking yield of more than 50 %. Forward voltage drops of 3.50 V @ 25 °C and 3.25 V @ 150 °C were measured. Practically no degradation could be observed after applying 100 A/cm^2 for 1 h demonstrating the suitability of the 4° off-oriented substrates for the production of high-voltage PiN-diodes. Different methods to influence the basal plane dislocation density in the epitaxial layer like e.g. wafer pretreatments and changes of growth parameters are currently under investigation in order to effectively improve the stability of high-voltage PiN-diodes.

ACKNOWLEDGMENTS

The authors thank Dr. D. Peters for processing of Schottky Barrier Diodes, W. Bartsch for processing of PiN-diodes, Dr. R. Schörner for electrical measurements, and Epigress AB (Sweden) for technical assistence with the 6×100mm upgrade.

REFERENCES

1. A.A. Burk, M.J. O'Loughlin, M.J. Paisley, A.R. Powell, M.F. Brady, R.T. Leonhard, and D.A. McClure, *Mat. Sci. Forum* **527-529**, 159 (2006).
2. B. Thomas, C. Hecht, R. Stein, and P. Friedrichs, *Mater. Sci. Forum* **527-529**, 135 (2006).
3. C. Hecht, B. Thomas, and W. Bartsch, *Mater. Sci. Forum* **527-529**, 239 (2006).
4. B. Thomas, C. Hecht, R. Stein, and P. Friedrichs, *Mater. Res. Soc. Symp. Proc.* **911**, 37 (2006).

5. B. Thomas and C. Hecht, *Mater. Sci. Forum* **483-485**, 141 (2005).
6. J. Zhang, U. Forsberg, M. Isacson, A. Ellison, A. Henry, O. Kordina, and E. Janzen, *J. Cryst. Growth* **241**, 431 (2002).
7. C. Hecht, B. Thomas, R. Stein, and P. Friedrichs, *presented at ICSCRM 2007, Otsu, Japan* (2007) (in press).
8. J.P. Bergman, H. Lendenmann, P.A. Nilsson, U. Lindefelt, and P. Skytt, *Mat. Sci. Forum* **353-356**, 299 (2001).
9. J.J. Sumakeris, M.K. Das, S. Ha, E. Hurt, K. Irvine, M.J. Paisley, M.J. O'Loughlin, J.W. Palmour, M. Skowronski, H. McD. Hobgood, and C. H. Carter Jr., *Mat. Sci. Forum* **483-485**, 155 (2005).
10. J.J. Sumakeris, J.P. Bergman, M.K. Das, C. Hallin, B.A. Hull, E. Janzen, H. Lendenmann, M.J. O'Loughlin, M.J. Paisley, S. Ha, M. Skowronski, J.W. Palmour, and C. H. Carter Jr., *Mater. Sci. Forum* **527-529**, 141 (2006).
11. B. Kallinger, B. Thomas, and J. Friedrich, *presented at ICSCRM 2007, Otsu, Japan* (2007) (in press).
12. W. Bartsch, B. Thomas, H. Mitlehner, B. Bloecher, and S. Gediga, *presented at EPE 2007, Aalborg, Danmark*, (2007) (in press).

Mater. Res. Soc. Symp. Proc. Vol. 1069 © 2008 Materials Research Society   1069-D05-01

Improved SiC Epitaxial Material for Bipolar Applications

Peder Bergman[1,2], Jawad ul Hassan[1], Alex Ellison[2], Anne Henry[1], Philippe Godignon[3], Pierre Brosselard[3], and Erik Janzén[1]

[1]Department of Physics, Chemistry and Biology, Linköping University, IFM, Linköping, SE-581 83, Sweden

[2]Norstel AB, Ramshällsvägen, Norköping, S-60116, Sweden

[3]CNM-IMB-CSIC, Campus UAB, Bellaterra, Barcelona, ES-08193, Spain

ABSTRACT

Epitaxial growth on Si-face nominally on-axis 4H-SiC substrates has been performed using horizontal Hot-wall chemical vapor deposition system. The formation of 3C inclusions is one of the main problem with growth on on-axis Si-face substrates. In situ surface preparation, starting growth parameters and growth temperature are found to play a vital role in the epilayer polytype stability. High quality epilayers with 100% 4H-SiC were obtained on full 2" substrates. Different optical and structural techniques were used to characterize the material and to understand the growth mechanisms. It was found that the replication of the basal plane dislocation from the substrate into the epilayer can be eliminated through growth on on-axis substrates. Also, no other kind of structural defects were found in the grown epilayers. These layers have also been processed for simple PiN structures to observe any bipolar degradation. More than 70% of the diodes showed no forward voltage drift during 30 min operation at 100 A/cm^2.

INTRODUCTION

The superior physical properties of SiC, like wide band gap, high thermal conductivity and high breakdown electric field, make it potentially useful semiconductor material for high power, high temperature and high frequency electronics devices. SiC exists in many different polytypes and the difference between polytypes lies in the stacking sequence of Si-C bilayers along the c-axis. In order to replicate the substrate polytype into the epilayer, off-axis substrate with several degree off-cut typically along [$11\bar{2}0$] direction is used. The steps produced on the off-axis surface reveal the substrate polytype and act as a continuous source of polytype information of the crystal during epitaxial growth [1].

However, the major problem with the growth on off-cut substrate is the replication of the basal plane dislocations (BPDs) from the substrate into the epilayer. In the case of high power bipolar electronic devices the BPDs are reported to act as a source of the formation of the stacking faults on the basal plane during bipolar injection. These stacking faults result in an increased forward voltage drop and is known as bipolar degradation [2,3]. Bipolar degradation has in recent years been extensively studied. The increase in the forward voltage drop in bipolar PiN diodes due to the expansion and propagation of the stacking faults along the basal-plane throughout the active area has been well understood. Different techniques have also been proposed to reduce the number of basal-plane dislocations that are the main source of the stacking faults [4,5]. These techniques increase the conversion rate of the basal plane dislocations to threading dislocations usually at the substrate-epilayer interface.

In this work we have investigated another possible way to reduce the bipolar degradation by using on-axis epitaxial layers. The main advantage of growth on on-axis is that the basal plane dislocation originally in the substrate will not propagate into the epitaxial layer or into the active region of a PiN diode, where the presence of electron-hole plasma provides the energy for the stacking fault expansion. Therefore, the total number of defects can also be reduced in the epilayer. Several groups have reported successful homoepitaxial growth on C-face, on-axis 4H-SiC substrates [6] owing to the fact that C-face is generally used for the bulk growth of 4H-SiC while bulk growth on Si-face results in switching of the polytype to 6H-SiC [7] therefore, C-face has polytype stability for 4H-SiC. However, epitaxial growth on C-face is of less interest, especially for high power devices, due to higher background doping of the active device layer and bad control on high doping of the p-type contact layer. Also, processing such as etching and implantation is more difficult and less established than on Si-face [8].

The main challenge with on-axis epitaxial growth is the difficulty to maintain the polytype replication from the substrate [9,10], which is better controlled by the off-axis step flow growth conditions. It has been shown previously that the polishing scratches on the substrate surface are the preferential sites for the growth of 3C inclusions [11]. Therefore, surface preparation is a crucial step for the homoepitaxial growth. In-situ surface preparation in hydrogen ambient, in the temperature range of 1400-1800 °C, is already know to improve the surface morphology through removing polishing scratches on off-cut substrate. In situ surface preparation under different ambient conditions may have different effect on on-axis substrate. In this study we report on in-situ surface preparation and homoepitaxial growth on full 2" Si-face on-axis wafers. Furthermore, we will show our preliminary results on the electrical characterization of the PiN diodes fabricated on these layers.

EXPERIMENT

The samples used in this study were commercially obtained nominally on-axis, Si-face polished 4H-SiC. The samples were either full 2" wafers or 1 cm^2 pieces cut from 3" wafers. Hot-wall chemical vapor deposition [12] was used for in-situ etching of the samples and homoepitaxial growth. Hydrogen purified through heated pladium membrane was used as carrier gas. Silane and propane gases were used as the sources of Si and C respectively while nitrogen and trimethylealuminum were used as n- and p-type dopants respectively. Etching was performed under C-rich conditions, Si-rich conditions and in pure hydrogen. Optical microscopy with Nomarski interface contrast and Atomic Force Microscopy (AFM) in tapping mode were used to observe the surface morphology, step structure and surface roughness on all samples before and after etching. The growth was performed at a temperature of 1620 °C with a growth pressure of 200 mbar with C/Si = 1. A series of samples with epilayer thickness of 10-40 μm were grown with a growth rate of 3 μm/h. Doping of the n-type layers was between 5×10^{14} cm^{-3} to 5×10^{18} cm^{-3} while for p-type was between 1×10^{15} cm^{-3} to 5×10^{19} cm^{-3}. The off angle of the substrates were measured by x-ray diffraction (XRD) using 'X-Pert' XRD setup. The doping of the epilayer was measured either through CV measurements on mercury probe station. The polytype identification was done using high resolution X-ray diffraction (HRXRD) or illumination of the epilayer under UV laser light at 77K. Also PL mapping at liquid helium temperature using either 244nm or the 351 nm line of an Ar$^+$ laser was performed to identify the presence of any other polytype. Synchrotron white beam X-ray topography (SWBXT) was performed using the synchrotron radiation source HASYLAB-DESY Hamburg. SWBXT was made in back reflection mode on some selected areas of the sample.

DISCUSSION

In-situ surface preparation

As received SiC samples are known to have polishing related damages as shown in Fig. 1. In-situ etching, under c-rich conditions, prior to the epitaxial growth is a well known process to reduce the surface damages and to avoid the formation of Si droplets on the surface [13,14]. We have performed in-situ etching using different ambient conditions such as etching in pure hydrogen, in a mixture of hydrogen and propane (carbon rich condition) and in a mixture of

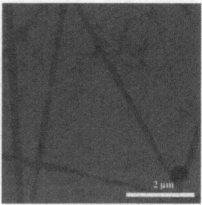

Fig. 1 AFM image taken from as received substrate. Polishing related damage is visible.

hydrogen and silane (silicon rich conditions) detail of which can be found somewhere else [15]. In-situ etching, under all conditions, results in the removal of all polishing related damages and also the surface appears with step bunching which is attributed to the high surface energy on Si-face.

Etching under pure hydrogen resulted in the formation of zigzag shaped macro-steps. C-rich conditions resulted in the formation of broken, non-linear and irregularly spaced macro-steps while under Si-rich conditions macro-steps, on the sample surface, appeared as linear, continuous and regularly spaced as shown in Fig. 2a. AFM images taken with tapping mode showed a small and uniformly distributed macro-steps on the sample etched under Si-rich conditions as compared to that etched under pure hydrogen and under C-rich conditions also the surface roughness on the sample etched under Si-rich conditions was small. AFM Analysis of the whole surface of all samples showed that all of the screw dislocations opened up in the form of shallow hexagonal etch pits with typical hexagonal spiral geometry, as shown in Fig. 2b. The micro-steps originating from the screw dislocations cover the entire surface and the height of the micro-steps is equal to unit cell height i.e., 1nm in the case of 4H-SiC. These high density micro-steps act as a sink for the incoming atoms to the surface and replicate the polytype information from the substrate into the epilayer.

Under pure hydrogen ambient conditions, Si-droplets are know to form on the surface of off-cut substrate during in-situ etching and even during temperature ramp-up. No such behavior was found on the surface of on-axis substrate even under etching in Si-rich conditions. Similar behavior was found on both Si and C-face substrates regarding Si-droplets. This could be attributed to a low partial pressure of Si on on-axis substrate surface as compared to that on off-cut substrate under same conditions. The experimental evidence showed that in-situ etching under Si-rich conditions effectively removed the surface damages and produced homogeneous step structure with lowest surface roughness on Si-face on-axis substrate surface. Therefore, in situ etching under Si-rich condition was employed during all of the growth experiments.

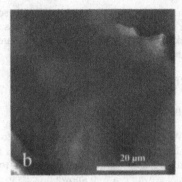

Fig. 2   Si-face etched in (a) Si-rich conditions b) AFM image taken from the sample etched in Si-rich conditions.

Homoepitaxial growth of high quality epilayers

Most of the defects, including 3C inclusions, nucleate at the substrate-epi interface and continue to grow in the epilayer. Therefore, any abrupt change, at the beginning of the growth, in the gas phase may result in the formation of such defects. Therefore, after in-situ etching it is important to change the growth mode from negative growth (etching) to positive growth very slowly and gradually. A very low supersaturation at the beginning of the growth avoids the 2D nucleation of 3C-SiC on wide terraces and the polytype of the substrate replicates into the epilayer through preferential growth at the steps produced by the screw dislocations. The growth was started with a very low concentration of the precursors with C/Si = 1, at the inlet of the susceptor. The flow rate of the precursors was gradually increased, to avoid any abrupt change in the gas phase, to obtain a stable growth rate of 3 μm/h. After the substrate is covered with a few hundreds nm of high purity homoepitaxial layers, growth can be continued with a stable higher growth rate.

Epitaxial layers, on off-cut substrate, are normally grown in the temperature range of 1500 °C to 1580 °C. On off-cut substrates the step density is high and at the given temperature range the adatom diffusion length is high enough to incorporate at a suitable lattice site to reproduce the substrate polytype into the epilayer. In the case of on-axis substrate, depending on the small off-cut, the steps density produced by the minor off-cut is very low. Therefore, homoepitaxial layers grown on nominally on-axis Si-face substrates show a columnar growth following the spiral growth mode. Such features appear on the surface as a result of low adatom surface diffusion length. The adatom surface mobility can be increased by increasing the growth temperature. Growth at high temperature also enhances re-evaporation of the unstable 3C nuclei along with other defects formed at the epi-substrate interface. Increased adatom diffusion length at high temperature will on the other hand increase the probability of the adatom to reach at a suitable lattice sites.

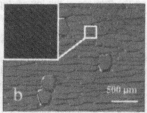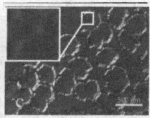

Fig. 3 Optical image taken from 12 μm thick epilayer with a) very high screw dislocation density with local surface orientation close perfectaly on-axis b) low screw dislocation density will localy tilted surface orientation c) low screw dislocation density with local surface orientation close to perfectaly on-axis.

Epitaxial layers with a thickness of 10-12 μm thickness were grown at a growth temperature of 1620 °C. The polytype identification based on illumination of the epilayer at 77K under UV light showed 100% 4H-SiC. Fig. 3a-c shows three different kinds of surface morphologies of the epilayer. In epitaxial growth, growth mechanism mainly depends on the surface structure of the substrate. In the case of high off-cut substrate the growth is mainly followed through the step-flow growth mode while in the case of ideally on-axis surface the growth is followed through the spiral growth mode via the micro-steps provided by the screw dislocations intersecting the surface. In the case of nominally on-axis substrate the situation could be entirely different. As it is not possible to get absolutely on-axis surface, there is always a small tilt. Especially in the case of SiC substrate, grown at very high temperature, there is always a bending of the lattice planes in the wafer which may result in local variation in the surface off-orientation. Therefore, on the same wafer we may have local areas with surface orientation close to perfectly on axis and local areas with surface orientation with a little off-cut. The typical specification from most suppliers is off-axis angle of $\pm 0.05°$ for nominally on-axis wafers. From our HRXRD measurements of the wafers used in this study we have observed small variations of $\pm 0.08°$. In the areas where surface is close to perfectly on-axis, spiral growth mechanism is dominant as can be seen in the high magnification optical images a and c in Fig. 3. The columnar growth in c is probably due to low dislocation density in this area as compared to in a. Also, the surface roughness, as measured with AFM in tapping mode, is higher in c as compared to that in a. The nearby interacting spirals meet each other perfectly and no defect or foreign polytype inclusion was found at the interface. AFM image taken from c, seen in the inset, shows that the surface is covered with micro-steps of unit cell height with typical spiral shaped geometry. In the high magnification image b step-flow growth is found to be dominant, probably due to the locally small off-axis in this area. AFM image taken from this area, seen in the inset, shows that the surface is covered with periodically arranged linear steps of unit cell height. The growth steps, at the beginning of the growth, are supplied by the micro-steps originated from threading screw dislocations intersecting the surface and uncovered during in-situ etching. Under very low supersaturation the incoming atoms incorporate preferentially at these micro-steps and the growth is started through the spiral growth mode. After the growth of a few tens of nm the micro-steps start to accumulate at the boundaries of macro-steps but the terraces of the macro step are still covered with micro-steps of unit cell height. The incoming atoms incorporate at the micro-steps and in this way the macro-steps keep advancing in the off-cut direction on the terraces of the lower steps while the micro-steps follow the crystallographic orientation of the original spiral and the growth mode switches to the step-flow growth mode.

A 40 μm thick epitaxial layer grown on Si-face, on-axis, 2" full wafer, under the growth conditions described above, also showed 100% replication of substrate polytype into the epilayer. The polytype identification was primarily done by the illumination of full wafer, immersed in liquid nitrogen, under UV light as shown in Fig. 4. No sign of 3C-SiC inclusions was seen including at the edges of the wafer. In order to examine the purity of the epilayer, low temperature PL mapping was performed on the entire wafer. PL spectrum, taken from the middle of the wafer, shown in Fig.4b, in the energy range of 3.3-2.1 eV does not show the presence of any foreign polytype. The PL spectrum is dominated by the NBG emission where the N-BE and the FE lines can be observed. The inset shows characteristic NBE spectrum, both N-BE and FE related peaks are visible. As the layer was intentionally doped the intensity of N-BE is higher then the FE peak. For low doped material with nitrogen concentration $< 3 \cdot 10^{16}$ cm^{-3} the relative intensity between the N-BE no-phonon line, such as Q_0 in 4H-SiC, see the inset in Fig. 4b, and one of FE line, such as I_{76} the strongest phonon replica of the FE, has allowed a quantitative estimation of the doping concentration [16]. The concentration of the nitrogen determined by this way is $3 \cdot 10^{15}$ cm^{-3} for this doped layer. AFM analysis of the epilayer showed that the surface roughness is increasing with the increasing thickness of the epilayer. The macro-step height, as found by AFM, could reach to over 100 nm for 40 μm thick epilayer without giving rise to the nucleation of 3C-SiC inclusions.

The HRXRD analysis of the epilayer also confirms the 100% replication of the substrate polytype into the epilayer and no 3C or any other foreign polytype related reflection was observed in 2θ-ω scan. In the case of homoepitaxial growth on off-cut substrate BPD conversion into TED, at epi-substrate interface, is a well known process. Though, in this process, the total

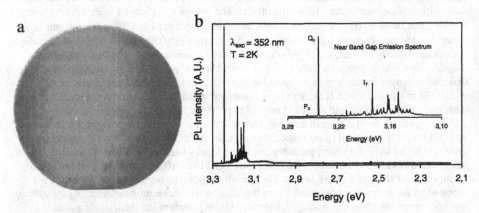

Fig. 4   (a) PL image taken under UV illumination at 77K from a full 2" wafer with 40 μm thick epilayer. (b) PL spectrum taken at 2K at close to the edge of the wafer, the inset shows the chracteristic NBG emission spectrum.

number of BPD in the epilayer is reduced but still the total number of dislocations is same as that in the substrate. One of the main reasons to grow homoepitaxial layers on on-axis substrate was to eliminate the BPD propagation form the substrate into the epilayer so in this way total number of defects can also be reduced in the epilayer. SWBXT images taken in the back reflection do not show basal plane dislocation related contrast, in any possible reflection, in the epilayer. Epilayers grown on off-cut substrate are also known to have some typical epi-defects like triangular defects, half moon, carrots and growth related pits on the surface. Preliminary studies based on optical microscopy and SWBXT show that the epilayers grown on Si-face on-axis substrate do not show such kind of surface or structural defects. In-grown basal plane stacking faults are another kind of structural defects which are often present in epilayers grown on off-cut substrate. CL study on on-axis grown epilayers do not show any basal plane stacking faults. On the other hand, the surface roughness is quite high as compared to the epilayers grown on the off-cut substrates which is a subject of further studies.

Processed PiN Diodes on On-axis grown epitaxial layers

The surface preparation described above was used to grow a complete structure for a PiN diode. The epitaxial layers ere grown in-situ and consisted of a n+ buffer layer, a 30 µm n-layer doped with nitrogen to 1-2 10^{15} cm^{-3}, a p+ -layer of 1 µm and Al doped to 2-3 10^{18} cm^{-3}, and finally a 5 µm p++ contact layer doped > 10^{19} cm^{-3}. The structures was processed to obtain mesa etched diodes but without any passivation. The diodes showed bipolar modulation even if the forward voltage drop was relatively high (Fig. 5). This is part to the high resistivity of the on-axis substrate, and possibly to a too low p++ doping in the epitaxial contact layer. The epitaxial layer grown for the device structures did also have a relatively low measured carrier lifetime. This is not related to the on-axis epitaxial growth since previously grown layers has shown no difference in carrier lifetime between on on-axis of off-axis substrates. It is instead more likely due to specific conditions during the growth, such as residual contamination or an ageing of the

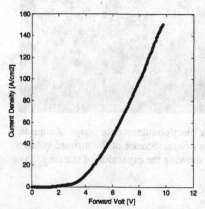

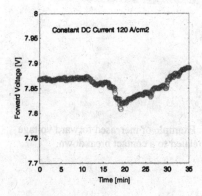

Fig 5. IV-characteristics of a PiN diode grown on-axis substrate.

Fig 6. Forward voltage at a constant current level, as a function of time.

susceptor coating. The diodes were made with a top contact including windows to be able to observe the electroluminescence from the bipolar plasma.

The diodes were tested by a DC forward current injection of 125 Acm-2 during 30 minutes (Fig. 6). The measurements were performed on-wafer using a single probe. During the test the electroluminescence image from the diodes was continuously recoded to be able to correlate any change in forward voltage drop with a change in electroluminescence image.

Most of the diodes exhibit a very stable forward voltage drop where the observed variations most likely are related to temperature variations or changes of the contact properties during the measurement time. More than 70 % of the diodes exhibited at change in forward voltage of less than 0.1 V. And no change in the electroluminescence image that could be related to the expansion of stacking faults was observed for these diodes.

For some other diodes we did observe a drastically increase of the forward voltage drop, as shown in Fig. 7, mainly occurring due to breakdown or destruction of the contact. In some cases when this happen we could also observe expanding faults that rapidly covered the entire diode area, seen in Fig. 8. From the sudden appearance and high density of these stacking faults we believe that they are expanding in the basal plane just below the top layer, and are created by the contact breakdown. It could be noted that the on-axis growth is used in order to minimize the propagation of BPD dislocations from the substrate, but if any dislocations are present or created in a BPD of the epitaxial layer, it will expand over a larger area than in the on-axis epitaxial layers.

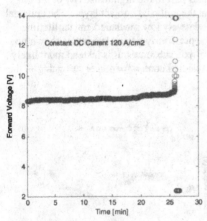 

Fig 7. Example of increased forward voltage drop related to a contact breakdown.

Fig 8. Electroluminescence image of a diode with a drastic increase of the forward voltage drop showing the expansion of stacking faults.

CONCLUSIONS

On-axis epitaxial growth has been obtained for a full 2" 4H-SiC wafer. In situ surface preparation along with improved starting growth parameters and high temperature growth has been shown to be a key process for 100% polytype stability in the epilayer on Si-face on-axis substrate. Si-rich conditions are found to be batter to remove the polishing related surface damages and to get uniform step structure on the surface with low surface roughness. Also, in situ etching under Si-rich conditions does not result in the formation of Si-droplets on the surface. The grown epilayers are of high quality with no structural defects, which appear in epilayers grown on off-cut substrates, like triangular defects, half moon, carrot or growth pits are seen. More importantly no basal plane dislocations are found the epilayer, which means that total number of dislocations are reduced in the epilaye.

PiN diodes has been produced on in-situ grown epitaxial layers on on-axis substrates. The diodes have been showed to be stable with respect to bipolar degradation, when tested during high current forward operation.

ACKNOWLEDGEMENTS

The authors are grateful for the financial support from Norstel AB, Swedish Governmental Agency for Innovation Systems, (Vinnova), and Swedish Energy Agency (STEM).

REFERENCES

[1] H. Matsunami, T. Kimoto, Mater. Sci. and Eng. R20, 125 (1997).

[2] J.P. Bergman, H. Lendenmann, P.Å. Nilsson, U. Lindefelt, P. Skytt. Mat. Sci. Forum 299, 353 (2001).

[3] H. Lendenmann, F. Dahlquist, N. Johansson, R. Söderholm, P.A. Nilsson, J.P. Bergman, P. Skytt. Mat. Sci. Forum 727, 353 (2001).

[4] J. J. Sumakeris, J. P. Bergman, M. K. Das, C. Hallin, B. A. Hull, E. Janzen, H. Lendenmann, M. J. O'Loughlin, M. J. Paisley, S. Ha, M. Skowronski, J. W. Palmour, C. H. Carter, J. Mater. Sci. Forum 527-529, 141 (2006).

[5] Z. Zhang, E. Moulton, T.S Sudarshan, Appl. Phy. Lett. 89, 81910 (2006).

[6] K. Kojima*, H. Okumura, S. Kuroda, K. Arai, J. of Crystal Growth 269, 367 (2004).

[7] G. Augustine, H.M.C.D. Hobgood, V. Balakrishna, G. Dunne, R.H. Hopkins, Phys. Stat. Sol. B 137, 202 (1997).

[8] A. Hallén, M. S. Jansona, A. Yu. Kuznetsov, D. Åberga, M. K. Linnarsson, B. G. Svensson, P. O. Persson, F. H. C. Carlsson, L. Storasta, J. P. Bergman, S. G. Sridhara, Y. Zhang, Nuclear Instruments & Methods in Physics Research, Section B (Beam Interactions with Materials and Atoms) 186-94, 186 (2002).

[9] S. Nakamura, T. Kimoto, H. Matsunami, J. of Crystal Growth 256, 341 (2003).

[10] C. Hallin, Q. Wahab, I. Ivanov, P. Bergman, E. Janzén, Mater. Sci. Forum 457-460, 193 (2004).

[11] J. Hassan, J. P. Bergman, A. Henry, H. Pedersen, P. J McNally, E. Janzen, Mater. Sci. Forum 556-557, 53 (2007).

[12] A. Henry, J. ul Hassan, J. P. Bergman, C. Hallin, E. Janzén: Chemical Vapor Deposition 12, 475 (2006).

[13] S. Yu. Karpov, Yu. N. Makarov, M. S. Ramm, Silicon Carbide and Related Materials 1995. Proceedings of the Sixth International Conference 177 (1996)

[14] E. Neyret, L. di Cioccio, E. Blanquet, C. Raffy, C. Pudda, T. Billon, J. Camassel, J. Mater. Sci. Forum 338-342, 1041 (2000).

[15] To be published in the Journal of Crystal Growth

[16] I.G. Ivanov, C. Hallin, A. Henry, O. Kordina, and E. Janzén; J. Appl. Phys. 80, 3504 (1996).

Mater. Res. Soc. Symp. Proc. Vol. 1069 © 2008 Materials Research Society   1069-D05-02

Lateral/vertical Homoepitaxial Growth on 4H-SiC Surfaces Controlled by Dislocations

Yoosuf N. Picard[1], Andrew J. Trunek[2], Philip G. Neudeck[3], and Mark E. Twigg[1]
[1]Electronics Science and Technology, Naval Research Lab, Code 6812, 4555 Overlook Ave. SW, Washington, DC, 20375
[2]Ohio Aerospace Institute, 21000 Brookpark Rd., Cleveland, OH, 44135
[3]NASA Glenn Research Center, 21000 Brookpark Rd., Cleveland, OH, 44135

ABSTRACT

This paper reports the influence of screw dislocations on the lateral/vertical growth behavior of chemical vapor deposited (CVD) on-axis homoepitaxial 4H-SiC films grown on patterned mesas. Electron channeling contrast imaging (ECCI) was utilized to image both atomic steps and dislocations while the film structure/orientation was determined using electron backscatter diffraction (EBSD). The presence and position of screw dislocations within the mesa impacted the resultant film thickness, lateral shape, and atomic step morphology. Mesa side walls that incline inwards due to faceting during screw-dislocation driven vertical film growth can intersect with the dislocation step sources near the side walls. If this occurs for all screw dislocations on a mesa, we observe a transition towards laterally dominated growth that produces webbed structures and films surfaces exhibiting significantly lower step densities. Transition from vertical to lateral dominated growth is consistent with ECCI imaged dislocation very near a mesa side wall.

INTRODUCTION

Homoepitaxial growth of 4H-SiC by chemical vapor deposition (CVD) is a critical processing step for both fabricating devices and producing high quality substrate surfaces. Many process parameters during CVD growth influence nucleation behavior, step-flow dynamics, and growth rates, but dislocations intrinsic to the original wafer substrate can also play a key role. For example, efforts to fabricate step-free 4H-SiC surfaces using patterned mesa structures show film nucleation, growth, and step morphology can be strongly affected by the presence of dislocations [1]. The influence of dislocations on the resulting morphology of on-axis homoepitaxial films grown on patterned mesas is investigated in this work using electron backscattering detection (EBSD) and electron channeling contrast imaging (ECCI). These combined analysis techniques allow structural/orientation determination of the epitaxial film [2] and dislocation/atomic step imaging of the film surfaces [2,3]. Various films grown on individual, isolated mesa structures are analyzed to correlate dislocations and atomic step morphologies to the observed vertical and lateral growth behavior.

EXPERIMENT

Isolated mesa structures were fabricated on 4H-SiC (0001) on-axis wafer surfaces prior to step-flow homoepitaxial growth by CVD. Specific mesa shapes were formed by photolithographic patterning followed by reactive ion etching [1] for two separate specimens, Sample A and B. Sample A consisted of mesas spanning up to 500 μm wide with a variety of complex shapes in order to encourage webbed formation [4]. Sample B consisted simply of various rectangular shaped mesas ranging from 20-200 μm in size. After etching these mesa shapes, specimens were loaded into a horizontal-flow, cold-wall reactor and subjected to a 2 minute in-situ pre-growth etch at 1000 mbar of H_2 at elevated temperature (~1600°C). Following the pre-growth etch, the pressure was reduced to 200 mbar and the step-flow growth process was initiated utilizing silane (SiH_4) and propane (C_3H_8) as the Si and C precursors in a hydrogen carrier gas. The parameters of the pre-growth etch and CVD growth process for Sample A and B are shown in Table 1.

Table 1. Pre-growth etch and CVD step-flow growth parameters

Sample ID	Pre-growth Etch (H_2 pressure/temp.)	Growth Time	Growth Temp.	Growth Si/C ratio
A	1000 mbar / 1620°C	330 minutes	1600°C	0.82
B	1000 mbar/ 1580°C	20 minutes	1600°C	1.0

Individual mesa structures were analyzed in a commercial scanning electron microscope equipped with a commercial EBSD system with forescatter diode detectors. Specimens were oriented with the [2$\underline{11}$0] parallel to and tilted ~75° relative to the incident 20 keV electron beam operating at 2.4 nA current. The structure and orientation of epitaxial SiC films were determined by indexing Kikuchi patterns recorded using the EBSD system. Atomic step morphologies and individual dislocations were imaged by ECCI via the forescatter diode detectors. Atomic step morphologies and step heights were confirmed by subsequent analysis using atomic force microscopy (AFM).

RESULTS AND DISCUSSION

Mesas subjected to longer growth duration

Figure 1 presents images of three films grown mesas at exactly the same conditions on three patterned mesas with identical pre-growth shape from Sample A. The film surface in Fig. 1(a) most closely resembles the original mesa profile prior to CVD growth. This mesa morphology in Fig. 1(a) can be attributed to vertically dominated screw-dislocation-assisted growth during the 330 minute long CVD process [1,4]. However, the films in Fig. 1(c) and 1(e) appear to have filled in the large gaps between the original mesa shape, indicative of enhanced lateral growth during the CVD process. Because the original RIE etched mesa side walls were vertical, the observed height of the inclined side walls should be close to the overall thickness of the epitaxial films (denoted by white lines in Fig. 1(a,c,e). Overall, the 4H-SiC epitaxial film in Fig. 1(a) appears to be roughly twice as thick as the film in Fig. 1(c).

Kikuchi patterns recorded by EBSD of the surface of each epitaxial film are also shown in Figure 1. The observed structure was consistent across the entire film surface and recorded patterns provided the structure and epitaxial relationship between film and substrate for each mesa: 4H-SiC (0001) ‖ 4H-SiC (0001) for Fig. 1 (a,b), 4H-SiC (0001) ‖ 4H-SiC (0001) for Fig. 1(c,d), and 3C-SiC (111) ‖ 4H-SiC (0001) for Fig. 1(e,f). The 3C-SiC film surface in Fig. 1(e) exhibits dished or depressed regions within the central areas of the mesa. It has been previously proposed that as SiC film growth converts from 4H to the 3C polytype on isolated mesa structures, growth becomes dominated by edge/corner nucleation and leaves a depleted region in the central area characterized as having a much lower growth rate [5]. Also, the formation of these surface undulations may result from relief of stresses due to the thermal/lattice mismatch between the 3C-SiC film and 4H-SiC mesa substrate.

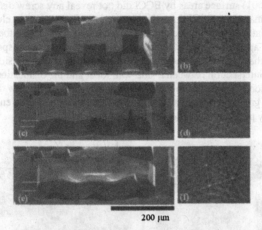

200 μm

Figure 1. SEM micrographs and corresponding Kikuchi pattern of mesas with epitaxial films exhibiting (a,b) vertically dominated growth and purely 4H-SiC structure, (c,d) laterally dominated growth and purely 4H-SiC structure, (e,f) laterally dominated growth and 3C-SiC structure. White lines denote the approximate thicknesses of the epitaxial films on each mesa.

The surfaces of the two 4H-SiC epitaxial films identified in Fig.1 were further investigated by ECCI with the results presented in Figure 2. Close inspection of the surface of the thicker epitaxial film indicated a high step density (~1.0 step/μm as measured by AFM) and the presence of two atomic spirals shown Fig. 2(b). These spirals are indicative of axial screw dislocations penetrating the (0001) surface. During the step-flow CVD growth process, these screw dislocations propagate through the growing epitaxial film and act as persistent step sources. Hence, vertical homoepitaxial growth of the 4H-SiC film in the <0001> direction is directly enabled by these screw dislocations [1,4]. However, no such screw dislocations or atomic spiral patterns were observed in the thinner 4H-SiC film surface, shown in Fig. 2(c). Instead a much lower step density (~0.15 step/μm) was observed across the film surface with an ~80 μm wide atomically flat region at the left edge. Closer inspection of this area by ECCI and AFM, shown in Fig. 2(d) confirmed this low step density where steps were 1 nm high and ~15 μm apart.

Combined ECCI, EBSD, and AFM analysis of other mesas confirmed similar results for other mesa shapes in Sample A. These results showed consistent trends between the estimated thickness of the epitaxial films, the film crystal structure (polytype), and the atomic step morphology. 3C-SiC films on mesas exhibited depressed regions in the central areas of the mesa and varying total film thicknesses, indicative that conversion from 4H-SiC to 3C-SiC for such mesas may have occurred at different times during growth. Thicker epitaxial 4H-SiC films exhibited atomic spirals while thinner epitaxial 4H-SiC films exhibited webbed growth, low step densities, and an atomically flat region adjacent to one or more edges of the mesa. These low step density morphologies consistently appeared concentric about the step-free region and indicated that the step source might have originally existed near one of the mesa edges. Since screw dislocations are the only clear step source, it is proposed that screw dislocations act to produce the steps observed in these low step density 4H-SiC films. However, close inspection in the atomically flat (0001) surface areas by ECCI did not reveal any screw dislocations present. Dislocations can be imaged by ECCI through characteristic fluctuations in electron intensity generated locally by lattice strain induced by the dislocation [3]. Because these low step densities consistently appear concentric about mesa side wall edges, we suspect that the screw dislocation acting as the original step source was intersected by an inward side-wall facet. We hypothesize that this intersection between a side-wall facet and the dislocation significantly inhibits the introduction of new steps to the (0001) top surface by the screw dislocation during growth. This leads to greatly reduced vertical growth rate accompanied by enhanced lateral growth that ultimately produces the observed webbed growth [4].

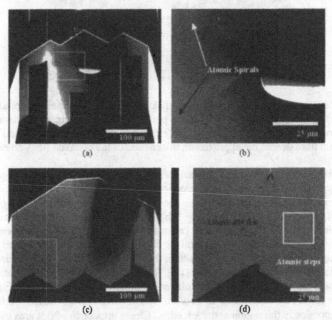

Figure 2. ECCI micrographs of (a) the mesa corresponding to Fig. 1(a) exhibiting (b) atomic spirals indicative of screw dislocations near the mesa center, and (c) the mesa corresponding to Fig. 1(c) exhibiting (d) an atomic flat area and low atomic step density. The inset in Fig. 2(d) is a 25 x 25 μm AFM image of the corresponding region.

Mesas subjected to shorter growth duration

In order to further investigate screw dislocation influence on 4H-SiC homoepitaxial growth behavior, ECCI analysis was performed for rectangular mesas subjected to shorter CVD growth durations (Sample B) and the results presented in Figure 3. Both of these mesas exhibited high step densities across the 4H-SiC epitaxial film surface. The mesa in Fig. 3(a) exhibited two atomic step spirals, one of which is presented at higher magnification in Fig. 3(b). At even higher magnification (Fig. 3(c)), an intensity fluctuation at the very center of the atomic spiral was observed. This intensity fluctuation is the characteristic feature denoting a threading screw dislocation when imaged by ECCI [3]. The mesa shown in Fig. 3(d) exhibited no atomic step spirals. Instead, an atomically flat region was observed at the bottom right corner of the mesa, shown at higher magnification in Fig. 3(e). This region appeared qualitatively similar to the thinner, lower step density 4H-SiC films described earlier.

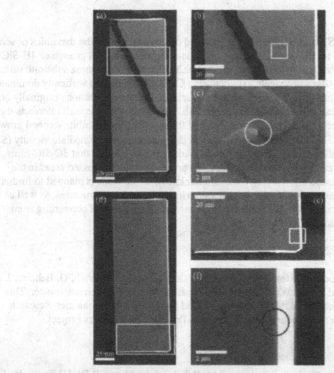

Figure 3. ECCI micrographs of (a) a mesa exhibiting (b) a spiral atomic morphology at the surface where (c) a screw dislocation at the center exhibits a channeling contrast by the characteristic light/dark feature, and (d) a mesa exhibiting (e) an atomically flat area in one corner where (f) a suspected screw dislocation is visible by channeling contrast within 100 nm of the mesa side wall. Circles denote screw dislocation position.

Higher magnification inspection of this atomically flat region near the mesa side wall edge revealed the presence of a channeling contrast feature indicative of a threading screw dislocation, shown in Fig. 3(f). This dislocation, qualitatively similar in size though weaker in intensity to the dislocation imaged in Fig. 3(c), is within 100 nm of the mesa side wall. It is proposed that this dislocation is an axial screw dislocation exhibiting weaker intensity due to partial strain relief brought about by its close proximity to the free surface of the mesa side wall. Hence, the step source for this entire mesa can be identified as a screw dislocation in the immediate vicinity of the mesa side wall edge. The step pattern of Fig. 3(e) illustrates that a clear transition in step density has occurred. Thus, a transition away from vertically dominated growth towards more laterally dominated growth appears to have already been initiated by the negligible distance between the sidewall facet and the screw dislocation. This would explain both the atomically flat region immediately surrounding the dislocation (similar to Fig. 1(c,d)) and the high step density observed across the rest of the mesa surface (similar to Fig. 1(a,b)).

CONCLUSIONS

Combined EBSD and ECCI analysis provided details concerning the dynamics of screw dislocation penetrated (0001) surfaces during CVD on-axis homoepitaxial growth of 4H-SiC films on patterned mesas. Screw dislocations originally penetrating the mesa substrate surface far from the side wall edges propagate axially during CVD growth lead to vertically dominated growth, thicker epitaxial films, and highly stepped surfaces. Screw dislocations originally near mesa side wall edges initially enhance vertical growth, but as inclined side wall intersects the dislocation, growth becomes laterally dominated and the resultant film exhibits webbed growth and low step densities. ECCI observation of a screw dislocation in the immediate vicinity (≤ 100 nm) of a mesa side wall edge confirmed this trend. Additionally, we find that 3C-SiC films spontaneously nucleated and grown during the CVD process exhibit depressed areas in the central regions of the mesa and varying total film thickness. More work is planned to further investigate the mechanisms for 3C-SiC growth behavior on these mesa structures, as well as to locate screw dislocations at the side walls for films exhibiting evidence of converting from vertically to laterally dominated 4H-SiC growth.

ACKNOWLEDGMENTS

The authors would like to thank L. Evans, B. Osborn, M. Mrdenovich, G. Beheim, J. A. Powell, G. Hunter, and L. Matus at NASA Glenn Research Center for their assistance. This work was funded by the Office of Naval Research and by the NASA Aeronautics Research Mission Directorate in the Fundamental Aeronautics Program, Supersonics Project.

REFERENCES

1. J.A. Powell, P.G. Neudeck, A.J. Trunek, G.M. Beheim, L.G. Matus, R.W. Hoffman, Jr., L.J. Keys, *Appl. Phys. Lett.* **77**, 1449 (2000).
2. Y.N. Picard, M.E. Twigg, J.D. Caldwell, C.R. Eddy, Jr., P.G. Neudeck, A.J. Trunek, and J.A. Powell, *J. Elec. Mater.*, in press (2008).
3. Y.N. Picard, M.E. Twigg, J.D. Caldwell, C.R. Eddy, Jr., P.G. Neudeck, A.J. Trunek, and J.A. Powell, *Appl. Phys. Lett.* **90**, 234101 (2007).
4. P.G. Neudeck, J.A. Powell, G.M. Beheim, E.L. Benavage, P.B. Abel, A.J. Trunek, D.J. Spry, M. Dudley, and W.M. Vetter, *J. Appl. Phys.* **92**, 2391 (2002).
5. P.G. Neudeck, H. Du, M. Skowronski, D.J. Spry, and A.J. Trunek, *J. Phys. D.* **40**, 6139 (2007).

Mater. Res. Soc. Symp. Proc. Vol. 1069 © 2008 Materials Research Society

Factors Influencing the Growth Rate, Doping, and Surface Morphology of the Low-Temperature Halo-Carbon Homoepitaxial Growth of 4H SiC with HCl Additive

Galyna Melnychuk, Huang De Lin, Siva Prasad Kotamraju, and Yaroslav Koshka
Mississippi State University, 216 Simrall Hall, Box 9571, Mississippi State, MS, 39762

ABSTRACT

In this work, a possibility to further suppress silicon vapor condensation and formation of Si clusters in order to improve the growth rate and morphology during the low-temperature halo-carbon epitaxial growth of 4H-SiC was investigated. While a pronounced dissociating of Si clusters was clearly demonstrated, the enhancement of the growth rate and morphology was less significant then expected. In addition, the homogeneity of the growth rate and doping along the gas flow direction indicated that a significant and non-equal depletion of Si and C growth species takes place at sufficiently high HCl supply. HCl flow-dependent formation of polycrystalline Si and SiC deposits in the upstream portion of the hot zone was shown to be the source of this depletion.

INTRODUCTION

Use of chlorinated precursors, including addition of HCl, in the epitaxial growth of SiC has been extensively investigated recently [1-2]. Significant improvements in epilayer morphology and the growth rate have been demonstrated. Most of the reports attribute the improvements caused by chlorinated precursors to enhanced dissociation of silicon clusters in the gas phase [1-3]. However, a possibility of a significant enhancement of the surface migration of adatoms due to HCl-assisted suppression of 2D nucleation was also suggested [4].

Most of the previous reports provide predominantly indirect evidences in support of the proposed mechanisms for the effect of HCl and other chlorinated precursors. The evidences are typically based on interpretation of the growth rate and morphology trends.

The incompleteness of our knowledge about the HCl effect is even more pronounced for the so called low-temperature halo-carbon epitaxial growth [5]. HCl addition during the halo-carbon growth at 1300^{0}C was successful for (1) improving the growth rate for the same growth conditions, and (2) making it possible to additionally enhance the growth rate by increasing the precursor flow rates without morphology degradation [6]. However, the observed trends did not allow making a simple conclusion that the improvements are merely due to increase of the Si supply to the growth surface caused by HCl-induced Si cluster dissociation or suppressed formation.

In this work, an attempt is made to provide direct experimental evidences for the mechanisms of cluster-etching by HCl during the low-temperature halo-carbon growth.

EXPERIMENT

Low-temperature epitaxial growth experiments were conducted in a hot-wall CVD reactor at 100-200 Torr with H_2 as the carrier gas, SiH_4 (3% in H_2) as the silicon source, and CH_3Cl as the carbon source. Different flow rates of HCl gas were used during the growth

A typical growth experiment utilized 10x10 mm square pieces of standard commercially available n-type 4H-SiC substrates. Substrates were vicinally cut 8^0 towards the [11-20] direction. A few pieces of the substrate were placed at different locations of the susceptor on the top of a 2"-diameter carrier wafer, which insured thermal conditions equivalent to the growth on a full-wafer substrate. In addition to 10x10 mm pieces, full 2"-diameter wafer growths were also successfully conducted at 1300^0C.

The epilayer thickness maps were obtained by reflective Fourier Transform Infrared Spectroscopy (FTIR). Surface morphology was examined by Nomarski optical microscopy. Capacitance-voltage (C-V) measurements were used to characterize unintentional doping.

RESULTS AND DISCUSSION

As was reported earlier [6], reduction and even complete disappearance of the so called "vapor trail" inside the hot-wall susceptor served as the first evidence of the suppressed Si clustering by HCl. Also, addition of HCl caused increase of R_g (at least for some ranges of HCl supply). More than two-times increase of R_g was observed at some growth conditions. However, the quantitative analysis of the R_g increase caused by HCl addition as well as the nature of the changes in doping and epilayer morphology suggested that the HCl-addition effect could not be explained by a simple model of the enhanced supply of Si growth species caused by Si cluster dissociation or their suppressed formation.

The effect of low HCl flow rate

While addition of HCl predominantly caused increase of R_g, this increase was found non-monotonous and also non-homogeneous from upstream to downstream of the susceptor (Fig. 1). For the particular carrier gas-flow conditions used, the R_g homogeneity was preserved up to HCl flow rate of 2 sccm, with almost 50% increase of the average value of R_g. Further, the HCl flow rate of 5 sccm caused almost two-times increase of R_g upstream of the hot zone; however the R_g increase was diminishing towards the downstream of the susceptor. Further changes of R_g were clearly non-monotonous, experiencing a temporary decrease when the HCl flow rate was changed from 5 to 8 sccm.

This behavior is illustrated in Fig. 2, where R_g at a fixed location inside the susceptor is shown as a function of the HCl flow rate. The corresponding changes in the net donor concentration are also shown, which in general should be expected to be the results of the change of the effective Si/C ratio caused by HCl supply as well as the result of the change of R_g.

Clearly, the improvements caused by HCl-addition could not be explained by a simple model of the enhanced supply of Si growth species released from Si clusters. Since the release of Si growth species from the clusters is proportional to the HCl flow rate, the HCl effect in the simple case would be a monotonous increase of R_g, which is not what follows from Figs.1 and 2. Also, the deterioration of the R_g non-homogeneity from upstream to downstream (Fig. 1) cannot

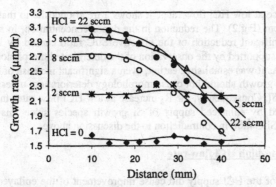

Fig. 1 Growth rate as a function of the distance from the leading edge of the substrate for different values of HCl flow rates.

be explained by a regular precursor depletion caused by epitaxial growth of SiC in the upstream regions of the substrate.

Moreover, an improvement in the epilayer morphology (as well as R_g) is expected when additional supply of Si causes an increase of the effective Si/C ratio above the growth surface. However, the increase in R_g at low HCl flow rates (i.e., the HCl flow of 2 sccm in Fig.2) was not accompanied with any obvious improvement of the epilayer morphology. Instead, the morphology actually degraded at low HCl flow rates (e.g., 1-2 sccm in Fig.2).

On the other hand, the expected increase of the effective Si/C ratio due to Si cluster dissociation is supposed to cause the increase in the net donor concentration due to the well known cite-competition mechanism. While the changes in the doping caused by HCl were observed to vary depending on the other growth conditions, the growth conditions of Fig. 2 show the changes in doping opposite to the expectations discussed above. Not only does the donor

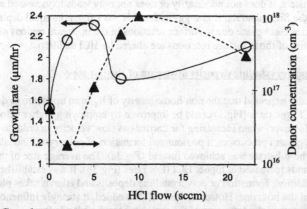

Fig. 2 Growth rate and the net donor concentration as a function of the HCl flow rate. For the particular growth conditions, HCl flow causes changes in the effective Si/C ratio that outweigh the effect of the growth rate on doping.

159

concentration not increase at low HCl flow rates, it shows a trend opposite to that of R_g at HCl flow rates below 2-3 sccm (Fig.2). The reduction in the donor concentration, in spite of the R_g increase, suggests a significant reduction of the effective Si/C ratio caused by low HCl flow rates. This conclusion is supported by the observation of the deteriorated morphology at low HCl flows described above. As it was established earlier, even a significant increase of R_g in the low-temperature halo-carbon growth should not cause morphology deterioration, unless the effective Si/C ratio is decreased [5]. Consequently, the R_g increase at low HCl flow rates in Fig. 2 cannot be predominantly caused by enhanced supply of Si growth species since this would cause increase of the effective Si/C ratio in contradiction to the discussion above.

The effect of moderate-to-high HCl flow rate

Further increase of the HCl supply did cause improvement of the epilayer morphology, thus suggesting that we get an increase of the effective Si/C ratio due to the enhanced supply of Si species. However, for the particular growth conditions of Figs. 1 and 2, R_g experiences a decrease when HCl flow rate increases from 5 to 8-10 sccm (Fig.2). Simultaneously, the increase in the net donor concentration with the HCl addition in this range of HCl flow rates clearly takes place. These two facts indicate that a significant increase of the effective Si/C ratio takes place in the range of the HCl flows from 5 to 10 sccm, and this increase is more due to a reduced supply of C species rather than enhanced supply of Si growth species caused by HCl addition.

Further increase of the HCl flow rates beyond 8-10 sccm caused a relatively modest increase of R_g (Fig.2). However, this increase could not be attributed only to an enhanced supply of Si growth species, which would increase the effective Si/C ratio. Instead, a moderate reduction of the effective Si/C ratio must be taking place since the net donor concentration is decreasing in this range of high HCl flow rates. Apparently, an additional supply of C growth species caused by HCl outpaces the supply of Si growth species, thus causing both the moderate increase of R_g and the decrease in the net donor concentration.

In summary, while it is clearly established that HCl addition causes etching or suppressed formation of Si clusters, it does not necessarily or does not only leads to increased supply of Si growth species to the SiC substrate. It is suggested in this work that non-equal depletion of Si and C growth species takes place due to surface reactions in the upstream regions of the hot zone, and the kinetics of those surface reactions are altered by HCl addition.

The effect of the polycrystalline deposits upstream of the hot zone

It was initially expected that the non-homogeneity of R_g from upstream to downstream caused by high HCl flow rates (Fig.1) could be improved by employing higher velocity of the carrier gas flow. However, when increasing the carrier gas flow velocity in order to delay the dissociation of the growth precursors, a pronounced formation of polycrystalline islands at the upstream edge of the substrate was achieved instead (Fig.3a). The average size of the polycrystalline islands increased at higher HCl flow rates (Fig.3b). It was established that even without the HCl addition, formation of polycrystalline deposits and islands takes place in the upstream portion of the hot zone. However, when HCl is added, it strongly influences the kinetics of the process and the extension of the region where the islands form.

Energy Dispersive Spectroscopy and X-ray diffraction allowed us to establish that the islands predominantly include polycrystalline Si and SiC. Since separate and non-equal fractions

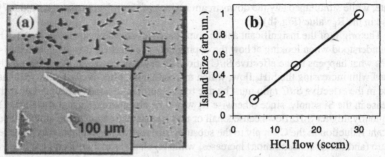

Fig. 3 (a) Polycrystalline islands formed at the upstream portion of the SiC substrate when the gas flow velocity was increased in an attempt to improve R_g homogeneity. (b) The average normalized size of the islands as a function of the HCl flow rate.

of Si and SiC phases are formed, it is suggested that the island deposition process serves as an additional mechanism for significant (and non-equal) precursors' depletion.

In order to further investigate the depletion of Si and C growth species and, consequently, variations of R_g and the effective Si/C ratio downstream as a result of the polycrystalline deposition upstream, the effects of SiH_4 and CH_3Cl flow rates were studied. A normal dependence of R_g on CH_3Cl flow rate was observed (Fig.4a). The increased supply of C growth species did not appear to significantly influence the precursor depletion. The value of R_g was increased similar to what was previously reported for low-temperature halo-carbon growth without HCl [5]. Correspondingly, the R_g homogeneity from upstream to downstream did not suffer when CH_3Cl flow was increased.

A qualitatively different behavior of R_g and its homogeneity was observed when the SiH_4 flow rate was changed (Fig.4b). No straightforward increase of R_g with increasing the SiH_4 flow was observed even though an abundant supply of Si species was expected due to Si cluster dissociation by HCl. Detailed inspection of the R_g distribution showed that more than 2.5-times increase in the SiH_4 flow rate caused less than 5% increase of R_g at the upstream portion of the

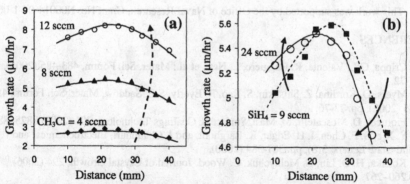

Fig. 4 Growth rate at higher gas flow velocity as a function of the distance from the leading edge of the wafer-carrier for (a) different values of CH_3C flow rates, and (b) different values of SiH_4 flow rates. The HCl flow rate used was 30 sccm.

substrate, while simultaneously the downstream regions of the substrate experienced 6-7% decrease in the R_g value (Fig.4b).

The origin of the insignificant and in some cases inverse dependence of R_g on SiH_4 flow can be understood when looking at how the doping depends on SiH_4 flow, which in turn indicates what happens to the effective Si/C ratio. The net donor concentration expectedly increased with increasing the SiH_4 flow, which means that the effective Si/C ratio increased. This increase in the effective Si/C ratio must be due to reduction in the carbon supply rather than due to increase in the Si supply, since otherwise R_g would be also increasing significantly. Consequently, higher SiH_4 flows cause small or no increase of Si supply and simultaneously a significant reduction in the C supply to the substrate surface, which explains why the effective Si/C ratio (and donor concentration) increases, while R_g changes insignificantly or even decreasing towards downstream. Premature dissociation (or suppressed formation) of Si clusters by HCl contributes to the depletion of Si and C supply by polycrystalline deposition.

CONCLUSIONS

A desirable pronounced etching of the Si clusters and/or suppression of the Si cluster formation by adding HCl during the low-temperature halo-carbon epitaxial growth was clearly confirmed. However, this etching did not result in an expected degree of increase of the growth rate or effective Si/C ratio. Instead, at certain values of the HCl flow rate, a reduction of R_g and/or reduction of the net donor concentration (i.e., the effective Si/C ratio) were observed. A premature release of Si growth species due to interaction of HCl with Si clusters enhanced another mechanism of growth precursor depletion – formation of polycrystalline deposits of Si and SiC, which was revealed as polycrystalline island formation on the SiC substrate at a sufficiently high velocity of the carrier gas flow. It was established that polycrystalline deposits in the upstream portion of the reactor caused non-equal depletion of Si and C growth species, which intern influenced R_g non-homogeneity and doping along the gas flow direction. In addition, the enhanced island formation in presence of the HCl-induced cluster dissociation may cause irregular dependence of R_g on SiH_4 flow rate, with R_g actually decreasing with SiH_4 flow increasing for some growth conditions.

This work was supported by the Office of Naval Research, Grant No. N00014-06-1-0637.

REFERENCES

1. D. Crippa, G.L. Valente, R. Ruggiero, L. Neri, et al.: Mater. Sci. Forum, 483-485 (2005), pp. 67-72.
2. R. Myers, O. Kordina, Z. Shishkin, S. Rao, R. Everly, S.E. Saddow, Mater. Sci. Forum, 483-485 (2005), pp. 73-76.
3. A. Fiorucci, D. Moscatelli, M. Masi, Surface & Coatings Technology 201 (2007) 8825–8829.
4. Z. Y. Xie, S. F. Chen, J. H. Edgar, K. Barghout, and J. Chaudhuri, Electrochemical and Solid-State Letters, 3(8), pp. 381-384 (2000).
5. Y. Koshka, H. D. Lin, G. Melnychuk, C. Wood, Journal of Crystal Growth, 294 (2006) pp.260–267.
6. Y. Koshka, B. Krishnan, H. D. Lin, and G. Melnychuk, Mater. Res. Soc. Symp. Proc. Vol. 911 p. 101 (2006).

Mater. Res. Soc. Symp. Proc. Vol. 1069 © 2008 Materials Research Society   1069-D07-09

Observation of Asymmetric Wafer Bending for 3C-SiC Thin Films Grown on Misoriented Silicon Substrates

Marcin Zielinski[1], Marc Portail[2], Thierry Chassagne[1], Slawomir Kret[3], Maud Nemoz[2], and Yvon Cordier[2]

[1]NOVASiC, Savoie Technolac, Arche Bât.4, BP267, Le Bourget du Lac cedex, 73375, France
[2]Centre de Recherche sur l'Hétéroépitaxie et ses Applications, CRHEA-CNRS, UPR10, rue Bernard Gregory, Valbonne, 06560, France
[3]Institute of Physics, Polish Academy of Sciences, Al. Lotnikow 32/46, Warsaw, 02668, Poland

ABSTRACT

We present an experimental study of asymmetric wafer deformation for 3C-SiC layers grown on deliberately misorientated silicon substrates. An asymmetric curvature has been observed both on (100) and (111) oriented layers. In this work we focus on the (100) oriented samples. The curvature of the wafers is studied as a function of wafer thickness and offcut angle. We look for the correlations between the observed asymmetric strain relaxation and the layer morphology and microstructure. We claim that different defect pattern, measured along [110] and [1-10] direction can be at the origin of almost complete relaxation of mismatch strain along the offcut direction.

INTRODUCTION

Cubic silicon carbide is considered as a potential candidate for development of power devices (performance of 3C-SiC MOSFET is better than that of 4H-SiC [1]). Its mechanical properties and chemical inertia make it an ideal candidate for MEMS applications working in harsh environment. 3C-SiC/Si is also successfully used as a template for the nitride growth, instead of more onerous 6H-SiC [2]. Since there is no native, bulk 3C-SiC substrate for homoepitaxial growth of cubic material (contrarily to hexagonal polytypes 4H, 6H, that can be obtained by sublimation methods), every deposition technique for 3C-SiC must include a stage of heteroepitaxial nucleation / growth. The problem of strain control, inherent of heteroepitaxy, has thus to be resolved in order to allow successful development of 3C-SiC based devices.

Another problem intimately connected to the heteroepitaxial growth of 3C-SiC is the formation of planar defects, like antiphase boundaries (APB) on (100) silicon substrates or double positioning boundaries (DPB) on (111) orientation. A (partial) solution of these problems consists in growing the layer on vicinal substrate. However, for both orientations, the use of vicinal substrates, allowing the reduction of APB or DPB densities, results in anisotropic strain relaxation of the layer and asymmetric bending along two perpendicular directions. This type of wafer deformation makes particularly difficult some technological steps like polishing or photolithography. The purpose of this study is to investigate the anisotropy of strain, determined from the curvature measurements of 3C-SiC/Si wafers, and to discuss the correlation between strain and defect density in 3C-SiC layer.

EXPERIMENT

The classical, low pressure, two stage growth process was applied [3] (carbonization at 1100°C followed by CVD step at 1350°C) using silane and propane as precursors and hydrogen as carrier gas. 3C-SiC films are grown on 2-inch vicinal Si (001) substrates, deliberately misorientated toward [110], with the offcut angles ranging from 1° to 6°. *On-axis* substrates were used for the curvature benchmarking. The grown layers were up to 16 μm thick.

The deformation of the wafers was measured using a surface profiler (Dek-Tak) along and perpendicular to the offcut axis. The curvature value was extracted as the second derivate of deformation profile. The crystalline quality of 3C-SiC was checked by X-Ray diffraction (Seifert 3003 PTS). The XRD technique has also been used to determine the lattice parameter of 3C-SiC layer and underlying Si substrate. Layer morphology was studied using Nomarsky microscope and, in more detail, using a Scanning Electron Microscope (JSM 7000f). The microstructure of crystalline defects was studied by means of Transmission Electron Microscopy. Cross-section samples were obtained along and perpendicular to the offcut axis by ion milling in Gatan PIPS at 3KV and incident angle of 8°. The bright field as well phase contrast images were taken with the Jeol 2000EX microscope operating at 200KV.

RESULTS AND DISCUSSION

Curvature measurements

In an extreme case of thick (111) oriented layers, for which the curvature can easily attain 9m^{-1}, the asymmetry in bending of vicinal wafer can be observed by naked eye (Fig.1a). For (100) oriented layers the curvature asymmetry can be deduced from stylus surface profiler measurements. Generally, this asymmetry depends on the growth conditions – an example is shown in Fig. 1b were the evolution of wafer curvature along two perpendicular directions was

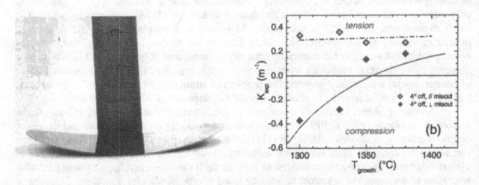

Fig. 1: a) giant bending of an off-axis (111) Si wafer with 2μm thick 3C-SiC layer. Note the asymmetry in wafer curvature. The bending of (100) oriented wafers is less pronounced, but the bending asymmetry still exists. b) curvature along and perpendicular to the offcut axis compared to on-axis results. 2.4μm thick layers are grown at various temperatures on (100) 4°off Si substrate

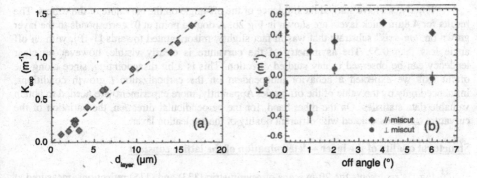

Fig. 2: a) curvature along the offcut axis versus epilayer thickness. The layers grown on 4°off (100) silicon wafers are from 1 to 16µm thick. b) curvature along and perpendicular to the offcut axis versus the off angle. The layers are 4.8µm thick

plotted as a function of growth temperature. In the direction perpendicular to the offcut axis the curvature depends on the growth temperature. The dependence is similar to that observed on *on-axis* wafers [4] represented by a continuous line on the plot. On the other hand, no significant variation of residual curvature is observed along the offcut axis. Furthermore, since its value is close to the calculated thermoelastic contribution that appears upon cooling down of the sample (dashed line on the plot) we suppose that the intrinsic (mismatch) contribution is almost entirely relaxed along the offcut axis [5].

This observation was confirmed on numerous samples. Fig. 2a shows the wafer curvature along the offcut axis as a function of layer thickness. The layers, up to 16µm thick, were elaborated under various conditions (carbonization step, growth rate, C/Si ratio...) on 4°off, 300µm thick silicon substrates. Despite the variety of growth conditions, a direct proportionality between layer thickness and measured curvature is observed. Calculated value of thermoelastic contribution is reported on the plot (dashed line).

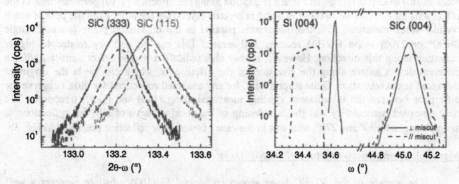

Fig. 3: a) 2θ-ω scans of asymmetric (333) and (115) reflections. b) ω scans of symmetric Si(004) and SiC(004) reflections, measured in two <110> directions.

Finally, we have studied the influence of the off-angle on the curvature of the wafers. The results for 4.8μm thick layers are shown in Fig 2b. A double point at 0° corresponds to the layer grown on *"on-axis"* substrate that was in fact slightly misorientated towards [1-10], with an off angle less than 0.5°. The asymmetry of the curvature is clearly visible, however no clear tendency can be observed in any studied direction. This is a bit disappointing, since along the offcut axis we expected a behavior independent on the carbonization / growth conditions, influenced only by the value of the off-angle. Apparently, more experiments are needed to obtain valuable data statistics. On the other hand, for the perpendicular direction, the variation of the curvature can be associated with different quality of the nucleation layer.

Structural quality of the layer – determination of the lattice constant

Fig. 3a represents the 2θ-ω scans of asymmetric (333) and (115) reflections, measured at azimuth angle $\Phi=0°$ (offcut axis [110] perpendicular to the diffraction plane) on a 16μm thick, 4°off sample. The different position of both maxima suggests that the symmetry of the 3C-SiC layer is no longer cubic (in a perfect cubic lattice $\theta_{115}=\theta_{333}$). The Bragg angle measured for each reflection is 66.605° and 66.675°, respectively. Interplannar spacing, calculated from the Bragg angle is 0.8393Å and 0.8388Å. Almost identical values were found at azimuth angle $\Phi=90°$, 180° and 270°. In order to determine the in-plane (parallel to the layer surface) and out of plane (normal to the layer surface) lattice parameters with more precision, we have analyzed 2θ-ω scans of other peaks of SiC: (002), (004), (111), (113), (115), (224) and (333). We find the in-plane lattice constant $a_1=4.3616$Å and the out of plane lattice constant $a_2=4.3583$Å. Supposing that the lattice parameter of an unstrained 3C-SiC material is $a_0=4.3596$Å, the 3C-SiC lattice is under tensile strain $\varepsilon=4.7\times10^{-4}$. This value is relatively close to the calculated value of thermoelastic strain $\varepsilon=6.4\times10^{-4}$.

Fig. 3b shows the ω scans of the symmetric SiC (004) reflection, measured in two <110> directions. It can be seen that the width and intensity of rocking curves are strongly dependent on the crystallographic direction. At azimuth angle $\Phi=0°$ and 180° the FWHM value is 250 arcsec. This is equivalent to the FWHM of similar layer grown on *on-axis* substrate and suggests that the mechanisms of defect formation / strain relaxation along the direction [1-10] perpendicular to the offcut axis may be similar to those existing in *on-axis* layers. On the other hand the peak is much broader when measured with the offcut axis parallel to the diffraction plane (azimuth angle $\Phi=90°$ or 270°) – the FWHM reaches 500 arcsec. This can be partially related to higher curvature along this direction. However, it may also indicate a higher defect density and/or a different defect nature along the direction of the offcut axis, which can be at the origin of enhanced stress relaxation. Quite surprisingly, for any analyzed reflections we didn't observe the difference between the interplannar spacing measured along [110] and [1-10] directions. The only observed particularity was the broadening of the rocking curve of all (hhl) reflections at azimuth angle $\Phi=90°$ and 270°, as it was in the case of symmetric reflection presented in Fig. 3b.

Morphology and microstructure of 3C-SiC layer

The surface of the 3C-SiC layer grown on vicinal Si (100) substrate presents a well known step bunching morphology, with terraces perpendicular to the offcut axis [110]. The terrace width decreases with increasing the off-angle, as illustrated in Fig. 4. The length of terraces taken along the [1-10] direction is limited to few micrometers.

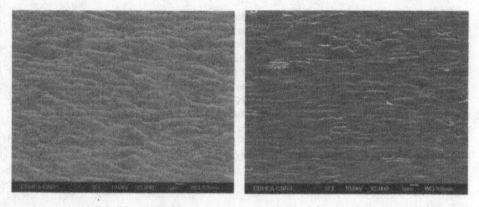

Fig. 4: Evolution of terrace width with off angle: from 2° off (left) to 6°off. 3C-SiC layers are 4.8μm thick

Fig. 5: TEM cross-sections of 2.5μm thick 3C-SiC layer grown on Si (100) 4°off substrate. Up-left: normal to the offcut axis. Up-right: along the offcut axis. Below: high resolution image of 3C-SiC/Si interface taken along the offcut axis.

The TEM cross-section images taken along two <110> directions on 2.5μm thick layer are shown in Fig. 5. The defects' pattern is strongly dependent on the crystallographic direction, especially in near surface region. On the first image, taken perpendicular to the offcut axis, the stacking faults (SFs) parallel to planes (1-11) and (-111) are clearly visible. They annihilate progressively through all the layer thickness. The second image shows the cross-section parallel to the offcut axis. All the SFs are parallel to the (-1-11) plane and their position coincides with terrace steps on layer surface. No defects in (111) plane are visible in the near surface region, even if they were present in the near interface region, as attested by the high resolution image of 3C-SiC/Si interface. Apparently the annihilation of SFs took place in the early stage of the growth. This behavior was already reported by Nagasawa *et al* [6] however no attempt was made to correlate this structural feature with the wafer curvature.

CONCLUSIONS

An asymmetric wafer deformation has been observed for the 3C-SiC layers grown on vicinal (100) Si surfaces, even for very low values of the off-angle (<0.5°). The curvature along the offcut axis [110] does not depend on any growth / carbonization conditions. Furthermore it is directly proportional to the wafer thickness and its value is very close to the calculated value of thermoelastic contribution. We conclude that along this direction the mismatch strain is totally relaxed during the growth and thus the residual curvature reflects the thermoelastic strain. Despite an obvious asymmetry in wafer curvature, the XRD measurements of lattice parameter didn't indicate any anisotropy of interplannar spacing along [110] and [1-10] direction. On the other hand, the TEM images reveal different defect patterns along two perpendicular <110> directions. The correlations between TEM and XRD results have to be studied in more detail. Similar effects are observed for (111) oriented layers.

ACKNOWLEDGMENTS

The authors would like to thank L.Nguyen for his help in SEM observations.

REFERENCES

1. A. Schöner, M. Krieger, G. Pensl, M. Abe, H. Nagasawa, Chem. Vap. Deposition, 12 (2006), 523
2. Y. Cordier, M. Portail, S. Chenot, O. Tottereau, M. Zielinski, T. Chassagne, Proceedings of ICSCRM 2008, *in press*, Materials Science Forum (2008).
3. S. Nishino, H. Suhara, H. Ono, H. Matsunami: J. App. Phys. 61, (1987) 4889
4. M. Zielinski, A. Leycuras, S. Ndiaye, T. Chassagne, Appl. Phys. Lett. 89 (2006) 131906.
5. M. Zielinski, M. Portail, T. Chassagne, Y. Cordier, Proceedings of ICSCRM 2008, *in press*, Materials Science Forum (2008).
6. H. Nagasawa, K. Yagi, T. Kawahara, N. Hatta, Chem. Vap. Deposition 12 (2006) 502.

Mater. Res. Soc. Symp. Proc. Vol. 1069 © 2008 Materials Research Society   1069-D07-10

Epitaxial Growth on 2° Off-axis 4H SiC Substrates With Addition of HCl

Jie Zhang[1], Swapna Sunkari[1], Janice Mazzola[1], Becky Tyrrell[1], Gray Stewart[1], R Stahlbush[2], J Caldwell[2], P Klein[2], Michael Mazzola[3], and Janna Casady[1]

[1]SemiSouth Laboratories, 201 Research Boulevard, Starkville, MS, 39759
[2]Naval Research Laboratory, 4555 Overlook Avenue SW, Washington, DC, 20375
[3]Mississippi State University, Department of Electrical and Computer Engineering, Starkville, MS, 39762

ABSTRACT

Epitaxial growth on 3-in, 2° off-axis 4H SiC substrates has been conducted in a horizontal hot-wall CVD reactor with HCl addition. The thickness of the epiwafers ranges from 3 μm to 11 μm and the growth rate is 7 – 7.5 μm/h. Although a rougher surface and a higher triangular defect density is observed using the standard process for 4° growth, an improved process has resulted in reduced triangular defect density down to around 4 cm^{-2} and a smoother surface with the roughness of 1.1 nm for a 3.7 μm thick epiwafer. Most interestingly, the basal plane dislocation density in the 2° off-axis epiwafers has been reduced to non-detectable levels, as confirmed by both the non-destructive UVPL mapping technique and the molten KOH etching on 2° epiwafers with thickness of around 10 μm.

INTRODUCTION

Silicon carbide is the material of choice in high power switches and rectifiers for critical applications such as hybrid or electric vehicles and electric power distribution. The use of SiC-based electronics will substantially contribute to energy efficient and environment friendly components and systems. The continuously increasing demand for the economy and performance of SiC-based components is driving the industry to develop lower off-axis SiC substrates and epilayers. The lower off-axis angle would reduce materials cost due to less materials waste upon slicing a boule. In addition, a lower density of basal plane defects, especially basal plane dislocations (BPDs), are expected to propagated from the substrate into the epilayer with a lower off-axis angle. The BPDs are responsible for the stacking fault formation in SiC bipolar devices, resulting in the drift of forward voltage drop during long-time forward operation [1]. Use of lower off-axis substrates and epiwafers thus presents an attractive means of overcoming the issue with forward voltage drift. For 4H SiC, 8° off-axis substrates have been used originally, followed by 4° off-axis angle, which has become the most common choice for large diameter substrates (3-in or larger). For improved economy and quality, further reduced off-axis angle is highly desirable. However, the epitaxial growth on 4H SiC substrates with an off-axis angle lower than 4° is a challenging task, as the epi surface is more prone to step-bunching and large morphological defects tend to form, which often leads to 3C formation with increased epi thickness. In this paper, we present epitaxial growth on 3-in, 2° off-axis 4H-SiC Si-face substrates in a horizontal hot-wall reactor with addition of HCl.

EXPERIMENTAL

The SiC epitaxial growth was conducted in a horizontal hot-wall CVD reactor with no gas foil rotation. The growth was performed in the chemical system of H_2-SiH_4-C_3H_8 with addition of HCl. Epiwafers with thickness in the range of $3 - 11$ μm were obtained at a growth rate of 7-7.5 μm/h. The substrates used in this study were 3-in, 2° off-axis Si-face 4H SiC, both the 2° and 4° substrates are from the same vendor. To our knowledge, the surface finish of the substrates with different off-axis angles is the same. The surface morphology was inspected in an optical microscope with Nomarski interference contrast. Defect count was conducted across a grid pattern corresponding to 14% of the full wafer area. The surface roughness was measured with Zygo using a white light interference technique. Basal plane dislocation distribution on one 2° epiwafer was mapped by UVPL imaging [3]. In addition, etching with molten KOH was conducted on another epiwafer, followed by manual counting under an optical microscope to reveal micropipes and dislocations. For the total etch pit density analysis, 9 points distributed along and perpendicular to the gas flow direction was manually counted with an area/ field being 0.44mm x 0.32mm. For basal plane dislocation density analysis, 37 points in a grid pattern across the whole wafer was manually counted with an area/field of 1.78mm x 1.26mm. The thickness of the epi layer was measured by an MKS Film Expert 2140 using Fourier Transformed Infrared spectroscopy (FTIR). The thickness uniformity was calculated as σ/mean. The doping concentration was obtained by Capacitance Voltage (CV) measurements using a SSM mercury probe system. The doping uniformity was calculated as (max-min)/(max+min).

RESULTS AND DISCUSSION

Surface morphology

The standard process at 7-7.5 μm/hr reproducibly produces a low triangular defect density of well below 1 cm^{-2} for 4° off-axis epiwafers with thickness up to 15 μm [2]. The surface roughness Ra on a 5 μm thick 4° epiwafer is around 1.1 nm as measured by Zygo. Using the same standard process at the same growth rate, a 4.6 μm thick epiwafer on a 2° off-axis substrate displays a considerably higher triangular defect density of around 7 cm^{-2}. A rougher surface is also observed in the optical microscope, and the roughness Ra value measured by Zygo is around 2.2 nm, shown in Figure 1 a) and 1b). The lower off-axis angle generates a wider terrace width in the initial substrate surface step structure, which would affect the surface mechanism during the subsequent etching and growth processes [3]. Consequently, a higher triangular defect density and a rougher surface are likely to occur under non-optimized conditions for growth on lower off-axis substrates.

Figure 1 a): Nomarski micrograph of a 4.6 μm thick epi grown on a 2° 3-in substrate with the standard process, showing step-bunching and high triangle density of 7 cm^{-2}. **b):** Same wafer as in figure 1a) showing Zygo surface roughness Ra of 2.2 nm.

Process optimization of both the etching and growth conditions have been investigated. An appropriate C/Si ratio (not too high or too low) and a relatively lower growth temperature seemed to improve the surface roughness and triangle count. However, the key to both reduced defect density and improved surface roughness has been the fine tuning of the initial growth process. A gradual introduction of precursor flows while minimizing excessive etching has shown great advantages. The surface roughness Ra is improved to 1.1 nm, and the triangle density is reduced to around 4 cm^{-2} for a 3.7 μm thick 2° 3-in epiwafer. A typical surface morphology is shown in Figure 2 a), and the Zygo surface roughness of the locations is depicted in Figure 2b). The process window to obtain a reasonable surface morphology becomes narrower when the epilayer thickness increases, as the initial defects tend to spread laterally with increasing thickness, while new defects may also occur during the longer growth time, as shown in Figure 3(a). With carefully optimized conditions, epiwafers with thickness of 10-11 μm have been grown on 3-in 2° substrates with reasonable surface morphology. As shown in the micrograph in Figure 4(a), smooth surface and a reasonable triangle defect density of 6 cm^{-2} was achieved in this 11 μm thick epiwafer.

Figure 2 a): Nomarski micrograph of a 3.7 μm thick epiwafter, showing smooth surface with a low triangle density of 4 cm^{-2}; **b):** Same wafer as in figure 2a) showing zygo surface roughness Ra of 1.1 nm.

Basal plane dislocations mapping

To investigate the basal plane dislocations (BPDs) penetrated from the substrate in the epilayer, full wafer mapping using UVPL has been conducted on a 10 μm thick 2°, 3-in epiwafer thickness uniformity of 9.6%, and doping levels of 1E16 cm^{-3} with a uniformity of 13%. As shown in Figure 3(b), the same epiwafer as in Figure 3(a) displays such a low basal plane dislocation density (BPD) that the basal plane dislocations are referred to as not detected, never before seen by the researcher at NRL. Noteworthy is that micropipes and in-grown stacking faults are observed in this wafer, although they have not been counted. For comparison, the measured BPD counts on substrates from a typical boule from this vendor are in the range of ~1000's/cm^2. No 4 epiwafers have been mapped with UVPL yet.

Figure 3 a): Nomarski micrograph of a 10 μm thick epiwafer, showing smooth surface and reasonable triangle defect density; **b):** UVPL Wafer-map from Naval Research Laboratory measurement. Commonly seen SiC wafer defects of micropipe dislocations and ingrown stacking faults are observed but _**basal plane defects were identified as not detected.**_

This very low BPD density has also been verified through molten KOH etching conducted on another epiwafer with thickness of 11 μm grown on a 3-in 2° wafer from the same boule, thickness uniformity of 7%, and doping levels of 6E15 cm^{-3} with a uniformity of 18%. As shown in Figure 4a), this 11 μm thick epiwafer has smooth surface and figure 4b) showing few triangle defect density of 6 cm^{-2}, despite the increased thickness. The inspection under an optical microscope of this KOH etched wafer has confirmed that the basal plane dislocation density is negligible, as illustrated in Table 1 below.

 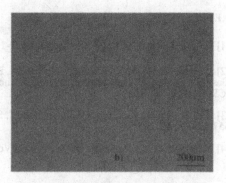

Figure 4 a): Nomarski micrograph of a 11 μm thick epiwafter, showing smooth surface **b):** Nomarski micrograph of a 11 μm thick epiwafer with a triangle defect density of 6 cm^{-2}

Table 1: Total Etch pit density and Basal plane dislocation density obtained on a virgin substrate and an epiwafer grown on a 2° off-axis substrate from the same boule. The values for average, max, and min are calculated from the 9pts manual counting for EPD and 37pts for BPD.

2° off-axis	Etch Pit Density (EPD) Avg / Min / Max (cm^{-2})	Basal Plane Dislocation (BPD) Avg / Min / Max (cm^{-2})
Virgin Substrate	4.61e4 / 2.15e3 / 1.72e5	1.27e3 / 0 / 7.15e3
Epi (11 μm)	6.66e4 / 3.58e3 / 3.12e5	NONE DETECTED

SUMMARY

Epitaxial growth on 3-in, 2° off-axis 4H SiC substrates with addition of HCl has been successfully demonstrated. Epiwafers with thickness ranging from 3 to 11 μm have been obtained with reasonable surface morphology. For thickness below 5 μm, the initial process optimization has resulted in 45% reduction in the triangle density to around 4 cm^{-2}, while the surface roughness Ra became close to that of the standard 4° epiwafers. UVPL and KOH etch investigations have verified negligible basal plane dislocation defects in the 2° off-axis wafer with thickness of around 10 μm.

ACKNOWLEDGMENTS

SemiSouth would like to gratefully acknowledge the support from scientists at SiCrystal, AG (www.sicrystal.de) including Mr. Peter Sasahara and Mr. Saber Bouloumi, for arranging and completing defect measurements. This work is partially supported by the Department of Energy Award # DE-FG02-07ER84693, monitored by Dr. Stan Atcitty. This work is also partially supported by the Unites States Air Force Research Laboratory under Contract # FA8650-06-D-2608 Delivery Orders 0001 and 0002, monitored by Dr. James D. Scofield.

REFERENCES

[1] J.P. Bergman, *ICSCRM 2007* abstract MO1-1

[2] J. Zhang, J. Mazzola, S.Sunkari, G. Stewart, P.B.Klien, R. Ward, E. Glaser, K.K.Lew, D.K. Gaskil, I. Sankin, V. Bondarenko, D.Null, D. Sheridan, and M.Mazzola, presented at *ICSCRM2007*, 2007

[3] R.E. Stahlbush, K.X. Liu, Q. Zhang, and J.J. Sumakeris, *Materials Science Forum* Vols. 556-557 (2007), pp. 295.

[4] Nakamura et al., *Materials Science Forum* Vols. 457-460 (2004), pp. 163-168.

Mater. Res. Soc. Symp. Proc. Vol. 1069 © 2008 Materials Research Society   1069-D07-11

High-frequency electron paramagnetic resonance study of the as deposited and annealed carbon-rich hydrogenated amorphous silicon-carbon films

A. V. Vasin[1], E. N. Kalabukhova[1], S. N. Lukin[1], D. V. Savchenko[1], V. S. Lysenko[1], A. N. Nazarov[1], A. V. Rusavsky[1], and Y. Koshka[2]

[1]V. E. Lashkaryov Institute of Semiconductor Physics, NASU, Pr. Nauki 45, Kiev, 03028, Ukraine

[2]Mississippi State University, 216 Simrall Hall, Box 9571, Mississippi State, MS, 39762

ABSTRACT

Three paramagnetic defects were revealed in amorphous hydrogenated carbon-rich silicon-carbon alloy films (a-SiC:H), which were attributed to the to silicon (Si) dangling bonds (Si DB), carbon-related defects (CRD), and bulk Si DB defect bonded with nitrogen atoms Si-N_2Si. The effect of thermal annealing on a-SiC:H films was studied. It was established that annealing at high temperatures leads to formation of graphite-like carbon clusters in a-SiC:H films and strong increase of the concentration of CRD, while EPR signal from Si DB disappeared. A dependence of the resonance positions for both Si DB and CRD signals on the orientation of the magnetic field relative to the a-SiC:H film plane was found at Q-band and D-band frequencies and was explained by the influence of demagnetizing fields, which is becoming significant at high spin density of the paramagnetic centers, higher microwave frequency, and low temperature. A demagnetizing field of 11 Gs was found in annealed a-SiC:H film at 140 GHz and 4.2 K.

INTRODUCTION

The hydrogenated amorphous carbon-rich silicon-carbon alloy films (a-$Si_{0.3}C_{0.7}$:H) has attracted great interest since they can be used for production of a large variety of materials with tuneable electrical, optical, and light emitting properties. However, the influence of preparation conditions and metastable paramagnetic defects on stability of electronic properties of a-SiC:H films are not yet well understood. The previous EPR study of the paramagnetic defects in nearstochiometric and carbon-rich a-SiC:H films was performed at X-band frequency (9.4 GHz). and EPR spectrum was represented by an isotropic single line with g = 2.0028. [1,2]. The investigation was mainly concentrated on the analysis of total concentration of paramagnetic defects and the width of the isotropic single line.

We present in this paper the results of high frequency (37 GHz and 140 GHz) EPR study of the carbon-rich a-SiC:H films and effects of thermal annealing. Our results show that in fact not one but three EPR lines contribute to the isotropic line observed at X-band frequency. A dependence of the resonance positions of EPR signals on the orientation of the magnetic field relative to the film plane was observed and attributed to the influence of demagnetizing fields.

EXPERIMENTAL DETAILS

Amorphous hydrogenated silicon-carbon alloy film (a-SiC:H) was deposited by reactive dc-magnetron sputtering of a monocrystalline silicon target using a mixture of Ar/CH₄ as working gas. Two-side polished Si (100) wafers (p-type, 40 Ohm·cm) were used as substrates. The substrate temperature during deposition was about 200 ^{0}C. The details of the deposition procedure have been reported elsewhere [1]. The thickness of the a-SiC:H layer measured by interferometer was found to be 500 nm. After deposition, samples were annealed in vacuum (10^{-6} Torr) for 15 minutes in the temperature range 400 – 950°C. The composition, local structure, and interatomic bonds were analysed by Auger-electron spectroscopy (JEOL, Jump 10s), Furrier-transform infra-red (FTIR) spectroscopy (2000FT-IR, Perkin Elmer), and Raman scattering spectroscopy (NR1800, JASCO, excitation with 514 nm line of Ar$^+$ laser). EPR measurements at 37 GHz were taken in a conventional Q-band EPR spectrometer equipped with a liquid helium cryosystem at temperatures between 4.2 and 300 K. EPR measurements at 140 GHz were performed in a superheterodyne D-band EPR setup at 4.2 K.

RESULTS AND DISCUSSION

Composition

Chemical composition of the a-SiC:H were measured by Auger electron spectroscopy. Using relative sensitivity factors (RSF) obtained for crystalline SiC(6H) standard we estimated the composition as following: Si (25 at.%), C (65 at %), oxygen (7 at.%), and nitrogen (3 at.%).

FTIR

FTIR-transmission spectra of the as-deposited and annealed at 950 ^{0}C samples are presented in figure 1 (spectra 1 and 2). As the transmission spectrum after high-temperature annealing became weak and highly distorted by interference, we recorded the reflection spectrum as well (figure 1, spectrum 3). The spectra of as-deposited samples in the spectral range of 400-3500 cm^{-1} are composed of four main absorption bands: 800 cm^{-1} (Si–C, stretching), 1000 cm^{-1} (Si:C–H$_n$, bending), 2100 cm^{-1} (Si–H$_n$, stretching) and 2800-3000 cm^{-1} (C(sp^3)–H$_n$ stretching) (figure 1, spectrum 1) [3]. A weak absorption is seen at 1260 cm^{-1}, which can be also ascribed

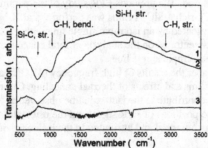

Figure 1. FTIR transmission spectra of the as deposited (1) and annealed at 950 ^{0}C (2) and reflection spectrum of annealed at 950°C (3) a-SiC:H film.

to carbon-hydrogen bending vibrations. The main effect of high-temperature annealing is breaking of Si–H and C–H bonds and effusion of hydrogen out of film. A strong narrowing of the 800 cm^{-1} Si-C stretching vibrations absorption band was also observed, which gives an evidence of nucleation of nano-crystalline SiC at 950 ^{0}C.

Raman scattering

No Raman scattering signal in the spectral ranges of 500-520 cm^{-1} (Si–Si) and 800 cm^{-1} (Si–C) was detected in all samples. As-deposited samples showed weak barely detectable Raman scattering in the range 1000-1800 cm^{-1} (disordered sp^2 C–C) (figure 2). Those samples also showed strong background of photoluminescence. Annealing at 950 ^{0}C resulted in strong enhancement of carbon related RS, while still no silicon and/or silicon carbide related signals were detected. Carbon related RS band after high temperature annealing was represented by broad band with two pronounced peaks at 1370 cm^{-1} and 1570 cm^{-1}, which are attributed to D- and G-band of disordered graphite-like local structure.

EPR results

Figure 3 shows Q-band EPR spectra of as deposited and annealed at 950 ^{0}C samples measured at room temperature. Deconvolution of the experimental EPR spectra was performed using Easy-Spin-2.6.0 toolbox program [4]. It was found that the EPR spectrum of as deposited a-SiC:H film is composed of three components labeled I_1, I_2, I_3. Two of them, I_1 and I_2 with g = 2.0059, and g = 2.0028, were previously assigned to silicon (Si) dangling bonds (SiDB) [5] and carbon-related defects (CRD) [6], respectively. The third broad EPR signal with g = 2.0033 and the peak-to-peak line width of $\Delta H_{pp} = 25 \pm 2$ Gs could be assigned to the so-called K-centers inherent for amorphous hydrogenated Si_3N_4 layers [7]. These defects are associated with threefold-coordinated SiDB where a Si atom is bonded with two nitrogen N atoms (Si-N$_2$Si). The linewidth of the K-center generally depends on the configuration of the defect and is determined by unresolved super-hyperfine interaction with ^{14}N (I=1) and ^{15}N (I = 1/2) nuclei.

Relative contribution of silicon- and carbon-related paramagnetic defects (N_{Si}/N_C) in as-deposited films was estimated to be about 1/5 with total concentration of spins about $5 \cdot 10^{19}$ cm^{-3}. The intensity of EPR lines due to Si-DB (g = 2.0059 and g = 2.0033) started to decrease at 400 ^{0}C annealing temperature and disappeared at 650^{0}C indicating saturation of the concentration of the silicon dangling bonds with increasing the temperature. From the other hand, FTIR

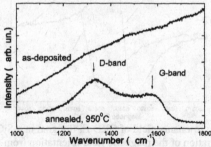

Figure 2. Raman scattering spectra of the as-deposited and annealed at 950 ^{0}C a-SiC:H film.

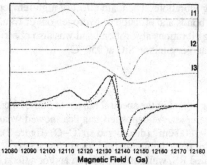

Figure 3. Q-band experimental and simulated (dashed lines) EPR spectra of as deposited (solid line) and annealed at 950^0C (dash dot-dot line, gain 0.03) a-SiC:H film. T = 300 K.

measurements showed no Si–H bonds in samples annealed at 650 ^{0}C, i.e. saturation of the silicon dangling bonds is not due to passivation by hydrogen. The explanation of this phenomenon is a challenge as the structure of a-SiC:H is still amorphous after annealing at 650 ^{0}C. Moreover this observation seems to be in discrepancy with an earlier suggestion about enhancement of silicon related EPR signal from nearstochiometric a-SiC:H films after annealing at 650 ^{0}C [1]. The intensity of the CRD EPR signal was increased by a factor of 30 at annealing temperature 950^0C and accompanied with a decrease of the line width (figure 3). The decrease in EPR line width of the CRD signal with increasing spin concentration demonstrates its exchange origin. Strong correlation of the increase of spin concentration and clusterization of graphite-like carbon allow us to suggest that CRD is mainly associated with sp^2-coordinated carbon clusters (figure 2). A dependence of the resonance positions for both Si DB and CRD signals on the orientation of the magnetic field relative to the a-SiC:H film plane was found at Q- and D-band frequencies, which was hidden in the natural line width at lower frequencies.

Figure 4 shows D-band EPR spectra of a a-SiC:H sample annealed at high temperature (950 ^{0}C) measured at 4.2 K with magnetic field orientation normal (B_0) and parallel (B_{90}) to the a-SiC:H film plane. It is seen that $B_0 > B_{90}$ with anisotropy of resonance field of about 16 Gs.

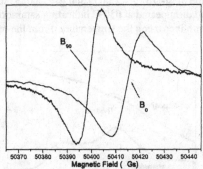

Figure 4. Anisotropy of the resonance field of CRD EPR signal in a-SiC:H film annealed at T=950^0C at D-band for variation of the magnetic field orientation from normal (B_0) to the film plane (B_{90}). T= 4.2 K.

Figure 5 shows the difference in resonance positions ($B_0 - B_{90}$) for the Si DB and CRD signals as function of the annealing temperature observed at Q- and D-band frequencies. The largest orientation effect of the resonance position for the CRD signal was observed in high temperature annealed a-SiC:H films having high spin density of the paramagnetic centers. This effect is explained by the influence of demagnetizing fields, which is increased linearly with the microwave frequency (figure 6). The observation of this effect in amorphous carbon at 9.4 GHz has been published earlier [8] and was explained by demagnetization factors of magnetized thin amorphous films. The influence of demagnetizing fields for a-C layers of about 100 Gs was also observed in [9] at 285 GHz. The origin of the demagnetizing fields is an impact of the macroscopic sample shape on resonant magnetic fields. In this case the internal magnetic field B_{int} of the sample will differ from the external field B_{ext}:

$$B_{int} = B_{ext} - N \cdot M \qquad (1)$$

Where M is the sample magnetization and N is the factor depending on the sample geometry. For a thin film, one can evaluate the sample magnetization M (demagnetizing field) from the following expressions [10]:

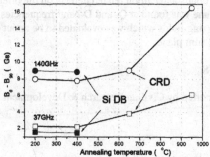

Figure 5. The difference in resonance positions ($B_0 - B_{90}$) of the Si DB and CRD EPR signals for the magnetic field orientation normal (B_0) and parallel (B_{90}) to the a-SiC:H film plane as function of the temperature annealing observed at Q- and D-band frequencies at 4.2 K.

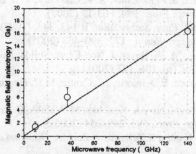

Figure 6. Microwave frequency dependence of the magnetic field anisotropy $B_0 - B_{90}$ for the CRD EPR signal observed in a-SiC:H film annealed at 950^0C. T = 4.2 K

$$B_{int}(0) = \sqrt{B_{ext}(B_{ext} - 4\pi M)} \text{ and } B_{int}(90) = B_{int} - 4\pi M \qquad (2)$$

Using experimental data it was found that the demagnetizing field at 4.2K in a-SiC:H film annealed at 950^0C were $4\pi M \cong 4$ Gs at 37 GHz and $4\pi M \cong 11$ Gs at 140 GHz.

CONCLUSIONS

High frequency EPR study of as deposited hydrogenated carbon-rich silicon-carbon alloy (a-SiC:H) films have shown that EPR spectrum consists of three EPR lines with g = 2.0059, g = 2.0028 and g = 2.0033, which were assigned to Si dangling bonds, carbon related paramagnetic defects, and silicon related defects associated with Si-N_2Si dangling defect, respectively. Annealing of a-SiC:H at high temperatures leads to the disappearance of the EPR signals caused by Si dangling bonds and strong increasing of the intensity of carbon related EPR signal. The mechanism of the saturation of the concentration of the silicon dangling bonds after 650 ^{0}C annealing is still unclear. The observed exchange narrowing of carbon-related EPR line with increasing of the corresponding spin concentration was explained by formation of graphite-like carbon clusters in the annealed a-SiC:H films. A dependence of the resonance positions for EPR signal of both Si dangling bonds and carbon related defects on the orientation of the magnetic field relative to the film plane was found at Q- and D-band frequencies and attributed to the influence of the demagnetizing fields which was evaluated to be about 4 Gs at 37 GHz and 11 Gs at 140 GHz in the a-SiC:H film plane.

ACKNOWLEDGMENTS

The work was supported by Civilian Research & Development Foundation (project No. UKE2-2856).

REFERENCES

1. A. V. Vasin, S. P. Kolesnik, A. A. Konchits, V. I. Kushnirenko, V. S. Lysenko, A. N. Nazarov, A. V. Rusavsky, S. Ashok, *Journal of Applied Physics*, 99, 113520-1 – 113520-8, (2006).
2. A.V. Vasin, A. A. Konchits, S. P. Kolesnik, A.V.Rusavsky, V. S. Lysenko, A. N. Nazarov, Y. Ishikawa, S. Ashok , *MRS Symp. Proc.*, 994, F03-13 (2007).
3. J. Bullot and M. P. Schmidt, *Phys. Status Solidi* B 143, 345, (1987).
4. Stoll S., Schweiger A, *J. Magn. Reson.*, 177, 390–403 (2005).
5. M. Stutzman. M. S. Brandi, M. W. Bayer, J. Non-Cryst. Solids, No. 266-269, P.1-22 (2000)
6. J. Robertson, *Phys. Stat. Sol. (a)*, 186, 177 (2001).
7. P.Aubert, H.J.von Bardeleben, F.Delmotte, J.L.Cantin, M.C. Hugon, *Phys.Rev B*, 59, 10677 (1999).
8. B. Druz, I. Zaritskiy, Y. Yevtukhov, A. Konchits, M. Valakh, S. Kolesnik, B. Shanina, V. Visotski, *Mat. Res. Soc. Symp. Proc.*, 593, 249–254 (2000).
9. H.J. von Bardeleben, J.L. Cantin, K.Zellama, A.Zeinert, *Diamond and Related Materials*, 12, 124-129 (2003).
10. C.Kittel, *Introduction to Solid State Physics*, Fourth Edition, J.Wiley and Son, Inc., NY, 1976.

Mater. Res. Soc. Symp. Proc. Vol. 1069 © 2008 Materials Research Society

Characterization and Growth Mechanism of B12As2 Epitaxial Layers Grown on (1-100) 15R-SiC

Hui Chen[1], Guan Wang[1], Michael Dudley[1], Zhou Xu[2], James. H. Edgar[2], Tim Batten[3], Martin Kuball[3], Lihua Zhang[4], and Yimei Zhu[4]

[1]Department of Materials Science and Engineering, Stony Brook University, Stony Brook, NY, 11794
[2]Department of Chemical Engineering, Kansas State University, Manhattan, KS, 66506
[3]H.H. Wills Physics Laboratory, University of Bristol, Bristol, BS8 1TL, United Kingdom
[4]Center for Functional Materials, Brookhaven National Laboratory, Upton, NY, 11973

ABSTRACT

A systematic study of the heteroepitaxial growth of $B_{12}As_2$ on m-plane 15R-SiC is presented. In contrast to previous studies of $B_{12}As_2$ on other substrates, including (100) Si, (110) Si, (111) Si and (0001) 6H-SiC, single crystalline and untwinned $B_{12}As_2$ was achieved on m-plane 15R-SiC. Observations of IBA on m-plane ($1\bar{1}00$)15R-SiC by synchrotron white beam x-ray topography (SWBXT) and high resolution transmission electron microscopy (HRTEM) confirm the good quality of the films on the 15R-SiC substrates. The growth mechanism of $B_{12}As_2$ on m-plane 15R-SiC is discussed. This work demonstrates that m-plane15R-SiC is potentially a good substrate choice to grow high quality $B_{12}As_2$ epilayers.

INTRODUCTION

Icosahedral boron arsenide ($B_{12}As_2$) is a wide band gap semiconductor (3.47eV) and is potentially useful for devices operating in high electron radiation environments. An attractive possible application for $B_{12}As_2$ is for beta cells, which are capable of producing electrical energy by coupling a radioactive beta emitter to a semiconductor junction [1-6]. The properties of $B_{12}As_2$ are associated with its unique crystal structure and stiff bonding. $B_{12}As_2$ is based on twelve-boron-atom icosahedra, which reside at the corners of an α-rhombohedral unit cell, and two-atom As-As chains lying along the rhombohedral [111] axis. Each boron atom occupies a vertex of an icosahedra, and is bonded to five other B atoms as well as either an As atom or another icosahedron [1, 4].

In the absence of native substrates of $B_{12}As_2$, it is necessary to grow $B_{12}As_2$ as heteroepitaxial films on substrates with compatible structural parameters. To date, this has been attempted on substrates with higher symmetry than $B_{12}As_2$ such as on Si and 6H-SiC [4, 5-9]. However, growth of a lower symmetry epilayer on a higher symmetry substrate often produces structural variants in the film that are related to each other by a symmetry operation that is present in the substrate but absent in the epilayer. A theoretical treatment of this phenomenon, which has been referred to as degenerate epitaxy [10,11]. For substrate crystals with a multi-atom basis, this list of symmetry operations must be extended to include glide planes with translation parallel to the surface. For the case of $B_{12}As_2$ grown on Si with (100), (110) and (111) orientation and (0001) 6H-SiC, rotational and translational variants are both predicted and observed [4, 9]. Single terrace behavior can be diminished by surface roughness or vicinality since the presence of the risers and terraces exerts an influence on the nucleation of variants. For example the use of offcut substrates potentially enables manipulation, to a limited extent, of the relative populations of the multiple variants although such studies on offcut (0001) 6H-SiC substrates have so far been largely unsuccessful [5, 7]. The effects of degenerate epitaxy can also be potentially used to advantage by choosing substrate orientations for which no variants are predicted. Single terrace, m-plane 15R-SiC potentially fulfils this requirement for $B_{12}As_2$. In this paper we show that $B_{12}As_2$ grown on the m-plane 15R-SiC is free from structural variants and is of high single crystalline quality. We discuss the factors that contribute to the difference in quality of the epilayers grown on these two kinds of substrate.

EXPERIMENT

$B_{12}As_2$ was deposited using chemical vapor deposition (CVD) onto m-plane 15R-SiC at 1200°C and 500 Torr of reactor pressure for 1 hour, using 1% B_2H_6 in H_2 and 2% AsH_3 in H_2 as sources. The epitaxial $B_{12}As_2$ film had a nominal thickness of 3 μm. The film/substrate orientations were determined by synchrotron white beam x-ray topography (SWBXT). The interfaces between the $B_{12}As_2$ and the SiC were examined by high-resolution transmission electron microscopy (HRTEM) using a JEOL 3000EX system at Brookhaven National Laboratory with an electron voltage of 300KeV.

DISCUSSION

Figs. 1(a) shows a SWBXT Laue pattern recorded from the $B_{12}As_2$ film on a *m*-plane 15R-SiC substrate. Compared to the previous study on twinned $B_{12}As_2$ grown on c-plane 6H-SiC [7-9, 11], the $B_{12}As_2$ in the current sample produced relatively strong diffraction spots indicating a (353) surface normal and low streaking indicating improved mosaicity. The Laue patterns also lacked any evidence for the existence of twins which would be manifested as (212) orientation $B_{12}As_2$. The improved crystal quality of $B_{12}As_2$ grown on 15R-SiC is also evident in the HRTEM micrograph shown in Fig. 1(b), with an abrupt, clean interface to the substrate. Detailed calculation based on the observed HRTEM images confirmed the film orientation as (353).

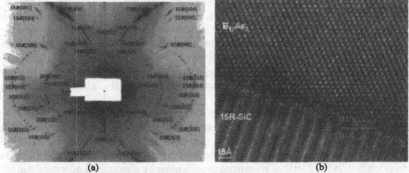

(a)   (b)

Fig. 1 (a) Laue pattern of $B_{12}As_2$ on 15R-SiC with the diffraction peaks marked with rhombohedral (*hkl*) indices; (b) Cross-sectional HRTEM image taken along [10$\bar{1}$] zone axis (equivalent to [11$\bar{2}$0]in hexagonal system) showing a sharp film/substrate interface.

The significantly improvement in microstructural quality of the films grown on m-plane orientated 15R-SiC can be understood by considering factors that may potentially contribute to the choice of surface orientation and/or the presence of structural variants in these heteroepitaxial films. These factors include: (1) the tendency to nucleate in (111) orientation on closed-packed atomic facets on the substrate surface with the attendant influence of symmetry on $B_{12}As_2$ nucleation; and (2) the tendency to minimize the in-plane lattice mismatch between $B_{12}As_2$ planes oriented approximately parallel to the SiC (0001) planes to minimize local strain in the film/substrate interface (similar tendency might be expected for the out-of-plane direction).

The hydrogen etched 15R-SiC substrate surface is expected to display a faceted structure featuring close-packed (474) and (212) facets that are successively three and two bilayers wide, respectively. If the surface comprised only these two types of facet, their asymmetry would result in a surface that would be offset by 3° from m-plane orientation. The fact that the crystal is "on-axis" m-plane orientation necessitates that other types of non-close-packed facet must appear periodically on the surface to compensate for the asymmetry of the close packed facets. A possible configuration which renders the surface "on-axis" is shown in Fig. 2 which shows the appearance of an additional (11·8·11) facet in a configuration that repeats following every fourth (474) facet. This Figure illustrates that this surface can be considered as vicinal with the (474) facets acting as terraces and the (212) and coupled (212) and (11·8·11) facets acting as step risers. Whatever the detailed configuration, the presence of such faceting breaks the aforementioned symmetry of the m-plane surface thus eliminating the associated advantage. Furthermore, the vast majority of the exposed facets are close-packed so that they each exhibit local symmetry which is actually the same as that of the $(111)_{B12As2}$ planes, *6mm*. This symmetry is expected to promote nucleation of $B_{12}As_2$ in (111) orientation which has *3m* symmetry. Growth of a *3m* film on a *6mm* substrate is predicted to lead to the creation of two rotational variants, contrary to our observations. In order to understand why the $B_{12}As_2$ adopts a highly crystalline, untwinned configuration on this substrate, it is instructive to consider the dangling bond configurations exhibited at the substrate/epilayer interface.

Fig. 2 Crystal visualization along [10-1] showing the closed-packed (474) atomic terraces and coupled (212) and (11·8·11) step risers exposed on m-plane 15R-SiC surface.

Straightforward consideration of lattice geometry reveals that if the $B_{12}As_2$ nucleates on either the (474) or (212) facets with the $(111)_{B12As2}$ planes aligned to the facets, the film would adopt $(353)_{IBA}$ orientation although if nucleation occurs simultaneously on both types of facet the film would be polycrystalline. However, the large dimension of the in-plane repeat unit of the $(111)_{B12As2}$ (twice that of the close-packed SiC planes) requires a certain minimum facet width for nucleation to occur. This width should be large enough to accommodate a nucleus of $B_{12}As_2$ which is at least two icosahedra wide. This minimum width is smaller than the width of the three bilayer-wide (474) facets but is larger than that of the two bilayer wide (212) facets, so that nucleation is expected on the former but not the latter. This type of nucleation is also not expected on the non-close-packed (11·8·11) facets. Therefore, it seems plausible that the choice of $(353)_{B12As2}$ orientation could be dominated by the influence of (1), the tendency to nucleate in $(111)_{B12As2}$ orientation on the broadest (474) closed-packed atomic facets on the substrate surface.

Based on the lattice geometry shown in Fig. 1(b) one can construct a schematic cross-sectional visualization of the $B_{12}As_2$/15R-SiC interface (see Fig. 3). In this orientation, which we define as "untwinned" a reasonable fit can be obtained between the dangling bonds exhibited by both the B triangles at the lower ends of the B icosahedra and the As dangling bonds and those exhibited by the triangular configurations of Si atoms on the (474) facets. For the vast majority of the (474) facets this can be achieved with nuclei which are at least two icosahedra wide. This long range fit is a consequence of both the in-plane and out-of-plane mismatch between the $B_{12}As_2$ and the stepped SiC surface. The in-plane lattice mismatch (~3.7%) can be well accommodated by the periodic appearance of interfacial dislocations (with one extra (555) half-plane on the SiC side) as shown in Fig. 3(a). Examples of such interfacial dislocations are indicated in Fig. 3(b). Furthermore, since the height difference between adjacent (474) facets is comparable to the spacing of the $(111)_{B12As2}$ planes (a mismatch of 15% measured perpendicular to (474) exists) a reasonable three-dimensional registry exists between the two structures. Help in accommodating the out-of-plane disregistry between the two structures is provided by the periodic presence of the deeper step riser comprising the coupled (212) and (11·8·11) facets. As such, the $B_{12}As_2$ is able to quickly self-adjust its own structure near the film/substrate interface to be compatible with that of the substrate and is then able to continue to grow with perfect structure. Thus both in-plane and out-of-plane mismatch is accommodated so that factor (2) favors nucleation in this orientation.

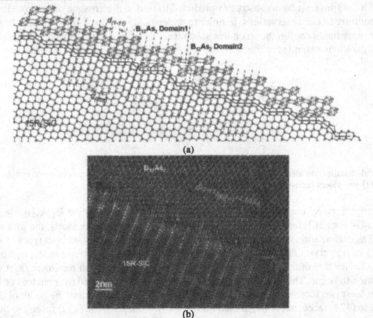

(a)

(b)

Fig. 3 (a) Cross-sectional crystal visualization of $B_{12}As_2$ grown on (474) facets with coupled (212) and (11·8·11) step risers expected on m-plane 15R-SiC surface, where the blue lines represent the extra half planes existing in the SiC; (b) HRTEM observation along [10$\bar{1}$] showing the presence of extra half planes in SiC, which are indicated by the blues lines.

The reason for the absence of twinning now becomes evident, since if the $B_{12}As_2$ nucleated on the (474) facets in twinned orientation, the long range interfacial order just described would no longer be possible. Defining a parallelogram with horizontal $(111)_{B12As2}$ planes and inclined $(\bar{1}11)_{B12As2}$ planes one can see how the combined in-plane and out-of-plane mismatch enables reasonable interfacial fit between the two structures for the untwinned orientation but not for the twinned orientation. For the untwinned structure, Fig. 4(a) (cross-sectional view) shows that the distance between the B triangles at corners O and Q of the parallelogram OPQR matches quite well (~2% mismatch) with the distance between the triangular configurations of Si dangling bonds labeled T1 and T2 indicated on the two (474) facets (located at corresponding positions to the right of the deeper step risers). On the other hand, for the twinned structure (Fig. 4(b)), the corresponding mismatch between the separation of the B triangles O' and Q' and the distance T1T2 increases to ~14% which is apparently large enough to preclude the appearance of twin oriented nuclei. In addition, in the twinned configuration the dangling bond configurations enable very few if the (474) facets to bond to nuclei which are more than one icosahedron wide. Therefore in summary, it appears that the growth is initiated by the coordinated nucleation of $(111)_{B12As2}$, in a single untwinned orientation with relaxed configuration on (474) closed packed facets on the m-plane 15R-SiC surface leading to $(353)_{B12As2}$ surface orientation.

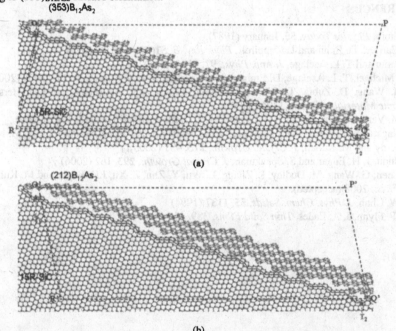

(a)

(b)

Fig. 4 Cross-sectional crystal visualization of $B_{12}As_2$ grown on m-plane 15R-SiC showing the combined in-plane and out-of-plane lattice mismatch in (a) $(353)B_{12}As_2$ and (b) its twinned orientation, $(212)\ B_{12}As_2$.

CONCLUSIONS

Epitaxial growth of $B_{12}As_2$ on m-plane 15R-SiC has been demonstrated. SWBXT and cross-sectional HRTEM revealed untwinned single-crystalline (353) orientated $B_{12}As_2$ in the film. The growth mechanism of the $B_{12}As_2$ on m-plane 15R-SiC has been discussed. It was found that the choice of film orientation on 15R-SiC resulted from the tendency to nucleate on closed-packed (474) atomic facets on the substrate and the tendency to minimize both in-plane and out-of-plane lattice mismatch between the film and the substrate.

ACKNOWLEDGMENTS

Financial support from the National Science Foundation under Grant No.0602875 and by the Engineering and Physical Science Research Council (EPSRC) under Grant No. EP/D075033/1 under the NSF-EPSRC Joint Materials Program is acknowledged. The SWBXT was carried out at Stony Brook Topography Facility (Beamline X19C) at the National Synchrotron Light Source (NSLS), Brookhaven National Laboratory, which is supported by the U.S. Department of Energy (D.O.E.) under Grant No. DE-AC02-76CH00016.

REFERENCES

1. D. Emin, *Physics Today*, **55**, January (1987).
2. M. Carrard, D. Emin and L. Zuppiroli, *Phys. Rev. B*, **51**(17), 11270 (1995).
3. D. Emin and T. L. Aselage, *J. App. Phys.*, **97**, 013529 (2005).
4. J.R. Michael, T. L. Aselage, D. Emin and P.G. Kotula, *J. Mater. Res.*, **20** (11), 3004 (2005).
5. R.H. Wang, D. Zubia, T. O'Neil, D. Emin, T. Aselage, W. Zhang and S.D. Hersee, *J. Electronic Materials*, **29** (11), 1304 (2000).
6. W.M. Vetter, R. Nagarajan, J. H. Edgar and M. Dudley, *Mater. Lett.*, **58**, 1331 (2004).
7. R. Nagarajan, Z. Xu, J. H. Edgar, F. Baig, J. Chaudhuri, Z. Rek, E. A. Payzant, H. M. Meyer, J. Pomeroy and M. Kuball, *J. Crystal Growth.*, **273**, 431 (2005).
8. X. Zhou, J. H. Edgar and S. Speakman, *J. Crystal Growth.*, **293**, 162 (2006)
9. H. Chen, G. Wang, M. Dudley, L. Zhang, L. Wu, Y. Zhu, Z. Xu, J.H. Edgar and M. Kuball, *J. Appl. Phys.*, 2008 (accepted).
10. S.W. Chan, *J. Phys. Chem. Solids*, **55**, 1137 (1994)
11. C. P. Flynn, J. A. Eades, *Thin Soild Films*, **389**, 116 (2001)

Device Processing and Characterization

Mater. Res. Soc. Symp. Proc. Vol. 1069 © 2008 Materials Research Society   1069-D09-01

Simultaneous Formation of n- and p-Type Ohmic Contacts to 4H-SiC Using the Binary Ni/Al System

Kazuhiro Ito[1], Toshitake Onishi[1], Hidehisa Takeda[1], Susumu Tsukimoto[1], Mitsuru Konno[2], Yuya Suzuki[2], and Masanori Murakami[1]

[1]Materials Science and Engineering, Kyoto University, Kyoto, 606-8501, Japan
[2]Hitachi High-Technologies Corporation, Hitachinaka, 312-0057, Japan

ABSTRACT

Fabrication procedure for silicon carbide power metal oxide semiconductor field effect transistors can be improved through simultaneous formation of ohmic contacts on both the n-source and p-well regions. We have succeeded in the simultaneous formation of Ni/Al ohmic contacts to n- and p-type SiC after annealing at 1000°C for 5 mins in an ultra-high vacuum. Ohmic contacts to n-type SiC were found when Al-layer thickness was less than about 5 nm while ohmic contacts to p-type SiC were observed for an Al-layer thickness greater than about 5 nm. Only the contacts with Al-layer thicknesses in the range of 5 to 6 nm exhibited ohmic behavior to both n- and p-type SiC, with specific contact resistances of 1.8×10^{-4} Ωcm^2 and 1.2×10^{-2} Ωcm^2 for n- and p-type SiC, respectively. An about 100 nm-thick contact layer was uniformly formed on the SiC substrate and polycrystalline δ-Ni_2Si(Al) grains were formed at the contact/SiC interface. The distribution in values for the Al/Ni ratio in the δ-Ni_2Si(Al) grains which exhibited ohmic behavior to both n- and p-type SiC was the largest. The smallest average δ-Ni_2Si(Al) grain size was also observed in these contacts. Thus, the large distribution in the Al/Ni ratios and a fine microstructure were found to be characteristic of the ohmic contacts to both n- and p-type SiC.

INTRODUCTION

Compared to Si, silicon carbide (SiC) is better suited to the next generation of high-powered devices due to such excellent intrinsic properties as a higher electric field breakdown strength and a higher saturation electron velocity[1]. However, several technical issues such as doping at high levels by ion implantation and development of formation techniques to create low resistance ohmic contacts[2] must be solved before the application of SiC in high-power devices can be realized.

For power metal oxide semiconductor field effect transistor devices with a double-implanted vertical structure, the ohmic contacts on both n+ -source region and p-well region are needed[3]. Nickel[4] and Ti/Al[5,6] contacts, where a slash symbol "/" indicates the deposition sequence staring with Ti, are well known as the low resistance ohmic contact to n- and p-type SiC, respectively. Thus, the ohmic contacts to n- and p-type SiC are currently fabricated using different contact materials and different annealing processes. Simultaneous formation of ohmic contacts to both n- and p-type SiC using the same contact material and in a one-step annealing process will simplify device fabrication processes and miniaturize the cell size.

Although a variety of ohmic contact materials to SiC have been developed, only a few materials are suitable for ohmic contacts to both n- and p-type SiC[7,8]. Fursin et al.[7] reported that a Ni contact showed ohmic behavior after annealing at 1050°C to both n- and p-type SiC in which N and Al were implanted at 10^{19} cm^{-3} and 10^{20} cm^{-3}, respectively. Tanimoto et al.[8] reported simultaneous formation of Ni-based contacts to heavily doped SiC (N>10^{20} cm^{-3}) after annealing at 1000°C. Tsukimoto et al.[9] in our group also reported that a Ni/Ti/Al ternary contact showed ohmic behavior after annealing at 800°C to both n- and p-type SiC in which N and Al were doped at 10^{19} cm^{-3} and 4.5 x 10^{18} cm^{-3}, respectively. However, the mechanisms responsible for the simultaneous formation of ohmic contacts to both n- and p-type SiC using the same material have not yet been clarified.

In the present study, to simplify the microstructures of the contacts to 4H-SiC, we deposited sequentially Ni and Al layers on 4H-SiC substrates, and investigated the electrical properties of Ni/Al contacts with various Al-layer thicknesses. Investigations of the microstructures at the contact/SiC interface were conducted using X-ray diffraction (XRD), cross-sectional transmission electron microscope (TEM) and energy dispersive X-ray (EDX) techniques to understand the mechanisms for the simultaneous formation of Ni/Al ohmic contacts to both n- and p-type SiC. Based on our present results, an optimized fabrication process for the simultaneous formation of Ni/Al ohmic contacts to both n- and p-type SiC was proposed.

EXPERIMENTAL PROCEDURES

The n- and p-type 4H-SiC epitaxial layers, doped with N at 1.3 x 10^{19} cm^{-3} and Al at 7.2 x 10^{18} cm^{-3}, were grown on 2-inch 4H-SiC wafers manufactured by Cree, Inc. (Durham, NC). The SiC substrates had 8° -off Si-terminated (0001) surfaces inclined toward the <$\bar{2}$110> direction. The 4H-SiC wafers were diced into small square pieces with about 8mm x 8mm. The square 4H-SiC substrates were used for measurements of electrical properties. For current-voltage (I-V) and specific contact resistance measurements, circular electrode patterns on the substrates were prepared using a photolithographic technique to remove portions of the SiO$_x$ layers. The substrates were cleaned by dipping in diluted hydrofluoric acid solution and rinsing in deionized water prior to metal depositions. Ni and Al layers were deposited sequentially by evaporation in a high vacuum below 1.5×10^{-4} Pa (the base pressure prior to deposition was approximately 4.0×10^{-6} Pa). Thicknesses of the Ni and Al layers are 50 nm and 0-10 nm, respectively. After lifting off the photoresist, the samples were annealed at 1000°C for 5 mins. in an ultra-high vacuum (UHV) chamber where the vacuum pressure was below 4.0×10^{-5} Pa (the base pressure prior to annealing was approximately 7.0×10^{-8} Pa).

The electrical property, microstructure and composition analyses were conducted using the annealed Ni/Al contacts. The I-V characteristics of Ni/Al contacts were measured by a two-point probe method using circular patterns with an interspacing of 8 μm. The specific contact resistances were measured using a circular transmission line model four-point probe method. The contact microstructures were analyzed by XRD and TEM. The Al/Ni ratio of the Ni/Al contacts was measured by EDX in a scanning electron microscope (SEM) and TEM, and by laser ablation inductively coupled plasma-mass spectrometry (LA-ICPMS) analyses. The LA-ICPMS measurements were conducted by JFE Techno-Research Corporation. To identify structure and composition of the contacts formed on the contact/SiC interfaces, the TEM samples with less than 100 nm in thickness and 10 μm in width were fabricated using focused ion beam (FIB) system (FB-2100, Hitachi High-Technologies Corporation).

RESULTS AND DISCUSSION

Simultaneous formation of ohmic contacts to both n- and p-type 4H-SiC

The I-V characteristics of the Ni/Al contacts to n- and p-type SiC are shown in Figs. 1(a) and 1(b), respectively. For n-type SiC (Fig. 1(a)), the contacts with Al-layer thicknesses in the range of 2.5 to 6 nm showed ohmic behavior, while the contact with 8 nm-thick Al layer showed rectifying (nonohmic) behavior. For p-type SiC (Fig. 1(b)), the contacts with Al-layer thicknesses in the range of 2.5 to 4 nm showed nonohmic behavior, while the contacts with Al-layer thicknesses in the range of 5 to 8 nm showed ohmic behavior. The contact properties of the Ni/Al contacts are given in Fig. 2. With increasing Al-layer thickness in the contacts, the contacts are found to change the ohmic behavior to from n-type SiC to p-type SiC. Only the contacts with Al-layer thicknesses in the range of 5 to 6 nm exhibited ohmic behavior to both n- and p-type SiC, with specific contact resistances of 1.8×10^{-4} Ωcm^2 and 1.2×10^{-2} Ωcm^2 for n- and p-type SiC, respectively. These values are higher than the specific contact resistance of the contacts which exhibited ohmic behavior to either n- or p-type SiC. This can be explained by noting that the area of n-type contact materials in the contacts which exhibited ohmic behavior to both n- and p-type SiC was smaller than that which exhibited ohmic behavior to only n-type SiC. This can be similar case for p-type contact materials.

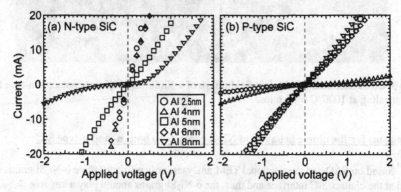

Figure 1. Current-voltage (I-V) characteristics of Ni(50nm)/Al(2.5-8nm) contacts to (a) n- and (b) p-type SiC after annealing at 1000°C for 5 mins.

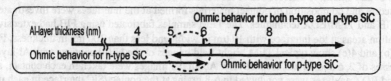

Figure 2. Schematic illustration of Al-layer thickness dependence of ohmic behavior to n- and p-type SiC.

Microstructure identification of Ni/Al contacts

XRD profiles of the Ni/Al contacts with various Al-layer thicknesses (not shown here) were in good agreement with a diffraction pattern of δ-Ni_2Si, indicating that the Ni/Al contacts reacted with the SiC substrates and that δ-Ni_2Si were formed. Figure 3 shows a cross-sectional TEM bright field image of a Ni(50nm)/Al(5nm) contact to n-type SiC after annealing at 1000°C for 5 mins. An electron incident beam with 300kV acceleration voltage was almost parallel to the <$\overline{2}$110> orientation of the SiC substrate. An about 100 nm-thick contact layer was uniformly formed on the SiC substrate. Based on elected area diffraction (SAD) analysis, polycrystalline δ-Ni_2Si grains with less than 50 nm in grain size and amorphous graphite grains with white contrasts (marked with arrows) were mostly observed. Only δ-Ni_2Si grains were found to form at the contact/SiC interface and thus, the δ-Ni_2Si grains should play a key role in determining electrical transport properties at the contact/SiC interface. Similar microstructures were observed in the Ni/Al contacts with other Al-layer thicknesses.

Figure 3. Cross-sectional TEM bright field image of a Ni(50nm)/Al(5nm) contact to n-type SiC after annealing at 1000°C for 5 mins.

Mechanisms for the ohmic behavior of Ni/Al contacts to both n- and p-type SiC

Based on SAD and conventional TEM analyses, only polycrystalline δ-Ni_2Si grains were formed at the contact/SiC interface and thus, the δ-Ni_2Si grains should play a key role in determining electrical transport properties between contacts and SiC. However, it is difficult to understand how only one material can exhibit ohmic behavior to both n- and p-type SiC. Thus, in order to obtain extensive microstructure information of the contacts at the contact/SiC interfaces, the structure and composition of contact materials formed at the interfaces were investigated using FIB and TEM/EDX techniques. The TEM samples fabricated using FIB had extensive observation areas at the interface with 10 μm in width and less than 100 nm in thickness. Figures 4(a), 4(b) and 4(c) show typical TEM bright field images of the Ni/Al contacts with Al-layer thicknesses of 2, 6 and 10 nm, respectively. About ten images were taken in each contact, and SAD and EDX analyses were conducted in 6-8 grains at the contact/SiC interface in each image.

Large grains with about 50 nm in grain size were found to form at the contact/SiC interface in the Ni/Al contacts with Al-layer thicknesses of 2 and 10 nm (Figs. 4(a) and 4(c)), while a fine microstructure was observed at the interface in the Ni/Al contacts with Al-layer thickness of 6 nm (Fig. 4(b)). This suggests that microstructure of the ohmic contacts to both n-

and p-type SiC is different from that to either n- or p-type SiC. All the SAD images (not shown here), which were taken from one grain, showed that the contact materials formed at the interface is δ-Ni$_2$Si. However, Al(at.%)/Ni(at.%) ratios of the δ-Ni$_2$Si(Al) grains at the interface varied with Al-layer thickness of the contacts. The Al/Ni ratios obtained in the Ni/Al contacts with Al-layer thicknesses of 2, 6 and 10 nm are plotted in Fig. 4(d). Note that the distribution in values of the Al/Ni ratio in the Ni/Al contacts with Al-layer thickness of 6 nm, where ohmic behavior was observed to both n- and p-type SiC, was the largest in the three samples. The Al/Ni ratios in some grains in the Ni/Al contacts with Al-layer thickness of 6 nm were lower than the ratios in the Ni/Al contacts with Al-layer thickness of 2 nm. Also the grain sizes of the contact materials formed at the interfaces tended to be the smallest in the Ni/Al contacts with Al-layer thickness of 6 nm (Fig. 4(e)). Thus, the large distribution of the Al/Ni ratios and fine microstructure were found to be characteristic of the ohmic contacts to both n- and p-type SiC. The reason why such the typical feature is formed has not been clarified yet. Figure 4(d) suggests that the δ-Ni$_2$Si(Al) grains with low and high Al concentrations exhibit ohmic behavior to n- and p-type SiC, respectively. The Al/Ni ratio in δ-Ni$_2$Si(Al) required for ohmic behavior to either n- or p-type SiC would be less than or larger than about 0.05 based on the results of Fig. 4(d).

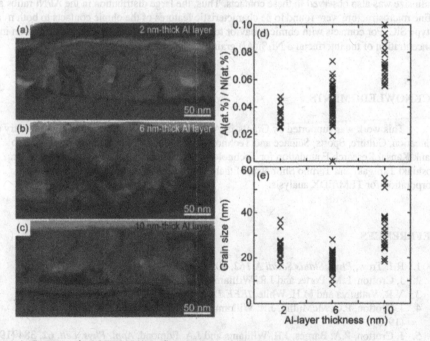

Figure 4. Cross-sectional TEM bright field images of (a) Ni(50nm)/Al(2nm), (b) Ni(50nm)/Al(6nm) and (c) Ni(50nm)/Al(10nm) contacts to n-type SiC after annealing at 1000°C for 5 mins. (d) Al/Ni ratio and (e) grain size of those contacts.

CONCLUSIONS

A technique was developed for the simultaneous formation of Ni/Al ohmic contacts to both n- and p-type SiC, where the doping level of N in the SiC substrates was 1.3×10^{19} cm^{-3} and the level of Al was 7.2×10^{18} cm^{-3}, after annealing at 1000°C for 5 mins in UHV. Ohmic contacts to n-type SiC were found when Al-layer thickness was less than about 5 nm while ohmic contacts to p-type SiC were observed for an Al-layer thickness greater than about 5 nm. Only the contacts with Al-layer thicknesses in the range of 5 to 6 nm exhibited ohmic behavior to both n- and p-type SiC, with specific contact resistances of 1.8×10^{-4} Ωcm^2 and 1.2×10^{-2} Ωcm^2 for n- and p-type SiC, respectively. An about 100 nm-thick contact layer was uniformly formed on the SiC substrate and consisted of polycrystalline δ-Ni$_2$Si(Al) grains and amorphous graphite. Only δ-Ni$_2$Si(Al) grains were found to form at the contact/SiC interface and thus, the δ-Ni$_2$Si(Al) grains play a key role in determining electrical transport properties at the contact/SiC interface. The average Al concentration increased lineally with increasing Al-layer thickness. However, the distribution in values for the Al/Ni ratio in the δ-Ni$_2$Si(Al) grains which exhibited ohmic behavior to both n- and p-type SiC were the largest. The smallest average δ-Ni$_2$Si(Al) grain size was also observed in these contacts. Thus, the large distribution in the Al/Ni ratios and a fine microstructure were found to be characteristic features of the ohmic contacts to both n- and p-type SiC. For contacts with ohmic behavior to both n- and p-type SiC, a wide distribution in Al concentration of the interfacial δ-Ni$_2$Si(Al) grains is required.

ACKNOWLEDGMENTS

This work was supported by Grants-in-Aid for Scientific Research from The Ministry of Education, Culture, Sports, Science and Technology (18360324). The authors would like to thank Kansai Research Foundation for Technology Promotion for financial support, and Dr. Toshiaki Tanigaki and Teruho Shimotsu, Hitachi High-Tech Manufacturing & Service Corporation for TEM/EDX analysis.

REFERENCES

1. R.J. Trew, *Phys. Status Solidi* A **162**, 409 (1997).
2. J. Crofton, L.M. Porter and J.R. Williams, *Phys. Status Solidi* B **202**, 581 (1997).
3. V.R. Vathulya and M.H. White, *IEEE Trans. Electron Dev.* **47**, 2018 (2000).
4. J. Crofton, P.G. McMullin, J.R. Williams and M.J. Bozack, *J. Appl. Phys.* **77**, 1317 (1995).
5. J. Crofton, P.A. Barnes, J.R. Williams and J.A. Edmond, *Appl. Phys. Lett.* **62**, 384 (1993).
6. O. Nakatsuka, T. Takei, Y. Koide and M. Murakami, *Mater. Trans.* **43**, 1684 (2002).
7. L.G. Fursin, J.H. Zhao and M. Weiner, *Electron. Lett.* **37**, 1092 (2001).
8. S. Tanimoto, N. Kiritani, M. Hoshi and H. Okushi, *Maer. Sci. Forum* **389**, 879 (2002).
9. S. Tsukimoto, T. Sakai, T. Onishi, K. Ito and M. Murakami, *J. Elec. Mater.* **34**, 1310 (2005).

Mater. Res. Soc. Symp. Proc. Vol. 1069 © 2008 Materials Research Society   1069-D10-04

Influence of Shockley Stacking Fault Expansion and Contraction on the Electrical Behavior of 4H-SiC DMOSFETs and MPS Diodes

Joshua David Caldwell[1], Robert E. Stahlbush[1], Eugene A. Imhoff[1], Orest J. Glembocki[1], Karl D. Hobart[1], Marko J. Tadjer[2], Qingchun Zhang[3], Mrinal Das[3], and Anant Agarwal[3]

[1]Naval Research Laboratory, 4555 Overlook Ave, S.W., Washington, DC, 20375
[2]Electrical Engineering Department, University of Maryland, College Park, MD, 20740
[3]Cree Inc., 3026 E. Cornwallis Rd, Research Triangle Park, NC, 27709

ABSTRACT

The increase in the forward voltage drop (V_f) observed in 4H-SiC bipolar devices due to recombination-induced Shockley stacking fault (SSF) creation and expansion has been widely discussed in the literature. It was long believed that the deleterious effect of these defects was limited to bipolar devices. However, it was recently reported that forward biasing of the body diode of a 10kV 4H-SiC DMOSFET, a unipolar device, led to similar V_f increases in the body diode I-V curve as well as a corresponding degradation in the majority carrier conduction characteristics. This degradation was believed to be due to the creation and expansion of SSFs. Here we report measurements comparing the influence of similar stressing, along with annealing and current-induced recovery experiments in DMOSFETs and merged pin-Schottky diodes with the previously reported results in 4H-SiC pin diodes. These experiments support the hypothesis that the majority-carrier conduction current degradation is the result of SSF expansion.

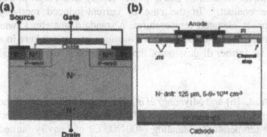

Fig. 1: Cross-sections of the (a) DMOSFETs and (b) MPS diodes.

INTRODUCTION

Silicon carbide is a desirable material for high power and temperature bipolar and unipolar electronic devices, such as high blocking voltage *pin* and Schottky diodes, respectively. However, electron-hole pair (ehp) recombination at basal plane dislocations (BPDs) in the drift layer of bipolar devices induces Shockley stacking fault (SSF) creation. Continued ehp injection causes the SSFs to expand, which induces an increase in the forward voltage drop (V_f) [1]. Recently, Agarwal et al. [2] determined that when the body diode of a DMOSFET was forward biased, a V_f drift of the body diode was observed. Furthermore, this drift was coupled with a reduction in the majority-carrier conduction current and an increase in the leakage current when the DMOSFET was operated in blocking mode. From these results, the authors implied that this degradation was due to the creation and expansion of SSFs via ehp injection from the body diode forward-bias stressing. If this implication is true, this is the first evidence of SSFs having a negative impact on the behavior of a majority-carrier SiC device. In order to determine the validity of this hypothesis, we report electrical stressing and annealing studies from both SiC high voltage DMOSFETs and merged pin-Schottky (MPS) diodes, studying the characteristic features of this degradation in comparison to those of the better understood degradation and the annealing-induced and current-induced V_f drift recoveries reported in 4H-SiC pin diodes by Caldwell et al. [3, 4].

EXPERIMENT

Two types of devices were studied: 10 kV, 0.1545 cm^2 4H-SiC DMOSFETs (Fig. 1a) and 10 kV, 0.04 cm^2 4H-SiC MPS diodes (Fig. 1b), both of which were fabricated by Cree Inc. A description of the fabrication details and the pertinent operating features of the DMOSFET structure may be found in the literature [2]. For the MPS diodes, a full-wafer of alternating MPS and pin diodes was fabricated. The wafer consisted of an epitaxially-grown 125 μm thick, N_D=5-6X10^{14} cm^{-3} drift layer, followed by ion implanted p$^+$ junctions, JTEs and channel stops. Prior to the implantation, but following a shallow RIE etching of the zero-level mask, a photoluminescence (PL) image map was acquired using the technique reported by Stahlbush et al. [5]. These illustrate the BPD and threading dislocations within the diode device regions. This allowed direct comparison between the individual device electrical characteristics and their response to electrical stressing with the BPD density within the device.

Pulsed I-V measurements were recorded at 30, 60, 100, 150 and 200^{0}C following each successive stressing or annealing procedure to monitor any changes in the electrical characteristics that were induced. In the case of the DMOSFET, the I-V traces were collected for both the body diode (minority-carrier conduction) and as a function of gate voltage (V_g, majority-carrier conduction). All of these measurements were performed using a Tektronix 371B high-power curve tracer, operated in pulsed-current mode. This instrument provides short (250 μs) current pulses at a duty cycle of 0.75%, simultaneously monitoring the voltage at each applied current. In this article, each trace represents the average of three consecutive I-V traces.

Stressing of the DMOSFETs was completed using the pulsed-current output of a Tektronix 371B, by negatively biasing the drain in reference to the mutually-grounded source and gate contacts. A 15A (97 A/cm^2) forward-bias current was maintained. The MPS diodes were stressed at 3A DC (75 A/cm^2) using an HP 6024A 200W power supply and a high power pogo-pin probe for the current-source contact. In the case of current-induced recovery measurements that involve ehp injection at elevated temperatures, a standard hot chuck was used. For annealing experiments, the samples were heated within a nitrogen gas atmosphere in a Centurion VPM NEY vacuum furnace. A further discussion of these procedures may be found in the literature [3, 4, 6].

DISCUSSION

Prior to the reports of Caldwell et al. [6] and Miyanagi et al. [7], it was believed that the 3C-SiC SSFs were the thermodynamically favorable state of the native 4H-SiC lattice. However, the authors showed that low temperature annealing (400-700^{0}C) of heavily-faulted (degraded) 4H-SiC pin diodes and epitaxial layers, respectively, induced a contraction of the expanded SSFs, illustrating that the SSFs are not energetically favorable. Caldwell et al. [4] later reported that this annealing behavior was coupled with a complete and repeatable recovery of the V_f drift in degraded pin diodes. Furthermore, it was also observed that if a current was maintained within the device at an elevated temperature (~250^{0}C), a current-induced recovery of the V_f drift occurred [3]. This recovery was drastically enhanced, both in the magnitude and in the speed of the recovery, in comparison to the annealing-induced recovery at that same temperature in the absence of injected ehps. These two phenomena, most especially the latter, are indicative of the SSF-induced degradation and such recovery mechanisms are not readily observed via other processes in such devices. Therefore, the observation of these two effects in electrically-degraded unipolar devices would provide substantial evidence supporting the Agarwal's hypothesis [2], further establishing the role that SSFs play in this degradation process.

DMOSFET RESULTS

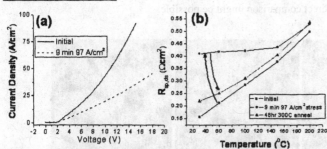

Fig. 2: (a) DMOSFET body diode forward-bias I-V curve prior to (solid curve) and following the 20 hour pulsed (9 min total injection time), 97 A/cm² current injection (dashed curve). (b) Temperature dependence of the specific on-state resistance of the DMOSFET prior to (squares), following the current injection (circles) and after 48 hours of annealing at 300°C (triangles). The lines are a guide to the eye and the arrows indicate the direction of the changes.

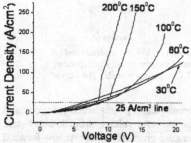

Fig. 3: I-V traces collected at temperatures as marked from the moderate BPD density MPS diode prior to any forward bias operation. The horizontal line indicates 25 A/cm².

Presented in Fig 2 (a) are I-V curves collected at 30°C from the body diode of a 10kV, 4H-SiC DMOSFET prior to (solid line) and following 20 hrs (9 min total injection time) of pulsed 97 A/cm² biasing of the body diode (dashed line). This graph shows the degradation that occurs following ehp injection into the body diode of the device, similar to that reported by Agarwal *et al.* [2] and to that observed in 4H-SiC pin diodes [8, 9]. Also consistent with Ref. [2], is the observed degradation of the majority-carrier characteristics, in this case presented in Fig. 2 (b) as an increase in the specific on-state resistance ($R_{sp,on}$) of the device at V_g=20V and I_{DS}=5A that was observed following the body diode stressing. Here we see that prior to ehp injection (squares, lines are guides to the eye), $R_{sp,on}$ increases linearly with increasing temperature. However, following electrical stressing (circles), not only was an increase in the V_f drift of the body diode observed, but there was an increase in the temperature-dependent $R_{sp,on}$. This increase is more predominant at the lower temperatures, which is consistent with SSF-induced electrical degradation. SSFs are known to act as 'structure-only' quantum wells [10, 11] within the 4H-SiC lattice, acting as electron traps at $\Delta E_c \sim$ -0.3eV. Similar to the behavior observed here, as the temperature is increased, the probability of trapping an electron within the 2D density of states associated with the SSF is reduced and, therefore, so is its impact upon the electrical characteristics of the device.

While the temperature dependence of $R_{sp,on}$ following electrical stressing is consistent with the presence of SSFs, it is by no means definitive. Therefore, the device was annealed at 300°C for two separate 24 hr periods and the device characteristics were recorded following this process (triangles). From this data, it was clear that the device recovered considerably following this process, with a significant reduction in the measured $R_{sp,on}$ being observed, along with a relaxation of its temperature dependence towards the prestress linear behavior. This recovery again supports the hypothesis that SSFs are the cause of the observed degradation in the majority-carrier electrical characteristics. However, continued experimentation with this and other DMOSFET devices was difficult due to gate reliability issues and the associated limitation on post-stress annealing temperature. Therefore a less complex device was desired for further investigation where a more direct comparison might be possible.

MPS DIODE RESULTS

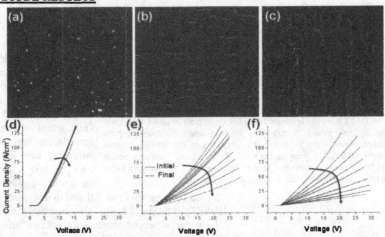

Fig. 4: Photoluminescence images collected from MPS diodes with (a) low, (b) moderate and (c) high BPD densities. The images were collected using a 780 nm long-pass filter. Corresponding I-V traces collected following each subsequent 75 A/cm² DC stressing period. The curved arrows indicate the direction of V_f drift.

MPS diodes utilize a single contact to both the interdigitated p⁺ implanted junctions and the n⁻ epitaxy to create a hybrid device providing a low on-state voltage drop, low off-state leakage, fast switching capabilities and good high temperature characteristics. At low forward biases and low temperatures, they exhibit minimal minority-carrier injection, however, as the temperature and/or forward bias is increased, the minority-carrier injection efficiency improves. The temperature dependent I-V traces from one of the MPS diodes is presented in Fig. 3, illustrating this behavior.

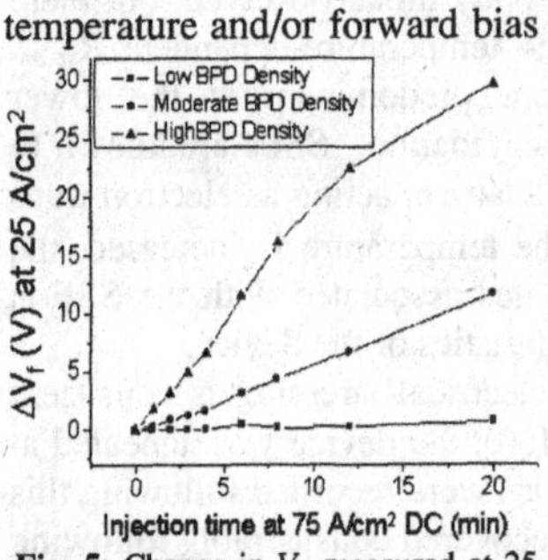

In order to decipher what role SSFs play in the forward bias degradation of MPS diodes, three diodes from the PL-mapped wafer were chosen to provide devices with low, moderate and very high BPD densities. Since BPDs are the initial epitaxial growth defects which lead to SSF creation, it would be expected that the devices with higher BPD density would exhibit significantly increased forward-bias degradation if SSFs are the cause. PL images of the three diodes chosen are presented in Fig. 4 (a)-(c), respectively. The BPDs may be identified as long white lines and the densities for the three diodes can be estimated as <25, >1000 and >10,000 cm⁻², respectively. The white dots are threading dislocations (both screw and edge), which are benign in terms of the forward-bias-induced electrical degradation.

Following device fabrication, pulsed I-V traces were

Fig. 5: Change in V_f, measured at 25 A/cm² on each I-V trace, as a function of injection time in MPS diodes with the low (squares), moderate (circles) and high (triangles) BPD densities. The lines are a guide to the eye.

collected from all three diodes at 30, 60, 100, 150 and 200 °C prior to any forward bias operation. The diodes were then stressed at 75 A/cm² DC at 30°C for periods of 1, 1, 1, 1, 2, 2, 4, and 8 mins (20 min total), with pulsed I-V(T) curves collected following each successive period. Presented in Fig. 4 (d)-(f) are the corresponding 30°C I-V curves after each stressing period for each of the diodes from the regions presented in Fig. 4 (a)-(c), (the arrows denote the direction of the Vf drift with continued electrical stressing). As the BPD density in the device increased, so too did the magnitude of the V_f drift. Presented in Fig. 5 is the change in V_f as a function of injection time for the three diodes. The V_f was measured at 25 A/cm² along each successive I-V curve. This value was chosen as the devices exhibited low minority-carrier injection, as shown in Fig. 3 (a dashed line designates the 25 A/cm² threshold). It is evident that there is a strong dependence of the magnitude of the V_f drift on the initial BPD density, which is indicative of a direct dependence of this degradation on the creation and expansion of SSFs.

In an effort to test this SSF-induced degradation further, the MPS diode wafer was annealed for 288 hours at 300°C, followed by an additional 96 hours at 400°C, with I-V(T) measurements being recorded following the first 96 hours and after an additional 192 hours of 300°C annealing and then again following the final 96 hour, 400°C anneal. Presented in Fig. 6 are I-V traces recorded at 30°C for the diode with moderate BPD density prior to electrical stressing (solid line), following 20 min of stressing at 75 A/cm² DC (dashed line) and following the entire annealing cycle (dotted line). Similar to the DMOSFETs, this extended period of annealing induced a significant, and in this case very close to a complete, recovery of the forward-bias-induced electrical degradation. The time dependence of the V_f drift is presented in Fig. 7. From this figure, it is clear that as the annealing temperature was increased the V_f-drift recovery rate also increased. This again is quite similar to the case of the SSF-induced V_f drift in 4H-SiC *pin* diodes, where an activation energy for this recovery was reported as 1.3 +/- 0.3 eV by Caldwell *et al.* [12].

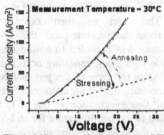

Fig 6: I-V curves of the moderate BPD density MPS diode prior to (solid line) and following 20 min of electrical stressing at 75 A/cm² (dashed line), and following a 244 hours 300°C anneal and a subsequent 96 hour 400°C anneal (dotted line). The arrows indicate the direction of the V_f drift.

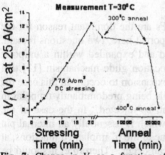

Fig. 7: Change in V_f as a function of stressing (squares) and annealing (circles). The lines provide a guide to the eye.

While all of the previous data presented is very supportive of the hypothesis put forth by Agarwal *et al.* [2], the data is still not definitive of a direct correspondence between SSFs and the observed degradation. Recently, it was reported that during forward-bias operation of 4H-SiC pin diodes, that continued electrical stressing led to a saturation of the V_f drift [3]. Furthermore, if the device temperature was increased and the current injection was continued at the same injection level, the V_f drift was actually *recovered* along with a SSF *contraction*. A plot of the V_f drift as a function of stressing time during such a measurement in a pin diode is presented in Fig. 8 (a). Initially, the 10kV *pin* diode was stressed at 13.9 A/cm² for an extended period, with the I-V characteristics being monitored following each successive stress. Following this, the device was heated to 242^0C and the same current was injected into the device for a series of prescribed time periods, with the I-V characteristics being measured following each period. In this case, a 22% recovery of the V_f drift was recorded. The observation that such a small temperature change ($\Delta kT \sim 19$ meV) could have such a dramatic effect on the SSFs and on their electrical consequences is something that does not occur via another known mechanism in similar devices. Therefore, the observation of this current-induced recovery effect in these MPS diodes would strongly supports Agarwal's hypothesis.

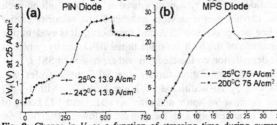

In order to test if this phenomenon were to occur, following the initial room temperature stressing of the MPS diodes, but prior to the aforementioned annealing cycle, the wafer was heated to 200^0C on the probe station hot chuck and allowed to equilibrate for 20 min. The wafer was then cooled to room temperature and the I-V trace was recorded to ensure that no significant annealing occurred.

Fig. 8: Change in V_f as a function of stressing time during current-induced recovery measurements performed on a (a) pin diode and (b) the moderate BPD density MPS diode. Initially, the diodes were stressed at 25^0C at 13.9 and 75 A/cm^2, respectively, (circles). The diodes were then stressed at the same current at 242 and 200^0C, respectively, (squares).

As in the case of the pin diodes, no change in the V_f drift was observed from such low temperature annealing attempts. The device was then heated to 200^0C again and this time 75 A/cm^2 was injected for periods of 1, 2 and 10 minutes at this temperature, with the device being cooled to room temperature for an I-V measurement following each successive period. The change in the V_f as a function of stressing time from this MPS diode is presented in Fig. 8 (b), where a recovery of 28% is observed. Note the strong similarities in the behavior between the current-induced recovery response in the pin diode and that which was observed in the MPS diode. The large variations in time-scales for this phenomenon between the two devices can be attributed to similar variations in the BPD densities and the large difference in current injection levels between the two studies. This observation leaves very little doubt that SSFs are the predominant cause for the observed degradation in the majority carrier electrical characteristics of primarily unipolar devices.

CONCLUSIONS

While from the above experiments it is clear that SSFs are the dominant reason for the electrical degradation of the DMOSFETs and MPS diodes reported here, two questions remain. The first involves the method in which the SSFs are formed and expanded within a unipolar device when it is clear from the recombination-induced dislocation glide mechanism [13] that bipolar carrier recombination at the faults is required for this expansion to occur. In this case the answer is rather straightforward. In both of these devices, while being predominantly unipolar in character, a mechanism exists by which minority carriers can be injected. In the case of the DMOSFETs, this occurs by forward biasing the body diode, which is essentially an internal pin diode to the device, whereas in the case of the MPS diodes, due to the p$^+$ implanted junctions, at high injection currents significant minority carrier injection can occur. Therefore, it should be realized that in any 4H-SiC device, bipolar injection into the active area of any device with BPDs, will result in the formation of SSFs, and that these faults will have negative implications even on the majority-carrier conduction characteristics. This is one of the reasons for the superior reliability of truly unipolar devices such as Schottky and JBS diodes. The second question involves the mechanism by which the SSFs themselves induce this change in the majority carrier characteristics. One could speculate that the SSFs effectively provide local areas of increased resistivity, thereby adding an additional series resistance to the current conduction path. Such a resistance could induce the observed change in the electrical characteristics of the device.

REFERENCES

[1] M. Skowronski and S. Ha, J. Appl. Phys. 99, 011101 (2006).
[2] A. Agarwal, H. Fatima, S. Haney and S.-H. Ryu, IEEE Electron Device Lett. 28, 587 (2007).
[3] J. D. Caldwell, O. J. Glembocki, R. E. Stahlbush and K. D. Hobart, Appl. Phys. Lett. 91, (2007).
[4] J. D. Caldwell, R. E. Stahlbush, K. D. Hobart, O. J. Glembocki and K. X. Liu, Appl. Phys. Lett. 90, 143519 (2007).
[5] R. E. Stahlbush, K. X. Liu, Q. Zhang and J. J. Sumakeris, Mater. Sci. Forum 556-557, 295 (2007).
[6] J. D. Caldwell, K. X. Liu, M. J. Tadjer, O. J. Glembocki, R. E. Stahlbush, K. D. Hobart and F. Kub, J. Electron. Mater. 36, 318 (2007).
[7] T. Miyanagi, H. Tsuchida, I. S. Kamata, T. Nakamura, K. Nkayama, R. Ishii and Y. Sugawara, Appl. Phys. Lett. 89, 062104 (2006).
[8] J. P. Bergman, H. Lendenmann, P. A. Nilsson, U. Lindefelt and P. Skytt, Mater. Sci. Forum 353-356, 299 (2001).
[9] R. E. Stahlbush, M. Fatemi, J. B. Fedison, S. D. Arthur, L. B. Rowland and W. Wang, J. Electron. Mater. 31, 370 (2002).
[10] O. J. Glembocki, M. Skowronski, S. M. Prokes, D. K. Gaskill and J. D. Caldwell, Mater. Sci. Forum 357-359, 347 (2006).
[11] K.-B. Park, Y. Ding, J. P. Pelz, M. K. Mikhov, Y. Wang and B. J. Skromme, Appl. Phys. Lett. 86, 222109 (2005).
[12] J. D. Caldwell, O. J. Glembocki, R. E. Stahlbush and K. D. Hobart, J. Electron. Mater. in press, (2008).
[13] K. Maeda and S. Takeuchi, in *Dislocations in Solids*, edited by F. R. N. Nabarro and M. S. Duesbery (North-Holland Publishing Company, Amsterdam, 1996), p. 443.

Mater. Res. Soc. Symp. Proc. Vol. 1069 © 2008 Materials Research Society

Performance of SiC Microwave Transistors in Power Amplifiers

S. Azam[1], R. Jonsson[2], E. Janzen[1], and Q. Wahab[1,2]

[1]Department of Physics, Chemistry and Biology (IFM), Linkaoping University, Linkaoping, 58183, Sweden

[2]Swedish Defence Research Agency (FOI), Linkaoping, 58111, Sweden

ABSTRACT

The performance of SiC microwave power transistors is studied in fabricated class-AB power amplifiers and in physical simulations of class-C switching power amplifier using the physical structure of an enhanced version of previously fabricated and tested SiC MESFET. The results for pulse input in class-C at 1 GHz are; efficiency of 71.4 %, power density of 1.0 W/mm. The switching loss was 0.424 W/mm. The results for two class-AB power amplifiers are; the 30-100 MHz amplifier showed 45.6 dBm (~ 36 W) output powers at P_{1dB}, at 50 MHz. The power added efficiency (PAE) is 48 % together with 21 dB of power gain. The maximum output power at P_{1dB} at 60 V drain bias and Vg= -8.5 V was 46.7 dBm (~47 W). The typical results obtained in 200-500 MHz amplifier are; at 60 V drain bias the P_{1dB} is 43.85 dBm (24 W) except at 300 MHz where only 41.8 dBm was obtained. The maximum out put power was 44.15 dBm (26 W) at 500 MHz corresponding to a power density of 4.3 W/mm. The PAE @ P_{1dB} [%] at 500 MHz is 66 %.

INTRODUCTION

Due to superior physical properties of SiC such as high saturated carrier velocity, high thermal conductivity and high breakdown field, SiC transistors have the capability of high power density and due to high transistor impedance, making matching easier. These qualities have been applied in the development of different generations of power amplifiers for use in digital audio and video broadcasting [1], [2], aerospace and military systems [3], [4] and also in UHF broadband amplifiers [5], [6]. The high breakdown field makes them suited for higher operation voltage. SiC transistors offer high output impedance, while Si LDMOS and GaAs based devices offers lower output impedance, which is more complicated to match. An experiment analyzing memory effects in the IQ plane showed that SiC-MESFET has less to no memory effects compared to Si LDMOS devices [7].

In this paper we are presenting the performance of SiC microwave power transistors based on simulation and fabrication of class-AB power amplifiers using Advance Design System (ADS) Software. The switching response of an enhanced version of a physical structure of a previously fabricated and tested SiC MESFET [8] is studied using an active load-pull simulation technique [9] in TCAD. The performance of these amplifiers is comparable to other power amplifiers (PAs) reported in the recent literature [10], but we achieved with high power, PAE and gain for these particular frequency bands. The frequency range considered in this work is a part of our Power amplifier research work for EW-system.

RESULTS AND DISCUSSION

A). Measured results of fabricated 30-100 MHz class-AB power amplifier

This amplifier design is based on S-parameters of the transistor in ADS software. Power measurements were performed between 30-100 MHz with 20 MHz intervals except between 90 MHz and 100 MHz. The output power at 1 dB gain compression (P_{1dB}) at 50 MHz was 45.6 dBm (~36 W) with power added efficiency (PAE) of 48 % and 21 dB gain. The maximum output power measured was 46.1 dBm (~41 W) at 2 dB gain compression. For the frequencies where measurements were performed, the output power is above 44.3 dBm (~27 W), the PAE is above 44 % and the gain is above 20 dB, all values taken at P_{1dB}. The maximum values for PAE, output power and gain are achieved at 30 MHz. At 30 MHz the PAE is 61.7 %, output power is 46.2 dBm (~41.7 W) and a gain of 21.1 dB as shown in Fig. 2. One power sweep was also done at higher drain bias (V_g= -8.5 V and V_d = 60 V) at 30 MHz, at this bias point the P_{1dB} was 46.7 dBm (~47 W). A picture of the fabricated PA is shown in Fig. 1.

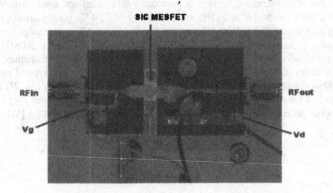

Fig. 1: A picture of the fabricated power amplifier for 30-100 MHz

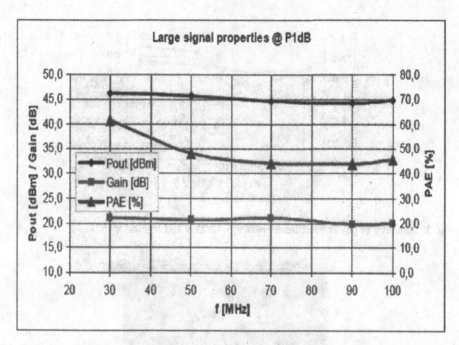

Fig. 2: Summary of power measurements at $V_g = -8.5$ V and $V_d = 50$ V.

B). Measured results of fabricated 200-500 MHz class-AB power amplifier

This amplifier is also designed using S-parameters of the transistor in ADS software. Power measurements were carried out between 200-500 MHz with 100 MHz intervals. All measurements were made at a gate bias of -2.25 V and a drain bias of 50 V and 60 V (Class AB). At 50 V, P1 dB was reached at an output power of 43.4 dBm (around 20 W) except at 200 MHz where only 39.7 dBm was measured. The maximum out put power measured was 43.4 dBm (21.88 W) at 400 MHz corresponding to a power density of 4.5 W/mm of gate periphery. PAE @ P1 dB [%] at 400 MHz is 60 %.

At 60 V drain bias and -2.25 V gate bias. The P1 dB was reached at an output power 20 W except at 300 MHz where only 15 W was measured. The maximum out put power measured was 44.15 dBm (26 W) at 500 MHz corresponding to a power density of 4.3 W/mm of gate periphery. The PAE @ P1 dB [%] in this design at 500 MHz is 66 % as shown in Fig. 3. A picture of the fabricated power amplifier is shown in Fig. 4.

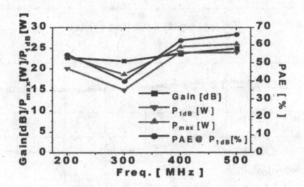

Fig. 3: Summary of power measurements at $V_g = -2.25$ V and $V_d = 60$ V.

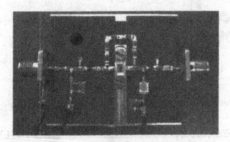

Fig. 4: A picture of fabricated power amplifier for 200-500 MHz

A). Class-C switching power amplifier results

In our simulations instead of sine wave we applied square pulses at the gate, while the RF source at the drain delivered a sine wave at the fundamental frequency thereby acting as a short at higher harmonic frequencies. We applied a gate pulse of constant duty cycle at 1 GHz PRF. While applying V_{ac} peak-to-peak signal of 90 V_{pp} together with V_{dc} of 55 V at the drain side. In order to calculate power added efficiency (PAE), power density, switching loss and gain of the amplifier, the time domain resulting current and voltage signals are then Fourier transformed into frequency domain using Fast Fourier transformation (FFT) in MATLAB. In this technique, there is no need of other matching than the active load.

We applied a 5 % duty cycle pulse of constant amplitude (-15 V) at the gate and Vac peak-to-peak signal of 90 V and Vdc of 55 V at the drain side. Typical values are efficiency of 71.4 %, power density of 1.0 W/mm. The switching loss was 0.424 W/mm.

The device show resistive loss behavior, which is clear from Fig. 5, where hatched area 'A' (which is resistive area) is more than the hatched area 'B' (which is capacitive area). And the difference of 'A-B' gives us a net resistive area. The resistive behavior is also clear from the electron current density of $1.6E^{+05}$ A/cm^{-2} in the channel at peak resistive area at the turn off time as shown in Fig 6. The reason is that, the drain current take a longer time to turn off and current and voltage have simultaneously high values in this region. This simultaneous high current and voltage increases the switching loss, which in turn reduces the efficiency.

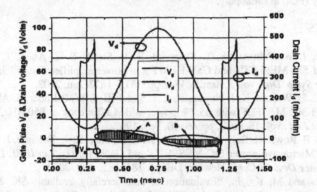

Fig. 5: Waveforms for the drain current and voltage and the gate pulse at 1 GHz. The hatched areas represent the integrated resistive and capacitive part in drain current during turn off time.

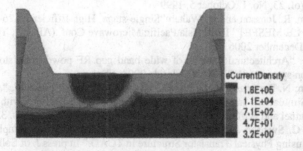

Fig. 6: Tech plot picture of SiC MESFET structure showing electron current density in the channel at peak resistive area at 1 GHz

CONCLUSIONS

The maximum PAE of 66 % and 61.7 % and maximum power of 47 W and 26 W achieved for class-AB power amplifiers using single transistors at corresponding frequencies and

PAE of 71.4 % and the power density of 1.0 W/mm for class-C power amplifier achieved in TCAD simulations of a SiC MESFET transistor structure indicates that although SiC based PAs presently can not compete with GaN and other conventional devices like GaAs in terms of frequency but in terms of power and efficiency, they could be the strong competitors and future devices for, RADAR, Electronic Warfare (EW), Wireless Communications and base stations applications.

ACKNOWLEDGMENTS

The authors wish to acknowledge the support of Microwave group at Swedish defense research agency (FOI) at Linköping.

REFERENCES

[1]   W. L. Pribble, J. W. Palmour, S. T. Sheppard, R. P. Smith, S. T. Allen, T. J. Smith, etal "Application of SiC MESFET's and GaN HEMT's in power amplifier design," in *IEEE MTT-S Int. Microwave Symp. Dig.*, vol. 3, Seattle, WA, pp. 1819–1822, Jun. 2002.

[2]   K. E. Moore, C. E. Weitzel, K. J. Nordquist, L. L. Pond III, J. W. Palmour, S. Allen, and C. H. Carter, Jr., "4H-SiC MESFET with 65.7% power added efficiency at 850MHz," *IEEE Trans. Electron Device Lett.*, vol. 18, no. 2, pp. 69–70, Feb. 1997.

[3]   A. P. Zhang, L. B. Rowland, E. B. Kaminsky, J. W. Kretchmer, R. A. Beaupre, J. L. Garrett, and J. B. Tucker, "Microwave power SiC MESFET's and GaN HEMTs," in *IEEE Lester Eastman High Performance Devices Conf.*, Newark, DE, pp. 181–185, Aug. 2002.

[4]   M. G. Walden and M. Knight, "Evaluation of commercially available SiC MESFET's for phased array radar applications," in *IEEE Int. Electron Devices for Microwave and Optoelectronic Applications Symp.*, Manchester, U.K., pp. 166–171, Nov. 2002.

[5]   J. F. Broch, F. Temcamani,1 P. Pouvil, O. Noblanc and J. P. Prigent "POWER AMPLIFICATION WITH SILICON CARBIDE MESFET" *Microwave and Optical Technology Letters* / Vol. 23, No. 1, October 5, 1999

[6]   Sher Azam, R. Jonsson and Q. Wahab "Single-stage, High Efficiency, 26-Watt power Amplifier using SiC LE-MESFET" IEEE Asia Pacific Microwave Conf. (APMC), Yoko Hama (Japan), pp. 441–444, December 2006.

[7]   Fischer G. "Architectural benefits of wide band gap RF power transistors for frequency agile base station systems" *IEEE WAMI*, Florida 2004, FD-1

[8]   J. Eriksson, N. Rorsman, H. Zirath, R. Jonsson, Q. Wahab, S. Rudner, "A comparison between Physical Simulations and Experimental results in 4H-Si C MESFETs with Non-Constant Doping in the Channel and Buffer Layers", Material Science Forum, 2001, Vols. 353-356, pp. 699-702.

[9]   S. Azam, C. Svenson and Q. Wahab,"Pulse Input Class-C Power Amplifier Response of SiC MESFET using Physical Transistor Structure in TCAD." In press J. of Solid State Electronics.

[10]   S. Azam and Q. Wahab: "GaN and SiC Based High Frequency Power Amplifiers", Review article submitted to the book "Micro and Nano Electronics", which will be published by NAM S&T Center, New Dehli.

Mater. Res. Soc. Symp. Proc. Vol. 1069 © 2008 Materials Research Society   1069-D11-02

Long-Term Characterization of 6H-SiC Transistor Integrated Circuit Technology Operating at 500 °C

Philip G. Neudeck[1], David J. Spry[2], Liang-Yu Chen[2], Carl W. Chang[3], Glenn M. Beheim[1], Robert S. Okojie[1], Laura J. Evans[1], Roger D. Meredith[1], Terry L. Ferrier[1], Michael J. Krasowski[1], and Norman F. Prokop[1]

[1]NASA Glenn Research Center, 21000 Brookpark Road, M.S. 77-1, Cleveland, OH, 44135
[2]OAI, NASA Glenn, 21000 Brookpark Road, M.S. 77-1, Cleveland, OH, 44135
[3]ASRC, NASA Glenn, 21000 Brookpark Road, M.S. 77-1, Cleveland, OH, 44135

ABSTRACT

NASA has been developing very high temperature semiconductor integrated circuits for use in the hot sections of aircraft engines and for Venus exploration. This paper reports on long-term 500 °C electrical operation of prototype 6H-SiC integrated circuits based on epitaxial 6H-SiC junction field effect transistors (JFETs). As of this writing, some devices have surpassed 4000 hours of continuous 500 °C electrical operation in oxidizing air atmosphere with minimal change in relevant electrical parameters.

INTRODUCTION

Extension of the operating envelope of useful transistor integrated circuits (ICs) to temperatures well above the effective 300 °C limit of silicon-on-insulator technology is expected to enable important improvements to aerospace, automotive, energy production, and other industrial systems [1,2]. For example, extreme temperature ICs capable of T = 500 °C operation are considered vital to realizing improved sensing and control of turbine engine combustion leading to better fuel efficiency with significantly reduced pollution. The ability to place 500 °C ICs in engine hot-sections would reduce weight and reliability penalties (from added wires and liquid cooling plumbing) that arise because silicon combustion-control ICs are restricted to operating temperatures well below 300 °C. In general, competitive performance benefits to large systems enabled by extreme temperature ICs are recognized as quite substantial, even though most such systems require only a relatively small number of extreme temperature chips [1].

One critical requirement for all ICs, including extreme temperature ICs, is that they function reliably over a designed product lifetime. Previous reports of extreme temperature ($\geq$ 500 °C) transistor or IC operation enabled by wide bandgap semiconductors have focused on current-voltage (I-V) properties and gain-frequency performance with little or no mention of how long such parts operated at high temperature. Without thousands of hours of useful operating life, extreme temperature semiconductor ICs will not benefit (and will not be inserted into) the vast majority of important intended applications. Aside from work at NASA Glenn Research Center [3-8], we are unaware of any published reports claiming stable semiconductor transistor operation for more than 10 hours at temperatures at or above 500 °C.

This paper reports semiconductor transistors and small ICs that have achieved thousands of hours of stable electrical operation at 500 °C. In particular, this work updates and expands our initial reports of transistor and differential amplifier testing for up to 3000 hours at 500 °C [6,7].

The results establish a new technology foundation for realizing usefully durable, extreme temperature ICs to benefit important harsh-environment applications.

EXPERIMENTAL

Driven by the need to realize integrated circuits with prolonged 500 °C operational durability, an epitaxial n-channel 6H-SiC junction field-effect transistor IC technology, shown in schematic cross-section in figure 1, was selected for development. In particular, the epitaxial SiC pn-junction gate structure (with low operating gate current) is believed inherently more robust against high temperature degradation than other (insulated gate, Schottky gate, bipolar, heterojunction and/or III-N) transistor technology approaches that would otherwise offer frequency, power dissipation, and/or circuit design and performance benefits. Though it is non-planar, the mesa-etched p$^+$ epi-gate structure avoids defects and extreme activation temperatures associated with high-dose p-type implants in SiC [9]. Despite inferior mobility compared to 4H-SiC, 6H-SiC was selected as having demonstrated better structural stability during some thermal processing steps [10,11]. Even though SiC is known to be chemically near-inert and diffusion resistant compared to silicon and III-V/III-N semiconductors, thermally activated degradation mechanisms at interfaces (such as metal-SiC, and/or SiC-insulator interfaces) or materials outside the semiconductor (such as metals, insulators, and/or packaging) have previously limited extreme temperature (i.e., $\geq$ 500 °C) stability/durability [3-5,12,13]. However, recently developed durable high temperature contacts to n-type SiC and high-temperature SiC packaging have demonstrated prolonged 500 °C operational capability in oxidizing air atmosphere [4,14]. The successful integration of these high temperature technologies into the epitaxial 6H-SiC JFET process is believed to be critical to the greatly prolonged 500 °C transistor and IC operation reported in this work.

A quarter wafer of small-signal 6H-SiC JFETs and simple ICs (configured using a single a metal interconnect layer) was fabricated starting from commercially purchased [15] epilayer substrate. Most fabrication process details are described elsewhere [3-8,12,14]. On-chip resistors were formed from the JFET n-channel layer and implants/contacts with the overlying p$^+$ gate layer removed. Interconnects and wire bond pads were simultaneously formed by patterning TaSi$_2$/Pt on top of reactive-sputtered Si$_3$N$_4$ dielectric.

A few SiC chips from the saw-diced quarter-wafer were custom-packaged without any lids (i.e., exposed to air) and mounted onto two custom high-temperature circuit boards using the high temperature packaging approach detailed in [4]. Two 6H-SiC JFETs (one with 200 µm wide / 10 µm long gate and another with 100µm/10µm gate dimensions) were packaged and tested as discrete devices. Several integrated circuits were packaged and tested, including an inverting amplifier stage, a differential amplifier stage, a digital inverter (i.e., NOT logic gate) and a two-input NOR digital logic gate. The circuit boards with test chips

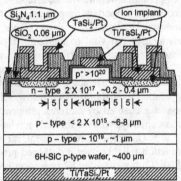

Figure 1. Simplified cross-sectional schematic of 6H-SiC JFET. Designed gate length (10 µm), gate edge to n$^+$ contact edge (5 µm) and n$^+$ contact edge to insulator via edge (5 µm) spacings are illustrated.

210

were placed in two laboratory ovens with ~ 30 cm long, 10 mil diameter unshielded Au wires running outside the ovens to nearby terminal strips connected to computer-controlled test instruments. The devices and circuits were operated continuously under electrical bias throughout the 500 °C test duration, with measurement data periodically (hourly at first, expanding to every 20 hours later) stored onto computer. The atmosphere inside the oven was ordinary room air (~ 21% O_2).

RESULTS

Discrete JFETs

All twenty one 100μm/10μm JFETs that were probe tested across the quarter-wafer prior to saw-apart demonstrated excellent room-temperature I-V behavior. However, threshold voltage (V_T) and saturated drain current (I_{DSS}) were respectively at maximum values around -14 V and 3.4 mA near the "top" of the wafer piece, and decreased across the wafer to minimum values around -7 V and 1.2 mA near the wafer piece "bottom". Until more-uniform commercial SiC JFET epilayers become available, SiC integrated circuit designs will need to account for such variability of these important transistor parameters.

Figure 2 compares drain current I_D versus drain voltage V_D characteristics of a packaged 200μm/10μm JFET measured by source-measure units (SMUs) during the 1st, 100th, and 4000th hours of 500 °C operation under $V_D = 50$ V and gate bias $V_G = -6$ V. Similar results were obtained for the 100μm/10μm JFET measured by a digitizing 60 Hz curve-tracer continuously operated with 50 V drain bias sweeps and -2V gate steps from $V_G = 0$ V to $V_G = -16$ V.

Figure 3 illustrates the measured variation of DC on-state I_{DSS}, transconductance g_m, drain-to-source resistance R_{DS}, and V_T for both packaged JFETs as a function of 500 °C operating time up to 4000 hours. The figure 3 plots are normalized to each transistor's measured value of I_{DSS0}, g_{m0}, R_{DS0}, and V_{T0} recorded at the 100 hour mark of 500 °C testing (i.e., after "burn-in"). Figure 3a shows I_{DSS} recorded at $V_D = 20$V, $V_G = 0$V. Figure 3b shows the g_m benchmarked at $V_D = 20$ V from $V_G = 0$V and -2V steps, and figure 3c shows the time evolution of R_{DS}. It is important to note that the y-axis scale limits for figure 3a-c plots are set to ±10% of each parameter's measured 100-hour value. With the exception of a few data points from the curve-tracer-measured 100μm/10μm JFET, the figure 3 data falls within this 10% parameter variation window. Figure 3d shows the precise time variation of V_T extracted from the computer-fit x-intercept of the SMU-measured $\sqrt{I_D}$ vs. V_G of the 200μm/10μm JFET. The measured V_T changes by less than 1%. This excellent stability reflects the fact that JFET V_T is determined by the as-grown 6H-SiC epilayer structure.

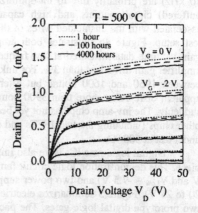

Figure 2. Drain I-V characteristics of packaged 200μm/10μm 6H-SiC JFET measured during the 1st, 100th, and 4000th hour of operation at 500 °C.

Integrated Circuits

The highly durable transistor properties summarized above enabled stable operation of prototype integrated circuits for thousands of hours at 500 °C. The inverting amplifier (inv-amp) and differential amplifier (diff-amp) are vital building blocks needed to realize more complex analog and digital circuits. Figure 4 summarizes the 500 °C durability data collected from packaged inv-amp and diff-amp ICs. The inv-amp consisted of two paralleled 40μm/10μm JFETs connected to a 20-square (516 kΩ at 500 °C) SiC load resistor. The diff-amp consisted of two source-coupled 20μm/10μm JFETs interconnected with three SiC epitaxial load resistors (see [6] for schematic). A 1 V peak-to-peak sine wave test input was applied (with -5 V DC bias for the inv-amp) and the V_{DD} supply voltage was 40 V with chip substrates grounded for both circuits. Figure 4a illustrates that there was less than 5% change in the 500 °C gain vs. frequency characteristics of these amplifiers measured at the beginning (black) and end (grey) of almost 4000 hours of 500 °C operation. The voltage gain drop-offs at higher frequency (above 10 kHz) are primarily due to un-optimized (un-buffered) circuit outputs and high capacitances associated with the unshielded wires of the oven-test setup. Figure 4b illustrates both amplifiers'

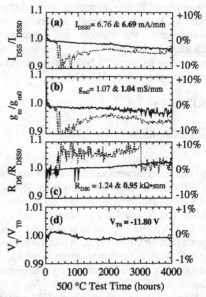

Figure 3. Normalized (see text) JFET parameters versus 500 °C test time for packaged devices with gate dimensions of 100μm/10μm (dashed, plain text, measured by digitizing curve-tracer) and 200μm/10μm (solid, **bold text**, measured by SMU).

gain vs. time data measured at low frequency (1 kHz or 2 kHz). Prior to failure, the inverting amplifier gain drifts less than 3%. For unknown reasons the diff-amp gain spiked some (~ 20%) between 1050 and 1600 hours, but afterwards, the amplifier gain stabilized with negligible apparent drift. After more than 3900 hours of 500 °C operation, the inverting amplifier failed suddenly. The inv-amp chip has not yet been removed from the oven for failure analysis because other chips (including the two JFETs and diff-amp IC) on the same printed circuit board remain under test at 500 °C.

Digital logic gate ICs were also implemented and packaged from the same 6H-SiC wafer piece. This demonstration logic circuit family features negative logic voltage levels (V_{High} ~ -2.5 V and V_{Low} ~ -7.5 V) and two power supplies (positive $+V_{DD}$ and negative $-V_{SS}$ in the range of 20 to 25 V). Figure 5 summarizes electrical results from prolonged 500 °C durability testing of two prototype digital logic gates. The packaged NOT gate successfully operated for over 3600 hours at 500 °C. Comparison of the figure 5a 1st hour and 3600th hour 500 °C output waveforms reveal that negligible change in NOT gate output signal occurred throughout the test duration. Figure 5b shows testing waveforms from the packaged two-input NOR gate collected during the 1st and 2405th hour of 500 °C operation. For a few days of this test, the setup was re-wired to

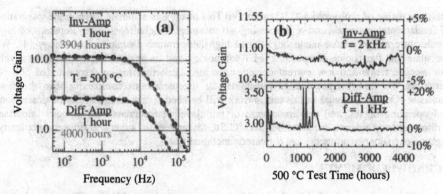

Figure 4. Analog amplifier circuit results recorded during prolonged 500 °C IC operation. (a) Gain vs. Frequency characteristics at beginning (black) and end (grey) of test. (b) Low-frequency voltage gain vs. 500 °C test time.

demonstrate the NOT gate output successfully driving the NOR gate inputs. Both logic gates failed suddenly soon after the maximum times shown in figure 5. The electrical failure of the NOR gate occurred in the form of an electrical near short-circuit, so failure of the insulating layer beneath a biased metal interconnect trace is suspected. However, detailed failure analysis is awaiting completion of 500 °C oven testing of other components on this board.

SUMMARY DISCUSSION

The increased 500 °C IC durability and stability demonstrated in this work is now sufficient for sensor signal conditioning circuits in jet-engine test programs. Although only a small number of devices have been packaged and tested for thousands of hours at high temperature, this demonstration establishes the feasibility of producing SiC integrated circuits

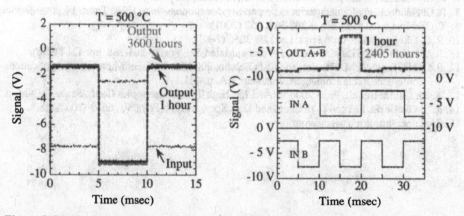

Figure 5. Digitized waveforms recorded near the beginning (black) and end (grey) of logic gate testing in 500 °C air-ambient oven. (a) Packaged NOT gate IC, and (b) NOR gate IC.

that are capable of prolonged 500 °C operation. This result was achieved through the integration of fundamental materials and/or processing advancements, including the development of high temperature n-type ohmic contacts [14] and high temperature packaging technology [4]. We speculate that the choice of epitaxial JFET technology and its designed operation at relatively low electric fields and low current densities are also important to the demonstrated 500 °C durability. For many envisioned applications, far greater circuit complexity than the few-transistor ICs demonstrated in this initial work will be needed. Shrinkage of device dimensions and operating biases, and implementation of multilayer interconnects are obvious important further steps towards realizing durable 500 °C SiC integrated circuitry with greater complexity, higher frequency performance, and increased functionality.

ACKNOWLEDGMENTS

This work was funded by the NASA Aeronautics Research Mission Directorate in both the Aviation Safety and Fundamental Aeronautics Programs under the Integrated Vehicle Health Management, Subsonic Fixed Wing, and Supersonics Projects. Work was carried out by NASA Glenn Research Center with the assistance of D. Lucko, K. Laster, J. Gonzalez, R. Lotenero, R. Buttler, M. Mrdenovich, B. Osborn, D. Androjna, J. Flatico, A. Trunek, G. Hunter, and L. Matus.

REFERENCES

1. P. G. Neudeck, R. S. Okojie, and L.-Y. Chen, Proc. IEEE **90**, 1065 (2002).
2. F. P. McCluskey, R. Grzybowski, T. Podlesak, *High Temperature Electronics*. (CRC Press, New York, 1997).
3. D. J. Spry, et al., Proc. 2004 IMAPS Int. High Temp. Electronics Conf., Santa Fe, NM, p. WA1.
4. L. Y. Chen, D. J. Spry, and P. G. Neudeck, Proc. 2006 IMAPS Int. High Temp. Electronics Conf., Santa Fe, NM, p. 240.
5. P. G. Neudeck, et al., Mater. Sci. Forum **556-557**, p. 831 (2007).
6. D. J. Spry, et al., Mater. Sci. Forum (2008) (in press).
7. P. G. Neudeck, et al., IEEE Electron Dev. Lett. (2008) (in press).
8. P. G. Neudeck, et al., manuscript in preparation for submission to IEEE Trans. Electron Devices.
9. T. Ohshima, et al., Physica B **308-310**, 652 (2001).
10. R. S. Okojie, et al., Appl. Phys. Lett. **79**, 3056 (2001).
11. J. A. Powell, D. J. Larkin, and A. J. Trunek, Mater. Sci. Forum **264-268**, pp. 421 (1998).
12. P. G. Neudeck, G. M. Beheim, and C. S. Salupo, 2000 Government Microcircuit Applications Conference Technical Digest, Anahiem, CA, p. 421.
13. A. M. Dabiran, et al., Proc. 2006 IMAPS Int. High Temp. Electronics Conf., Santa Fe, NM, p. 32?
14. R. S. Okojie, D. Lukco, Y. L. Chen, and D. J. Spry, J. Appl. Phys. **91**, 6553 (2002).
15. Cree, Inc., http://www.cree.com

Mater. Res. Soc. Symp. Proc. Vol. 1069 © 2008 Materials Research Society

Effect of SiC Power DMOSFET Threshold-Voltage Instability

A.J. Lelis[1], D. Habersat[1], R. Green[1], A. Ogunniyi[1], M. Gurfinkel[2], J. Suehle[2], and N. Goldsman[3]

[1]U.S. Army Research Laboratory, 2800 Powder Mill Rd, Adelphi, MD, 20783
[2]NIST, Gaithersburg, MD, 20899
[3]University of Maryland, College Park, MD, 20742

ABSTRACT

We have performed bias-stress induced threshold-voltage instability measurements on fully processed 4-H SiC power DMOSFETs as a function of bias-stress time, field, and temperature and have observed similar instabilities to those previously reported for lateral SiC MOSFET test structures. This effect is likely due to electrons tunneling into and out of near-interfacial oxide traps that extend spatially into the gate oxide. As long as the threshold voltage is set high enough to preclude the onset of subthreshold drain leakage current in the blocking state, then the primary effect of this instability is to increase the on-state resistance. For well-behaved power DMOSFETs, this would increase the power loss by no more than a few percent.

INTRODUCTION

It has been shown in the past few years that applying a gate-bias stress will shift the threshold-voltage of a SiC MOSFET, and that a subsequent bias of the opposite polarity will cause a shift back in the other direction [1-6]. This V_T instability is generally repeatable [2]. Relatively slow I-V measurements, which can take approximately 1 s or longer to complete, reveal threshold-voltage instabilities of around 0.25 to 0.33 V. Much faster I-V measurements reveal a much more significant amount of instability than previously realized, even for short stress times [3, 6]. This result is consistent with charge tunneling into oxide traps distributed spatially in the near-interfacial region, a region that has been shown to include excess carbon and other types of defects [7-9]. A tunneling mechanism would lead to a linear-with-log-time bias-stress response such that approximately half of the threshold-voltage instability would occur in the first microsecond for a 1-s bias stress. Likewise, a 1-s ramp-speed measurement will be highly influenced by the bias during the sweep. The faster the ramp speed, the more of the effect of the original bias stress can be observed. Of course, if the measurements are made fast enough the response of interface traps, which vary with energy, will also affect the response. Charge separation analysis suggests that there may be as many as 1×10^{12} cm^{-2} oxide traps present [10]. Fast I-V instability measurements suggest that a significant fraction of these traps can act as switching oxide traps [6]. It is important therefore to identify, understand, and reduce these oxide traps, while simultaneously keeping interface traps relatively low.

This work shows a direct correlation of the threshold-voltage instability in fully processed 4H-SiC vertical power DMOSFETs with results on lateral SiC MOSFET test structures, in which the instability has been shown to increase as a function of increasing gate-bias stress magnitude, gate-bias stress time, and temperature. Recent fast I-V measurements have also shown that the gate-voltage ramp speed of the measurement is directly related to how much instability is recorded. This greater-than-realized instability can potentially affect the performance and reliability of a power SiC DMOSFET in at least two ways. One, by increasing the drain leakage in the off-state due to a negative shift of the subthreshold I-V characteristic much further than slow DC measurements would indicate, if the threshold voltage margin is not set high enough. Two, by increasing the resistance in the on-state by increasing the threshold voltage and thus reducing the value of V_{GS}-V_T. While it is critical not to increase the drain leakage in the blocking state, it is also important not to set the threshold too high, especially since increased threshold voltage may be coupled with lower effective mobility. Thus, it is important to accurately estimate the magnitude of the actual threshold-voltage instability so as to find the optimum threshold voltage to achieve the best combination of low off-state leakage with low on-state resistance. Of course, the effects of temperature, which can also cause a change in the threshold voltage and on-state resistance, need to be taken into account as well.

EXPERIMENT

The basic experimental procedure has been described previously [2, 6]. In essence, we are applying a given gate-bias stress for a certain bias-stress time, measuring the effect by ramping the gate bias, and then applying a gate-bias stress of opposite polarity for the same bias-stress time and then re-measuring. This cycle is repeated several times, and the average shift in the threshold voltage, the bias-stress induced instability, is reported. Our measurements were made using an Agilent 4155 parameter analyzer, with a gate-ramp speed of about 1 s and with 50 mV applied to the drain. It has been previously reported that the V_T instability observed with a 20-μs gate-ramp speed can be more than three times as large [6]. This study was made with recent state-of-the-art 4-H SiC power DMOSFETs, which are three terminal devices. The source and drain were both grounded during the gate-bias stress. Under normal operating conditions, when the device is in the on state, a positive gate bias of 15 or 20 V is applied. In the blocking state, although the gate bias is zero, the large voltage on the drain leads to a negative field across the gate oxide. We have simulated this effect by applying a negative bias on the gate and grounding the drain.

Fig. 1 gives an example of the typical instability in the I_D-V_{GS} characteristics, plotted on both a logarithmic and a linear current scale, on the left and right, respectively. The gate is ramped down in voltage following a positive bias stress and ramped up following a negative bias stress. This reduces the influence of the gate bias during the measurement to better reveal the effect of the gate-bias stress itself. Fig. 1 shows the results of applying +20 V to the gate for 100s and then applying –20 V for the same amount of time to the gate of a 20-A DMOSFET for three full stress-and-measure cycles ($\pm$20 V), with an individual bias-stress time of 100 s for each half cycle. The effect is very repeatable. Fig. 2 plots the average V_T instability versus stress time calculated using two different methods: linear extrapolation to zero current and the voltage corresponding to a set current, in this case 90 μA ~ 0.1 mA. The two methods generally give similar results. Fig. 2 shows the results for the same 20-A DMOSFET whose I-V characteristics were shown in Fig. 1.

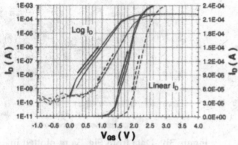

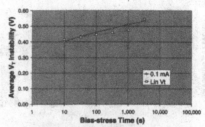

Figure 1. Instability of the I_D-V_{GS} characteristics of a 20-A SiC DMOSFET plotted on both a log (left) and linear (right) current scale, for three full cycles of bias stress (100s at +20 V and 100s at –20 V at room temperature) and measurement.

Figure 2. Average threshold-voltage instability for the 20-A SiC DMOSFET from Fig. 1 as a function of stress time, calculated by finding the voltage shift for a constant current of 90 μA and the shift in V_T using the linear extrapolation method.

RESULTS AND DISCUSSION

Fig. 3a shows the average gate-bias stress induced threshold-voltage instabilities, calculated by finding the voltage shift corresponding to a drain current of about 0.1 mA, for DMOSFETs of various sizes and vintages, ranging in size from 5-A to 20-A devices (rated for a current density of about 100 A/cm^2). In each case, the gate bias led to an oxide electric field of about 3 MV/cm by applying 15 V to the gate for devices with a gate-oxide thickness of about 500 Å, and 23 V for devices with a 750-Å thick gate oxide. As can be seen, all the devices tend to experience a linear-with-log-time increase in the V_T-instability versus individual bias-stress time. This is the same effect that was observed previously in as-processed SiC lateral MOSFETs and that was attributed to electrons tunneling to and from near-interfacial oxide traps. Figure 3b shows that when correcting for the thicker gate oxide of the more recent vintage 20-A power devices, the average number of oxide traps switching charge state is the lowest yet observed for the DMOSFETs—about 1.7×10^{11} cm^{-2} charge traps, based on these slow I-V measurements.

Fig. 4 shows the effect of the magnitude of the gate-bias stress. The bottom two curves were of two devices stressed with ±15 V on the gate (±2 MV/cm). Stressing with ±23 V (top diamond curve, ±3 MV/cm) applied to the gate leads to an increase in the instability of between 0.1 and 0.2 V. Following this stress, gate-bias stresses of ±20 V and ±15 V were applied, but showed similar results to the ±23 V case. (We note that lateral MOSFETs with similar gate oxides were stressed in a similar sequence and did not show lasting effects from the higher gate-bias stress.) Therefore, when the 20-A DMOSFET is stressed with ±15V on the gate, an even smaller value of only 1.2×10^{11} cm^{-2} oxide traps are switching charge state, based on these slow I-V measurements. Of course, the actual number switching charge state during the bias stress may be three times higher, which likely would be revealed with a faster I-V measurement.

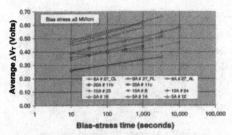

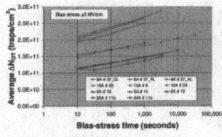

Figure 3a. Average threshold voltage instability for 5, 8, 10, and 20-A SiC DMOSFETs at room temperature as a function of stress time, calculated by finding the voltage shift for a constant current of 90 µA.

Figure 3b. Data from Fig. 3a re-plotted in terms of the number of switching oxide traps calculated to cause the V_T shifts.

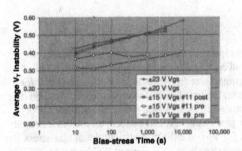

Figure 4. Average threshold voltage instability for the 20-A SiC DMOSFET from Fig. 1 vs stress time as a function of gate-bias stress, calculated by finding the voltage shift for a constant current of 90 µA. The instability with ±15 V increases to about the same level as that with ±23 V following the latter stress.

Figs. 5a and 5b show the effects of temperature on two different 5-A DMOSFETs. The room-temperature instability is the largest in each case. Fig. 5b clearly shows the instability increasing with temperature from its minimum value at $T = 100°C$. This effect is similar to what was observed in lateral test devices of the same vintage [4]. It should be noted that the threshold voltage did decrease with increasing temperature as well, with the onset of subthreshold leakage in these samples decreasing from about 3.5 V to less than 3.0 V.

Fig. 6 shows a comparison in the gate-bias stress-induced V_T instability between 5-A DMOSFETs and lateral MOSFETs with similar gate oxides. The as-processed epi lateral MOSFET has a slightly lower average instability, but the lateral MOSFET with an implanted epi region that simulates the implanted channel of the DMOSFET shows an average instability versus bias-stress time that is right in line with the response of the DMOSFETs. This result gives added confidence that we are looking at the same effect in the fully processed high-power DMOSFETs as in the low-power lateral test structures whose results have been previously reported [1-4, 6].

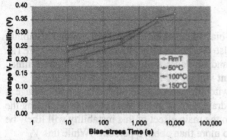
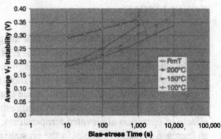

Figure 5a. Average threshold voltage instability for a 5-A SiC DMOSFET vs stress time as a function of temperature, calculated by finding the shift in the linearly extrapolated V_T.

Figure 5b. Average threshold voltage instability for another 5-A SiC DMOSFET vs stress time as a function of temperature, calculated by finding the shift in the linearly extrapolated V_T.

Fig. 7 shows the on-state characteristics for a 20-A DMOSFET. The arrow points to the gap between the $V_{GS} = 14$ V and $V_{GS} = 16$ V curves. The results reported here show DMOSFET V_T instabilities generally between 0.25 and 0.5 V. This correlates very well with the instabilities observed in similarly processed lateral MOSFETs. We know that the actual instability (difficult to measure because of the counter-acting effect of the gate bias during the measurement) can be three or more times larger. But suppose that the actual shift in V_T is about 1 V. If V_T was originally set to preclude the chance of turning on the subthreshold current in the blocking state, then the practical effect of a threshold-voltage shift would be to reduce the effective gate voltage applied in the on state. But Fig. 7 shows that even a shift of 1 V would only lead to an increase in V_{DS} of about 0.1 V, from about 1.9 to 2.0 V. This would increase the on-state losses about 5 percent, which should be manageable.

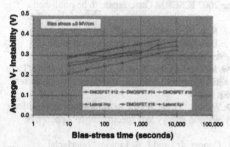
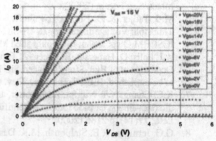

Figure 6. Comparison of the average threshold voltage instability of several 5-A SiC DMOSFETs with that of lateral MOSFETs with and without an implanted channel vs stress time, calculated by finding the shift in the linearly extrapolated V_T.

Figure 7. On-state I_D-V_{DS} as a function of V_{GS}. The effect of the V_T instability would be to increase the on-state resistance only by a few percent

CONCLUSIONS

In this work we have shown that the gate-bias stress-induced threshold-voltage instability previously reported in lateral SiC MOSFETs is also present in fully processed SiC DMOSFETs, and that the shifts are comparable in magnitude, and similar in their response to bias-stress time, gate oxide field, and temperature. The most recent vintage DMOSFETs tested showed the smallest number of near-interfacial oxide traps switching charge state. If the threshold voltage is set properly to prevent increases in subthreshold drain current in the blocking state, including the effects of operation at elevated temperatures, then the main effect of this instability will likely be an increase in resistive losses in the on state of no more than about 5 percent. While this V_T instability measurement remains a good diagnostic for finding a lower limit to the number of oxide traps present, and should help in the analysis of separating out the different types of interfacial charge present: interface traps, fixed charge, and oxide traps; and their effects on V_T, channel mobility, and overall device reliability; the effect of this instability on power devices for power converter applications should be manageable.

REFERENCES

1. A.J. Lelis, D. Habersat, G. Lopez, J.M. McGarrity, F.B. McLean, and N. Goldsman, "Bias Stress-Induced Threshold-Voltage Instability of SiC MOSFETs," Mater. Sci. Forum vols. 527-529, p 1317 (2006).
2. A.J. Lelis, D. Habersat, F. Olaniran, B. Simons, J.M. McGarrity, F.B. McLean, and N. Goldsman, "Time-Dependent Bias Stress-Induced Instability of SiC MOS Devices" Mater. Res. Soc. Symp. Proc., vol. 911, article 0911-B13-05 (2006).
3. M. Gurfinkel, J. Suehle, J.B. Bernstein, Y. Shapira, A.J. Lelis, D. Habersat, and N. Goldsman, "Ultra-Fast Characterization of Transient Gate Oxide Trapping in SiC MOSFETs," Intl. Rel. Physics Symposium Proceedings, vol. 45, p 462 (2007).
4. A.J. Lelis, D. Habersat, R. Green, and N. Goldsman, "Temperature-Dependence of SiC MOSFET Threshold-Voltage Instability" presented at the 2007 ICSCRM Otsu, Japan; to be published Mater. Sci. Forum (2008).
5. M. J. Tadjer, K. D. Hobart, E. Imhoff, and F. J. Kub, "Temperature and Time Dependent Threshold Voltage Instability in 4H-SiC Power DMOSFET Devices" presented at the 2007 ICSCRM Otsu, Japan; to be published Mater. Sci. Forum (2008).
6. A.J. Lelis, D. Habersat, R. Green, A. Ogunniyi, M. Gurfinkel, J. Suehle, and N. Goldsman, "Time Dependence of Bias-Stress Induced SiC MOSFET Threshold-Voltage Instability Measurements" to be published, IEEE Trans. Elec. Dev. (2008).
7. K.C. Chang, N.T. Nuhfer, L.M. Porter, and Q. Wahab, "High-carbon concentrations at the silicon dioxide–silicon carbide interface identified by electron energy loss spectroscopy,"Appl. Phys. Lett., 77(14), p 2186 (2000).
8. G.G. Jernigan, R.E. Stahlbush, M.K. Das, J.A. Cooper, Jr., and L.A. Lipkin, "Interfacial differences between SiO₂ grown on 6H-SiC and on Si(100)," Appl. Phys. Lett., 74(10), p 1448 (1999).
9. T. Zheleva, A. Lelis, U. Lee, G. Duscher, F. Liu, M. Das, and J. Scofield, "Transition Layers at the SiO2/SiC Interface," presented at the 2007 ICSCRM in Otsu, Japan and submitted to Appl. Phys. Lett. (2008).
10. D. Habersat, A.J. Lelis, G. Lopez, J. McGarrity, and F.B. McLean, "On Separating Oxide Charges and Interface Charges in 4H-SiC Metal-Oxide-Semiconductor Devices" Mater. Sci. Forum vols. 527-529, p 1007 (2006).

Mater. Res. Soc. Symp. Proc. Vol. 1069 © 2008 Materials Research Society   1069-D12-01

SiC-Based Power Converters

Leon Tolbert[1,2], Hui Zhang[1], Burak Ozpineci[2], and Madhu S. Chinthavali[2]

[1]Electrical Engineering and Computer Science, The University of Tennessee, 414 Ferris Hall, Knoxville, TN, 37996-2100
[2]Power Electronics and Electric Machinery Research Center, Oak Ridge National Laboratory, 2360 Cherahala Blvd., Knoxville, TN, 37932

ABSTRACT

The advantages that silicon carbide (SiC) based power electronic devices offer are being realized by using prototype or experimental devices in many different power applications ranging from medium voltage to high voltage or for high temperature or high switching frequency applications. The main advantages of using SiC-based devices are reduced thermal management requirements and smaller passive components which result in higher power density. An overview of the SiC research effort at Oak Ridge National Laboratory (ORNL) and The University of Tennessee (UT) is presented in this paper.

INTRODUCTION

Silicon carbide power electronic devices are expected to have better characteristics than their silicon counterparts once processing and packaging issues have been solved. SiC devices have higher blocking voltages, lower on-state resistance and switching losses, and higher thermal conductivity and operating temperatures. Use of these devices will impact several application areas including hybrid electric vehicles, electric grid interface with distributed energy sources, and high temperature environments.

In pursuit of mass production of hybrid electric vehicles, the automotive research industry has set goals such as reducing the size and weight of the power electronics and cooling systems and increasing their efficiency. The U.S. Department of Energy's research goals for the year 2020 include hybrid electric vehicle inverters that have power densities of more than 14.1 kW/kg and 13.4 kW/L, and efficiencies greater than 98 % at a cost less than $3.3/kW.

SiC devices are capable of operating at higher voltages, higher frequencies, and higher junction temperatures than comparable Si devices, which result in significant reduction in weight and size of the power converter and an increase in system efficiency. The objectives of research efforts on SiC based device applications at ORNL and UT are

- Assessing the impact of replacing silicon (Si) power devices in transportation applications such as hybrid electric vehicles (HEVs) or plug-in hybrid electric vehicles (PHEVs) with wide-bandgap (WBG) semiconductors, especially silicon carbide (SiC). Researchers have also examined what impact high voltage SiC devices would have in power electronics for electric utility applications.
- Developing temperature-dependent device and power converter models for various system simulation studies and analyzing the impact of SiC devices on the system performance.
- Building SiC-based prototype converters to validate the performance of SiC devices.
- Building high temperature packages and gate drives for SiC power devices to operate at 200°C ambient.

DEVICE CHARACTERIZATION AND MODELING

The static and dynamic characteristics of SiC Schottky diodes and JFETs are obtained experimentally. The experimental data are analyzed to extract device parameters to develop behavioral models. The behavioral models are integrated into system level models to study their impact on the overall system performance. The models of SiC Schottky diodes are briefly summarized below. For more information, see [1-4].

SiC Schottky Diode Models

Static characteristics: The forward characteristics of a 600 V, 4 A SiC Schottky diode obtained at different temperatures from 25 °C to 150 °C temperature are shown in Fig. 1(a). It can be approximated with the following equation [4]:

$$V_d = V_D + R_D \cdot I_d \tag{1}$$

where V_d and I_d are the diode forward voltage and current, V_D is the voltage drop at zero current, and R_D is the on-state resistance, which is modeled using equations (2)-(7).

$$R_D = \frac{4V_B^2}{\varepsilon E_c^3 \mu_n} \tag{2}$$

$$\mu(E) = \frac{\mu_0}{\left\{1 + \left|\frac{\mu_0 E}{v_s}\right|^\beta\right\}^{1/\beta}} \tag{3}$$

$$\mu_0 = \mu_{min} + \frac{\mu_{max} - \mu_{min}}{1 + \left(N_{tot}/N_{ref}\right)^\alpha} \tag{4}$$

$$\mu_{max} = A_{\mu_{max}} \times \left(\frac{T}{300}\right)^{-B_{\mu_{max}}} \tag{5}$$

$$\mu_{min} = A_{\mu_{min}} \times \left(\frac{T}{300}\right)^{-B_{\mu_{min}}} \tag{6}$$

$$N_{ref} = A_{N_{ref}} \times \left(\frac{T}{300}\right)^{-B_{N_{ref}}} \tag{7}$$

Table 1. Parameters used in the device models [5-6]

Property	4H-SiC
Breakdown electric field, E_c (kV/cm)	2200
Relative dielectric constant, ε_r	10.1
Doping coefficient of μ, α	0.76
Electric field exponent of μ, β	1
Coefficient of μ_{max}, $A_{\mu max}$	950
Exponent of μ_{max}, $B_{\mu max}$	2.4
Coefficient of μ_{min}, $A_{\mu min}$	40
Exponent of μ_{min}, $B_{\mu min}$	0.5
Coefficient of N_{ref}, A_{Nref}	2×10^{17}
Exponent of N_{ref}, B_{Nref}	1
Maximum saturated velocity, v_{smax} (cm/s)	4.77×10^{17}

Temperature impact is also considered in the model. For the diode shown in Figure 1, V_D is modeled as a linear function of temperature, T:

$$V_D = -0.0013T + 0.9573. \tag{8}$$

The characteristics of the SiC Schottky diode obtained from simulations are compared to those from tests in Figure 1(b) and (c).

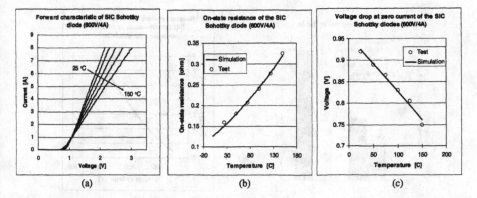

Figure 1. (a) Forward characteristics of a 600 V, 4 A SiC Schottky diode at different temperatures, (b) comparison of on-state resistance of the diode from simulations and tests, (c) comparison of voltage drop at zero current of the diode from simulations and tests.

Dynamic characteristics: The same SiC Schottky diode was also tested in a chopper circuit to observe its dynamic characteristics. The reverse recovery current waveforms are obtained at different temperatures (same reverse voltage and forward current) and at different forward currents (same temperature and forward current), which are shown in Figure 2. As seen in the figure, the turn-off behavior of the diode does not change with temperature and forward current. However, it changes with the different forward voltages. Since the depletion layer capacitance dominates the reverse recovery behavior and considering equations (9) and (10), the relationship between the energy loss during the diode reverse recovery and voltage can be estimated by equation (11).

$$C_d = \frac{\varepsilon}{w_d} = \sqrt{\frac{qN_d\varepsilon}{2(V_R + \phi_B)}} \tag{9}$$

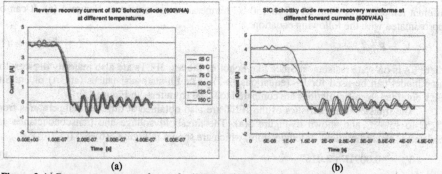

Figure 2. (a) Reverse recovery waveforms of a 600V/4A SiC Schottky diode at different temperatures, (b) reverse recovery waveforms of the SiC Schottky diode at different forward voltages.

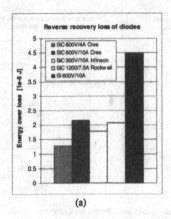

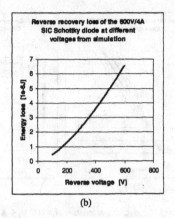

(a)   (b)

Figure 3. (a) Reverse recovery energy loss of diodes at reverse voltage of 200V and room temperature, (b) reverse recovery energy loss of the 600V, 4A SiC Schottky diode at different reverse voltages obtained from simulation.

$$E_{D,sw} = \frac{1}{2} C_d V_R^2 \tag{10}$$

$$E_{D,sw} = \left(\frac{V_R}{V_{RO}}\right)^{1.5} E_{D,sw0} \tag{11}$$

where C_d is depletion layer capacitance, V_R is reverse voltage, $Ed_{D,sw0}$ is the energy loss at voltage V_{RO}. The reverse recovery energy loss of the 600 V, 4 A diode is shown in Fig. 3(a) at a reverse voltage of 200 V, and it is expanded for other voltages using equation (11) in Fig. 3(b).

SiC JFET Models

Static characteristics: The forward characteristics of a 600 V, 5 A SiC JFET obtained at different temperatures from 25 °C to 150 °C temperature are shown in Figure 4 (a). It can be approximated with the following equation:

$$V_d = R_D \cdot I_d \tag{12}$$

where R_D is on-state resistance. Since like Schottky diodes, JFETs are also majority devices, the same model can be used for R_D (equations (2-7)). The on-state resistances (at 3A) of the SiC JFET obtained from simulations are compared to those from tests in Figure 4 (b).

The transfer characteristics of the JFET are also obtained at different temperatures from 25 °C to 150 °C and shown in Figure 5(a). The threshold voltage and transconductance are modeled as linear functions of temperature, which are shown in Figure 5(b) and (c).

$$V_{th} = -0.0091T + 3.0167 \tag{13}$$

$$g_m = -0.0072T + 1.9308 \tag{14}$$

224

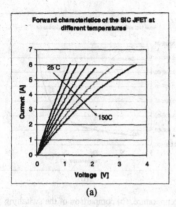
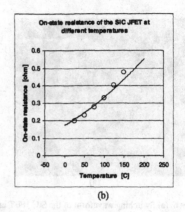

(a)   (b)

Figure 4. (a) Forward characteristics of a 600V, 5A SiC JFET at different temperatures, (b) comparison of the on-state resistances of the SiC JFET from simulations and tests.

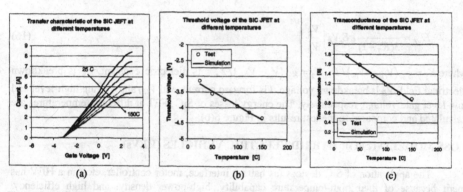

(a)   (b)   (c)

Figure 5. (a) Transfer characteristics of a 600 V, 5 A SiC JFET at different temperatures, (b) comparison of the threshold voltages of the JFET from simulations and tests, (c) comparison of transconductances of the JFET from simulations and tests.

Dynamic characteristics: The same SiC JFET was also tested in a chopper circuit to observe its dynamic characteristics. The switching waveform obtained at room temperature is shown in Figure 6(a). Assuming the gate drive resistance, R_G, is 0, and the applied gate voltage, V_G, is a step signal changing from V_{GL} to V_{GH} and vice versa. Then the Miller effect is minimized in the JFET, and the current rises and falls very quickly. The switching loss of the JFET can be modeled as the charging and discharging of the drain-to-source and drain-to-gate capacitance under the control of the gate drive current [5]. The equations of turn-on energy loss and turn-off energy loss are shown as follows:

$$E_{on} = \frac{1}{3(K_1 - 1)} \varepsilon E_c V_R \left(\frac{V_R}{V_B} \right)^{1/2} \tag{15}$$

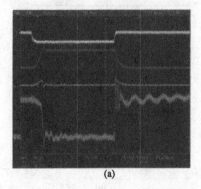
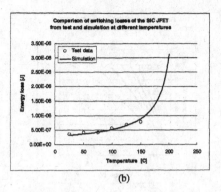

(a)   (b)

Figure 6. (a) Switching waveform of the SiC JFET at room temperature, (b) comparison of the switching energy losses of the SiC JFET at different temperatures obtained from simulations and tests.

$$E_{off} = \frac{1}{3(K_2 + 1)} \varepsilon E_c V_R \left(\frac{V_R}{V_B} \right)^{1/2} \tag{16}$$

where $K_1 = g_m (V_{GH} - V_{th})/I$, $K_2 = g_m (V_{th} - V_{GL})/I$, ε is dielectric constant, E_C is breakdown electrical field, V_B is breakdown voltage, I is forward current, and V_{GH} and V_{GL} are high-level and low-level gate voltage, respectively. The energy losses of SiC JFET at different temperatures are calculated and compared to the test results in Figure 6(b).

LOSS MODELING FOR HYBRID ELECTRIC VEHICLES (HEVs)

The application of SiC devices (as battery interface, motor controller, etc.) in a HEV has merit because of their high-temperature capability, high-power density, and high efficiency. Moreover, the light weight and small volume will affect the whole power train system in a HEV, and thus performance and cost.

An HEV system model based on 2004 Toyota Prius HEV was developed in PSAT (Powertrain System Analysis Tool - vehicle simulation software). An inverter model based on the previous discussion is inserted in the built-in motor drive model in order to calculate the power loss and efficiency of different inverters. A 3-phase dc/ac SiC inverter composed of SiC JFETs and SiC diodes is applied to the primary motor to take the place of the conventional Si inverter. Simulations are run for a UDDS cycle (US EPA-Urban Dynamometer Driving Schedule, which represents city driving conditions of light duty vehicles) for both a HEV with a SiC inverter and one with a Si inverter. The device performance and vehicle performance are observed and compared.

Assume the two inverters have the same size heatsink and cooling conditions, and the switching frequency is 20 kHz. The initial SOC is equal to final SOC. As for the inverter itself, the benefits of the SiC devices are demonstrated in Figure 7. Due to the lower power losses of the SiC devices, the junction temperatures of the SiC devices are much lower than those of Si ones (see Figure 7(a) and (b)). As a result, the power loss of the SiC inverter is reduced, and its efficiency is

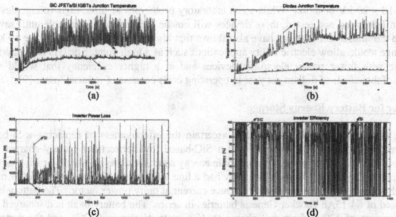

Figure 7. Comparison of the SiC and Si inverters in the 2004 Toyota Prius HEV: (a) SiC JFETs/Si IGBTs junction temperature, (b) diode junction temperature, (c) inverter power loss, (d) inverter efficiency.

much improved (see Figure 7(c) and (d)). Furthermore, the benefits of the SiC-based inverter are also seen at system level. For example, the system efficiency is improved from 31.3 % to 37.2 % (increased by 22.6 %, corresponding to 0.82 kWh) due to the energy saving in other powertrain components (such as engine, generator, mechanical accessories, and etc.) and the improvement in recuperating braking energy. As a result, the fuel economy is improved from 24.1 to 29.5 km/liter (increased by 18.8%).

A plug-in hybrid vehicle (PHEV) with the same powertrain architecture and components as the 2004 Toyota Prius HEV and 50 km all electric range (AER) is also studied. Compared to the conventional HEV, more benefits are observed at both inverter-level and system-level. The system efficiency is improved from 62.6 % to 79.6 % (increased by 27.2 %, corresponding to 1.9 kWh), and the electricity consumption is reduced from 124.4 to 83.8 Wh/km (decreased by 32.6%). In addition, the studies indicate that the optimized size of the battery bank for the plug-in vehicle with the SiC inverter and that with the Si inverter are 5.1 kAh and 7.8 kAh, respectively, compared to 1.1 kAh battery for 2004 Toyota Prius HEVs (assume initial SOC 90% and final SOC 30%). Thus, for this design, using a SiC-based inverter can reduce the size of the battery bank for a PHEV by 34.6%.

Since the junction temperatures of the SiC devices in the PHEV remained low (around 40 °C) when using the same heatsink as for a Si-based inverter, an additional study was done by reducing the size of the heatsink of the SiC inverter. The junction temperature is allowed to rise to the same level as the Si inverter (around 60 °C). It is found that the efficiency of the SiC inverter is lowered by only 0.5 % and at system level, the system efficiency is lowered by 0.3 % with the smaller heatsink. Thus, using SiC devices in such a system can simplify the thermal management system and improve system compactness. For more information, see [7].

MODELING FOR UTILITY APPLICATIONS

Because SiC is a wide-bandgap semiconductor, it inherently has higher blocking voltage. Companies are working on SiC switching devices (MOSFETs, IGBTs, thyristors) capable of blocking 10 – 20 kV or more. While some laboratory prototypes of these high-voltage devices have been fabricated and tested, these devices will not be commercially available until several years into the future. Simulations have also shown that high voltage SiC devices in the 10 kV – 20 kV range would allow electric utility applications such as a high voltage dc (HVDC) interface to operate with fewer power electronic devices and at a higher efficiency that would save hundreds of thousands of dollars per year in operating costs because of reduced losses [9].

Converter for Battery Energy Storage

A system model was developed to ascertain the effectiveness of replacing a Si-based inverter/rectifier for a battery charger with a SiC-based inverter/rectifier battery charger that would be used in combination with a renewable energy source such as solar cells or wind power. For the simulation, it was assumed the utility had a line voltage of 480 Vrms, a frequency of 60 Hz, and the converter was controlled to produce current at unity power factor. The battery bank is composed of 84 13Ah Hawker Genesis batteries in series. The battery bank is discharged at a constant current of 120 A from full charge to 40% state of charge (SOC), and the power is delivered to the utility through the three-phase, full-bridge converter. For its charging mode, it is charged at a constant voltage of 1109 V from 40% SOC to full charge, and the power is provided by the utility through the converter. In both discharge and charge processes, the converter works as an interface. They are expected to consume as little power as possible. The power losses and efficiencies of the SiC and Si converters were computed.

Based on the specific system information discussed previously, the battery system model shown in Figure 8(a) worked out the unknown system parameters required for power losses calculation. The lower on-state resistance of the SiC JFETs and lower switching losses of the diodes enabled the converter to operate at a lower temperature and have a higher efficiency than a comparable converter with Si devices as shown in Figure 8(b) for charging mode and Figure 8(c) for discharging mode.

Using MATLAB Simulink, two sets of simulations have been done for both the SiC-based and Si-based systems for a charge cycle and a discharge cycle, respectively. In the first set of simulations, the ambient temperatures and the heatsink sizes were the same for the Si and SiC systems. The junction temperatures are reduced by about 5-8 °C, and the average efficiency improvement is about 3.3-5.5% for the SiC-based converter. In the second set of simulations, the heatsinks were selected to limit the maximum junction temperature to 125°C for both the SiC and Si devices. The simulations indicate that the size of the heatsink required by the SiC converter is

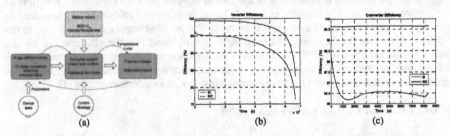

Figure 8. (a) Modeling methodology for battery-converter system, (b) converter efficiency during battery charging, (c) converter efficiency during battery discharging.

reduced to about 1/25 of that of the Si converter under the same cooling method (forced cooling with fan) or 4/5 when the SiC inverter is naturally cooled, and at the same time the efficiency is improved by 3.1% for discharge, 5.4% for charge. For more information, see [8].

55 kW HYBRID INVERTER TESTING

SiC Schottky diodes replaced Si diodes in a 55kW inverter. The superior switching performance of the diodes improves the efficiency of the inverter and also impacts the main power switches by reducing the stress on them and thus improving system performance.

SiC Schottky diodes are already commercially available at current ratings up to 50 A. These diodes are being used in niche applications such as power factor correction circuits or as antiparallel (free-wheeling) diodes in inverters. It is expected that the first impact of SiC devices on inverters will be as a result of SiC Schottky diodes replacing the conventional Si pn diodes.

ORNL collaborated with Cree and Semikron to build a hybrid 55 kW (Si IGBT–SiC Schottky diode) inverter by replacing the Si pn diodes in Semikron's automotive inverter with Cree's SiC Schottky diodes [10]. A performance comparison of the all-Si and the hybrid inverter is shown in Figure 9. The hybrid inverter efficiencies are higher than the all-Si inverter for all operating conditions. The results show up to 33.6% reduction in the losses.

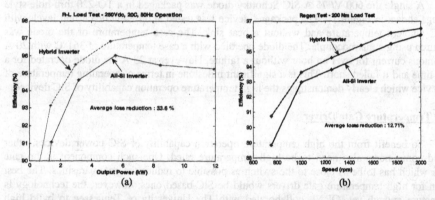

Figure 9. (a) R-L load test efficiency curves for one of the load conditions, (b) Dynamometer test – regeneration mode efficiency plots at 70°C.

HIGH TEMPERATURE APPLICATIONS

SiC power electronics are able to operate at ambient temperatures well over 200°C. However, for SiC devices to operate at these elevated temperatures, packaging that can survive these temperatures and the stresses from thermal cycling are needed.

High Temperature Packaging

One of the most important characteristics of SiC power devices is that they can operate at much higher junction temperatures (>300°C) than Si power devices. Presently, SiC devices use

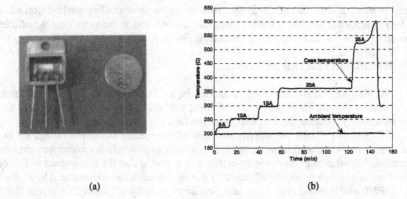

(a)   (b)

Fig. 10. (a) TO-220 through-hole high temperature package of 75 A SiC Schottky diode, (b) test results.

Si device packages that limit their operation temperature (~150°C). ORNL collaborated with the University of Arkansas and University of Idaho to build several high temperature packages that can operate at 200°C ambient [11].

A single die 600 V/ 75 A SiC Schottky diode was packaged in a TO-220 thru-hole style package shown in Figure 10. The packaged device was tested at different dc current levels with 200°C ambient temperature and without a heat sink. The case temperature of the diode was measured using a thermocouple. The diode operated with a case temperature of 361°C with 20 A continuous current for over an hour without a failure. However at 25 A the diode operated for a short time and it failed short. This is a significant milestone in terms of operating temperature of the device which clearly demonstrates the high temperature operation capability of SiC devices.

High Temperature Gate Driver

To benefit from the high temperature operation capability of SiC power devices, other related components also need to be high temperature rated. One such component is the gate driver which has to be as close to the switch as possible to reduce the gate parasitics. The best option for high temperature gate drivers would be SiC-based ones; however, the technology is not mature enough yet. ORNL collaborated with The University of Tennessee to build high temperature gate drivers using silicon-on-insulator (SOI) technology that can handle temperatures of up to 200°C.

A high temperature gate driver integrated circuit (IC) was designed and fabricated using 0.8 micron, 3-metal and 2-poly bipolar CMOS and DMOS (BCD) on silicon on insulator (SOI) process [12]. Figure 11(a) shows the fabricated chip's microphotograph. This circuit occupies an area of 3.6 mm^2 (2,240 μm × 1,600 μm) including pads. The two high-voltage NMOS devices of the half-bridge output stage occupy a major portion of the die area. They are sized (W/L = 24,000 μm/1.6 μm) to provide large peak current as needed to minimize the switching loss of the power FET. Each of these NMOS transistors is comprised of six hundred 45 V NMOS devices (W = 40 μm) connected in parallel. The high-voltage devices are well isolated from the low-voltage devices through a thick dielectric layer. Multiple pad connections are used for the power supply and output nodes to minimize the bond wire's parasitic inductances. All critical metal

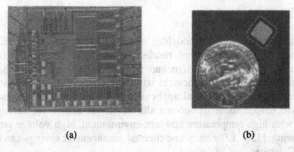

Figure 11. (a)Micrograph of the gate driver IC with bond wires. (b) Gate driver IC in CDIL44 package in comparison with a dime.

interconnects are made thick to avoid electromigration which is a potential failure mechanism at higher temperatures. Figure 11(b) shows a gate driver chip in QFN44 package placed next to a dime for visual comparison of the size of the chip.

A SiC power MOSFET developed by Cree was used to test the gate driver chip. A 10 ohm resistive load was connected between the 50 V power supply and the drain terminal of the MOSFET. Approximately 5 A switching current was flowing through the load and the MOSFET when it was turned ON by the gate driver chip. The test board was placed inside a temperature chamber and the SiC MOSFET was kept outside the chamber as it was not packaged for the high temperature operation. Starting from room temperature, the chip was tested up to 200°C.

One chip was successfully tested up to 175°C and at 40 kHz switching frequency (Figure 12(a)). Another IC successfully switched ON/OFF the MOSFET up to 200°C (Figure 12(b)). The gate driver was placed inside the temperature chamber without any cooling facility and the temperature was raised from room temperature to 200°C in five steps (85°C, 125°C, 150°C, 175°C and 200°C) and at each temperature step, it was kept for 15 minutes before taking any readings. At 200°C it was tested for more than 30 minutes. This way, the chip successfully operated over 150°C for more than 90 minutes without any cooling mechanism.

After the successful testing of the first version of the chip, a second version has been designed and fabricated to improve the gate drive reducing its temperature sensitivity and adding an on-chip voltage regulator.

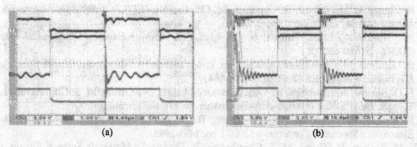

Figure 12. (a) Gate driver test results for 175°C and 40 kHz switching frequency, (b) Gate driver test results for 200°C and 20 kHz switching frequency.

231

CONCLUSIONS

The device characterization, system modeling, and application tests using SiC-based power electronics have revealed some important results, which were discussed with the device manufacturers to further improve the design and fabrication. The impact of SiC devices on performance parameters compared to Si devices will have a large impact on system efficiency and thermal management issues for several applications. The inherent properties of SiC-based power electronics will enable these to become the device of choice for certain high power applications - those with high temperature ambient environment, high voltage requirements, or high switching frequency [13]. Efficiency and thermal management savings can help to justify the higher cost of SiC devices.

ACKNOWLEDGMENTS

The authors wish to thank Dr. Mike Mazzola of SemiSouth and Dr. Anant Agarwal and Jim Richmond of Cree for providing SiC devices samples for testing and characterization. The authors wish to thank Dr. Fred Barlow of the University of Idaho for his high temperature packaging of SiC diodes. We also thank M. Aminul Huque, Dr. Ben Blalock, and Dr. S. Kamrul Islam at The University of Tennessee for their development of a high temperature gate driver.

REFERENCES

1. H. Zhang, L. M. Tolbert, B. Ozpineci, *IEEE Workshop on Computers in Power Electronics* (2006) pp. 199-204.
2. H. Zhang, L. M. Tolbert, B. Ozpineci, M. Chinthavali, *IEEE Industry Applications Society Annual Meeting* (2005), pp. 2630-2634.
3. B. Ozpineci, L. M. Tolbert, *IEEE Power Electronics Letters*, vol. 1 (2003), pp. 54-57.
4. B. Ozpineci, M. Chinthavali, L. M. Tolbert, *International Journal of High Speed Electronics and Systems*, vol. 16, (2006) , pp. 545-556.
5. B. J. Baliga, *Modern Power Devices*, 1987.
6. M. Roschke, F. Schwierz, "Electron mobility models for 4H, 6H, and 3C SiC," *IEEE Transactions on Electron Devices*, vol. 48, (2001), pp. 1442-1447.
7. H. Zhang, Ph.D. Dissertation of the University of Tennessee, Knoxville, December 2007.
8. H. Zhang, L. M. Tolbert, B. Ozpineci, M. Chinthavali, *IEEE Industry Applications Society Annual Meeting* (2006), pp. 346-350.
9. M. Chinthavali, L. M. Tolbert, B. Ozpineci, *IEEE Power Engineering General Meeting* (2004), pp. 680-685.
10. B. Ozpineci, M. Chinthavali, A. Kashyap, L. M. Tolbert, A. Mantooth, *IEEE Applied Power Electronics Conference* (2006), pp. 448-454.
11. F. D. Barlow, A. Elshabini, K. Vanam, B. Ozpineci, L. D. Marlino, M. S. Chinthavali, L. M. Tolbert, *IMAPS 40th International Symposium on Microelectronics* (2007).
12. M. Huque, R. Vijayaraghavan, M. Zhang, B. Blalock, L. Tolbert, S. Islam, *IEEE Power Electronics Specialist Conference (2007)*, pp. 1491-1495.
13. L. M. Tolbert, H. Zhang, M. S. Chinthavali, B. Ozpineci, *Materials Science Forum*, vols. 556-557 (2007), pp. 965-970.

Mater. Res. Soc. Symp. Proc. Vol. 1069 © 2008 Materials Research Society   1069-D12-02

3D Thermal Stress Models for Single Chip SiC Power Sub-Modules

Bang-Hung Tsao[1], Jacob Lawson[1], and James Scofield[2]
[1]University of Dayton Research Institute, 300 College Park, Dayton, OH, 45469-0074
[2]Air Force Research Labratory, WPAFB, OH, 45433

ABSTRACT

Three dimensional models of single chip SiC power sub-modules were generated using ANSYS in order to simulate the effects of various substrate materials, heat fluxes, and heat transfer coefficients on temperature and thermal stress contours. Silicon nitride, aluminum-nitride, alumina were compared as substrates with or without an additional layer of CVD diamond on either top or bottom of the surfaces. Simulated heat fluxes of 100 to 300 watts/cm^2 resulted in device junction temperatures in the range of 359 to 7289 K. With modest cooling, represented by a heat transfer coefficient (hconv) of 3350 watts/m^2-K, SiC chips operated at 300 watts/cm^2 power density maintained junction temperatures $T_j < 688$ K. In the applied heat flux range, the maximum Von Mises stress of a simulated single SiC device sub-module was between 767 MPa to 54.9 GPa. Whereas, the maximum shear stress was between 153 MPa and 8.1 GPa. Regardless of stacking configuration, the maximum chip temperature, Von Mises stress, and shear stress decreased with increasing heat transfer coefficient from 50 to 5000 watts/m^2-K. If consistent with simulation results, CVD diamond integrated substrates should be superior in most cases to those comprised of only AlN, Al$_2$O$_3$, and Si$_3$N$_4$. Experimental validation of ANSYS results and more extensive multiple-chip power module simulations will be explored.

INTRODUCTION

SiC is an excellent candidate for modern power electronics due to its superior breakdown voltage, thermal conductivity, and inherent resistance to radiation and chemical attack [1]. Discrete SiC devices have many advantages, most notably, reduced switching losses, high voltage, and high temperature capability [2]. As a result, the use of SiC devices can increase system efficiency and reduce volume and weight. Though a SiC switch module potentially requires less cooling and offers increased reliability, power module failure mechanisms tend to be thermally activated or enhanced [3]. These thermo-mechanical failure modes are expected to be significantly accelerated under some of the high temperature operating conditions projected for SiC power electronic devices. The objective of this study was to use ANSYS [4] finite element analysis (FEA) to predict temperature and stress distributions for notional SiC power module geometries under various loading and environmental conditions to be subsequently validated by experiment.

EXPERIMENT

This study began with the generation of highly simplified stacking configuration geometries representative of a partial power sub-module using ANSYS FEA software. As seen in Figure 1 three basic stacking configurations were chosen. All three geometries begin with SiC as the top layer, representative of a generic "device". The second material in the stack is a thin layer of copper, representative of a metallic contact pattern. In the case of the simplest geometry, the next layer is an insulating ceramic substrate comprised of either Al_2O_3, Si_3N_4, or AlN. The two more complicated geometries have a thin layer of diamond inserted either between the copper and the substrate, or added as a bottom layer below the ceramic.

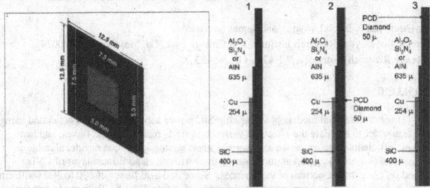

Figure 1. Basic geometry and 3 stacking configurations.

To simulate the effect of energizing the "device," a heat flux of either 100 or 300 W/cm^2 was applied to the top most surface of the SiC in each stacking configuration. To represent modest convective cooling, four different heat transfer coefficients were applied as constants to the bottom most surfaces: 50 W/m^2-K which is in the natural convention cooling range, 298 W/m^2-K which is slightly higher than natural convection, 3350 W/m^2-K which might result from single-phase forced convection, and 5000 W/m^2-K which is in the boiling regime. 25 W/m^2-K was applied to all other free surfaces. A reference temperature of 298 K was selected for all the simulations. Figure 2 depicts typical simulation results seen in this study.

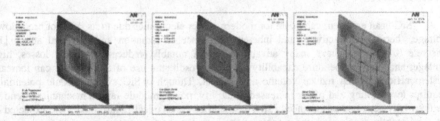

Figure 2. Typical simulation results for a Si_3N_4 stacking configuration with no diamond layer added, at an applied heat flux of 300 W/cm^2 and a hconv equal to 3350 W/m^2-K: temperature contour (left), Von Mises stress contour (middle) and shear stress contour (right hand side).

RESULTS AND DISCUSSION

Upon compiling all the simulation results, several general trends were clearly noticed. As represented by Figures 3 and 4 increasing heat flux from 100 to 300 results in a consistent increase in maximum temperature, Von Mises stress and shear stress. This seems to be true for each respective configuration. Furthermore, the change in magnitude of temperature or stress decreases with increasing hconv as seen in Figure 5; greater than 3350 W/m²-K, values tend to saturate and converge slightly. The addition of a layer of diamond either on the top or bottom of the ceramic results in a decrease in temperature compared to stacks without diamond. Due to its low coefficient of thermal expansion (CTE), however, diamond tends to increase Von Mises stresses at 100 W/cm². Similarly, if diamond is added to the bottom, the shear stress tends to increase due to CTE mismatching.

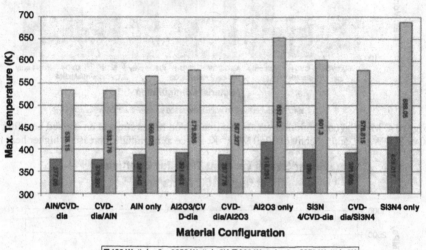

Figure 3. Material configuration versus maximum temperature at an applied heat flux of 100 or 300 W/cm² and a hconv equal to 3350 W/m²-K: notice a magnitude increase at higher heat flux.

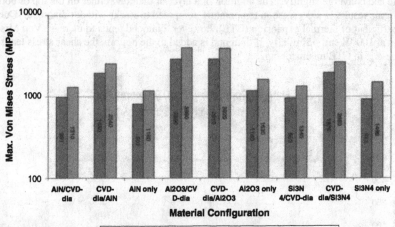

Figure 4. Material configuration versus maximum Von Mises stress at an applied heat flux of 100 or 300 W/cm² and a hconv of 3350 W/m²-K: notice a magnitude increase at higher heat flux.

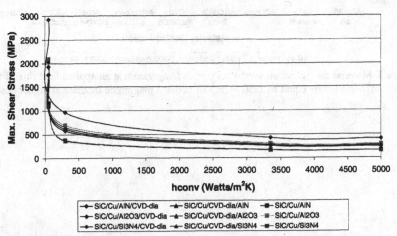

Figure 5. Maximum shear stress versus hconv for each stacking configuration: notice that the shear stress values tend to decrease and level out with increasing hconv.

The remainder of this section groups and discusses simulation results on the basis of each ceramic substrate. This is followed by relative comparison of the least and most optimal stacking configurations for this study in terms of maximum temperature and stress.

Al_2O_3

As mentioned, the addition of diamond tends to reduce the maximum temperature when compared to the respective Al_2O_3 configurations without diamond. The difference is relatively small, but the addition of a diamond layer to the top of a stack results in a lower temperature than if diamond is added to the bottom. Regardless of placement, the diamond increases the Von Mises stress at both 100 and 300 W/cm^2. The results are close, but it seems that adding diamond on the top yields a higher Von Mises stress than diamond on the bottom if hconv is less than 2000 W/m^2-K. If greater than 2000, the bottom diamond results in a higher Von Mises. Diamond on the top does cause a reduction in shear stresses, but the values are very similar in magnitude to those resulting from having no diamond layer.

AlN

Though the diamond layers reduce temperature overall, it is difficult to determine a configuration which is "best" as far as temperature reduction is concerned. Similar to the Al_2O_3 results, the diamond layers increase Von Mises stress in the AlN stacks at both applied heat fluxes. For this set of simulations, the plotted results are more widely spaced apart, so it is easier to see that: diamond on top causes the greatest increase in Von Mises stress, diamond on the bottom causes the second highest, and stacks without diamond result in the lowest stresses. Furthermore, the shear stress is lowered if diamond is on top, but the results are also close to those of the cases without diamond.

Si_3N_4

For this set of simulations, the diamond results are also very similar, but the placement of a diamond layer on top does yield an overall lower maximum temperature compared to the configurations without diamond. Usually a diamond layer increases the Von Mises stress, but for the Si_3N_4 stacks a decrease in stress is noticed at 300 W/cm^2 if diamond is placed on bottom and hconv is greater than 500 W/m^2-K. Otherwise, Von Mises stresses resulting from a bottom diamond layer are close to and only slightly higher than those of stacks without diamond. The addition of a top diamond layer consistently results in a higher Von Mises stress for each case. Similar to the results of the other two materials, if diamond is placed on top, shear stresses are close to those from cases where there is no diamond present; however it is difficult to say whether or not there is a clear decrease in magnitude as compared to configurations without diamond.

Comparison

The highest shear stresses occurred in simulations where Al_2O_3 was selected as a substrate material, and a diamond layer was added to the bottom. At 3350 W/m^2-K and 100 W/cm^2, this value was as high as 427 MPa, whereas at 3350 W/m^2-K and 300 W/cm^2, stresses as high as 566 MPa are noted. In contrast, the lowest shear stresses resulted from either Si_3N_4 without a diamond layer or AlN with a diamond layer on top. For example, at 3350 W/m²-K and 100 W/cm², Si_3N_4 shows the lowest shear stress values (162 MPa), but at the same hconv and 300 W/cm², AlN seems to be superior (211 MPa). The lowest Von Mises stresses correspond to cases where AlN is simulated without a diamond layer: 807 MPa at 3350 W/m²-K and 100 W/cm², and 1180 MPa at 300 W/cm². Again, it is difficult to determine the configuration which promoted the highest Von Mises stress. At 3350 W/m²-K and 100 W/cm², Al_2O_3 with diamond on top yielded a higher value (2910 MPa), but at 300 W/cm², placing diamond on bottom results in an inferior configuration (3990 MPa). Finally, the highest temperature values can be attributed to stacking configurations with Si_3N_4 and no diamond layer: 428.017 K at 3350 W/m²-K and 100 W/cm², and 688.05 K at 300 W/cm². The lowest temperature might be attributed to either AlN with diamond on top or with diamond on bottom, since the results tend to overlap considerably; at least at 3350 W/m²-K, AlN with diamond on the top results in the lowest junction temperatures (376.392 K at 100 W/cm² and 533.176 at 300 W/cm²).

CONCLUSIONS

In summary, if excessive device temperature and shear stresses greater than those allowable for a given material are considered as the primary failure criteria, stacking configurations of AlN as a bulk substrate with a layer of diamond placed between the copper and the ceramic should result in a superior overall performance. If consistent with simulation results, CVD diamond integrated substrates in general should be superior in most cases to those comprised of only AlN, Al_2O_3, and Si_3N_4. However, a greater understanding of initial boundary conditions and possible failure mechanisms needs to be obtained. Future studies will involve developing more sophisticated FEA models and verifying these results experimentally.

ACKNOWLEDGMENTS

This work was supported by the Propulsion Directorate, Air Force Research Laboratory, under contract no. FA8650-04-D-2403 (DO#13). The authors would like to thank Dr. Jeff Brown, Dr. Larry Scanlon, and Dr. Joe Fellner of AFRL for their assistance with ANSYS.

REFERENCES

1. International Energy Outlook, U.S. Dept. of Energy, Energy Information Administration, April 2004, pp7-107.
2. T.P. Chow. Handbook of Thin Film Devices, Vol.1 Hetero-Structures for High Performance Devices, chapter, 7, "Silicon Carbide Power Devices," Academic Press (2000).
3. S. G. Leslie. Electronic Cooling Magazine, November, 12, Number 4B (2006).
4. ANSYS 10.0

Mater. Res. Soc. Symp. Proc. Vol. 1069 © 2008 Materials Research Society   1069-D13-03

MOS Capacitor Characteristics of 3C-SiC Films Deposited on Si Substrates at 1270°C

Li Wang[1], Sima Dimitrijev[1], Leonie Hold[1], Frederick Kong[1], Philip Tanner[1], Jisheng Han[1], and Gunter Wagner[2]
[1]Queensland Microtechnology Facility and Griffith School of Engineering, Griffith University, 170 Kessels Road, Brisbane, 4111, Australia
[2]Institute of Crystal Growth, Max-Born-Str. 2, Berlin, 12489, Germany

ABSTRACT

SiC films were deposited on Si substrate by low pressure hot-wall CVD using C_3H_8 (5% in H_2)-SiH_4 (2.5% in H_2)-H_2 gas system at 1270°C and 1370°C. MOS capacitors were fabricated on the grown 3C-SiC films. In this paper, we compare the electrical characteristics of MOS capacitors fabricated on 3C-SiC films deposited at high and low temperatures, 1370°C and 1270°C, respectively. The cross-sectional TEM images indicate similar SiC/Si interface micro-structural quality for 3C-SiC films deposited at different temperatures, though a quicker elimination rate of stacking fault with increasing thickness at 1370°C, and rocking curves from XRD measurements indicate better crystalline perfection at 1370°C. The average surface roughness measurements performed by an atomic force microscope show that the surface roughness increases with elevated deposition temperature. The MOS capacitors were characterized by high-frequency capacitance-voltage (HFCV), conductance-voltage (G-V), and current-voltage (I-V) measurements at room temperature. The MOS capacitors fabricated on both films exhibit good and almost identical C-V characteristics. Measurements of current-voltage characteristics in accumulation region showed smaller leakage for the film deposited at 1270°C. It is concluded that the decrease of the deposition temperature from 1370°C to 1270°C does not bring any remarkable negative impact on the interface properties of fabricated MOS capacitors.

INTRODUCTION

The limitation imposed by Si and GaAs devices able to operate under extreme conditions has encouraged some research groups to investigate other semiconductors with wider energy gap. Hetero-epitaxial growth of 3C SiC films on silicon substrate has been attracting increasing attention [1, 2] because SiC-based devices can be integrated with the mainstream Si devices showing great potential in both electronic and micro/nano-mechanical applications. The conventional technique of deposition of 3C SiC films on silicon substrate is chemical vapour deposition (CVD) using separate precursors such as SiH_4 as Si source and C_3H_8 as carbon source [3]. There is a report showing that the best crystal quality and purity is usually obtained at relatively high temperature around 1350°C to 1400 °C for the deposition on Si substrate [4]. Ciobanu et al published a paper about the traps at the interface of 3C-SiC/SiO_2 MOS capacitor, in which the 3C-SiC film was deposited homoepitaxially by CVD at 1600 °C [5].

Heteroepitaxially grown 3C-SiC films on Si substrates by hot-wall CVD at 1370°C and 1270°C were used for the analysis presented in this paper. The electrical characteristics of metal-oxide-semiconductor (MOS) capacitors fabricated on these films were analyzed to investigate the influence of deposition temperature.

EXPERIMENT

SiC films were deposited on Si substrates using C_3H_8 (5% in H_2)-SiH_4 (2.5% in H_2)-H_2 gas system at 1270°C and 1370°C in a low pressure hot-wall CVD reactor. A detailed description of the CVD-system was already published elsewhere [6]. Ultra pure H_2 was used as the carrier gas, and the deposition of 3C-SiC films took place with the total pressure in the reactor of 150 mbar. MOS capacitors were fabricated on grown 3C-SiC films using the following process. The bare 3C-SiC/Si wafers were first cleaned in a mixture of H_2SO_4 and H_2O_2, followed by an RCA clean. This was then followed by 1% HF dip for 1 min. For better SiC/SiO$_2$ interface quality, the gate oxide layers were grown in 100% NO at 1160°C [5]. After that aluminum was thermally evaporated to form the gate electrodes and square capacitors with area of 0.0025 cm^2 were then defined by photolithography. The MOS capacitors were characterized by high-frequency capacitance-voltage (HFCV), conductance-voltage (G-V) and current-voltage (I-V) measurements at room temperature. The HFCV measurements were performed at the sweep rate of 0.1 V/s and the sweep range from 2V to -14V, using a computer-controlled HP4284A LCR meter. The I-V measurements were performed at the sweep rate of 0.1 V/s and the sweep range from 0V to 15V, using a computer-controlled HP4145 meter. All the measurements were performed under light-tight and electrically-shielded conditions. The electrical analysis was complemented by transmission electron microscopy, high resolution X-ray diffraction (HRXRD) rocking curve and atomic-force microscopy to obtain information on structural characteristics and average surface roughness of the deposited 3C-SiC films.

RESULTS AND DISCUSSION

It is usually believed that higher temperature helps to improve the film quality. The full-width at half-maximum (FWHM) of 3C-SiC (002) HRXRD rocking curve decreases from 670 arcsec for sample A (deposited at 1270°C, 4.6 µm) to 380 arcsec for sample B (deposited at 1370°C, 4.9 µm), as shown in figure 1. The cross-sectional TEM images indicate similar SiC/Si interface micro-structural quality for both samples (shown in figure 2), while sample B shows a quicker elimination rate of stacking fault with increasing thickness than sample A. No etch pits are observed in either sample. All these results indicate a better crystalline quality at higher deposition temperature.

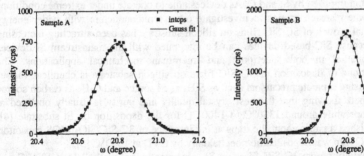

Figure 1 3C-SiC (002) HRXRD rocking curves of sample A (deposited at 1270°C) and sample B (deposited at 1370°C)

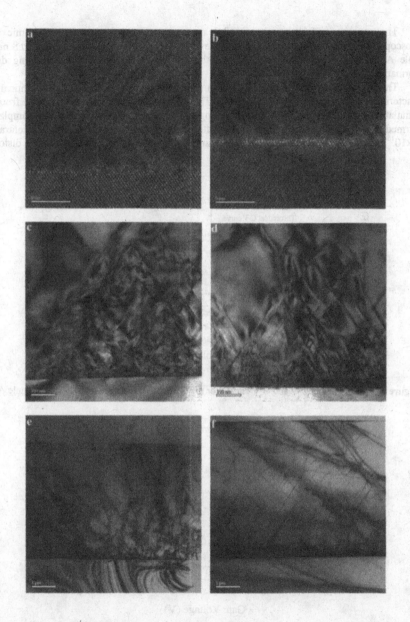

Figure 2 TEM micrographs of SiC films deposited at 1270 °C (a, c, e, for sample A) and 1370 °C (b, d, f, for sample B).

The average surface roughness (Ra) measurements performed by an atomic force microscope show that the Ra increases with elevated deposition temperature, from 10.8 nm for sample A to 34 nm for sample B. A smoother surface is preferred for improving device performance.

The MOS capacitors fabricated on both samples exhibit good and almost identical C-V characteristics (shown in Figures 3 and 4). The gate-oxide thickness, determined from the accumulation capacitance, is 28.3 nm. According to the measured *C-V* curves, both samples are confirmed to be *n*-type conduction and when fitted to theoretical curves, electron concentrations of 7×10^{15} cm^{-3} (sample A) and 5×10^{15} cm^{-3} (sample B) were determined. Minimum distortion

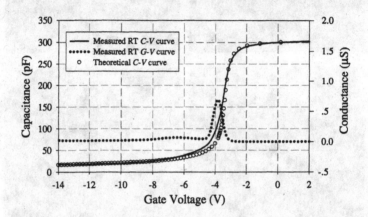

Figure 3 High frequency *C-V* and *G-V* curves for the 3C-SiC film deposited at 1270 °C (sample A).

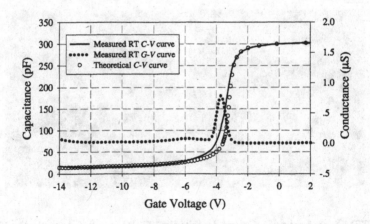

Figure 4 High frequency *C-V* and *G-V* curves for the 3C-SiC film deposited at 1370 °C (sample B).

and small stretch-out observed in the measured as compared to the theoretical C-V curves indicate low interface traps density at the SiO_2/SiC interface [7]. Similar level of conductance peaks also indicates similar level of interface-trap density for SiC films deposited at different temperatures. The C-V curves display a negative flat-band voltage shift for both capacitors. The negative flat-band voltage indicates the traps at SiO_2/SiC interface have an effective value of 2.05×10^{12} and 1.91×10^{12} electronic charges/cm^2 for samples A and B, respectively. These results show that good quality oxide can be grown on the SiC films that are deposited at lower temperature, compared to the conventional deposition temperature.

Current voltage measurements in the MOS capacitor accumulation region were also used to study the quality of the deposited SiC films on samples A and B. Typical current voltage characteristics of both samples are shown in Figure 5. A smaller leakage current is obtained for sample A, deposited at 1270°C. This agrees well with the average surface roughness measurement results as the gate leakage current is affected by surface roughness, a smoother SiC surface makes a smoother SiC/SiO_2 interface and SiO_2 surface, and thus a smaller leakage current. It is possible that for the MOS capacitor the surface roughness plays a more important role than the stacking fault density in the 3C-SiC films. The yield of the MOS capacitors fabricated on both of the films is about 85%.

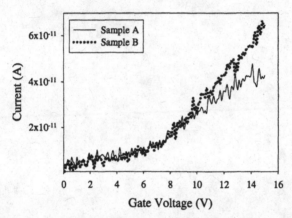

Figure 5 Current-voltage characteristics in the accumulation region for both samples

CONCLUSIONS

In summary, decreasing the deposition temperature of 3C-SiC films on Si substrates from 1370°C to 1270°C does not bring any remarkable negative impact on the interface properties of fabricated MOS capacitors, though the deposited 3C-SiC film has lower stacking fault density at 1370°C. Gate oxide grown by the NO process can form a high quality SiO_2/SiC interface on SiC films deposited at different temperatures. Current voltage measurement indicated smaller leakage current for the SiC film deposited at 1270°C. It is possible that for the MOS capacitor the surface roughness of SiC film plays a more important role than the stacking fault density in the deposited 3C-SiC film. It is practical to deposit MOS capacitor quality 3C-SiC films at 1270°C on Si substrates.

ACKNOWLEDGEMENTS

The authors would like to thank Dr. Martin Schmidbauer at Institute of Crystal Growth (Berlin, Germany) for the HRXRD rocking curve measurements on the samples.

REFERENCES

[1] X.M.H. Huang, C.A. Zorman, M. Mehergany, and M.L. Roukes, *Nature* **421** (2003).
[2] S. Dimitrijev, *Microelectron. Eng.* **83** (2006) 123.
[3] J. Yun, T. Takahashi, Y. Ishida, H. Okumura, *J. Cryst. Growth* **291** (2006) 140.
[4] J-H Boo, S. A. Ustin, W. Ho, *Thin Solid Film*, **343-344** (1999) 650.
[5] F. Ciobanu, G. Pensl, H. Nagasawa, A. Schoner, S. Dimitrijev, K-Y Cheong, V.V. Afanasev and G. Wagner, *Mat. Sci. Forum* **433-436** (2003) 551.
[6] G. Wagner, D. Schulz, D. Siche, *Progress in Crystal Growth and Characterization of Materials* **47** (2003) 139.
[7] D.K. Schroder, *Semiconductor Material and Device Characterization*, 3rd ed. (John Wiley & Sonsm Inc., Hoboken, New Jersey, 2006), pp. 350-351.

Mater. Res. Soc. Symp. Proc. Vol. 1069 © 2008 Materials Research Society   1069-D07-12

Impact Ionization in Ion Implanted 4H-SiC Photodiodes

Wei Sun Loh[1], Eric Z. J. Goh[1], Konstantin Vassilevski[2], Irina Nikitina[2], John P. R. David[1], Nick G. Wright[2], and C. Mark Johnson[3]

[1]Electronic and Electrical Engineering, The University of Sheffield, Mappin Street, Sheffield, S1 3JD, United Kingdom

[2]School of Electrical, Electronic and Computer Engineering, Newcastle University, Newcastle, NE1 7RU, United Kingdom

[3]School of Electrical and Electronic Engineering, The University of Nottingham, University Park, Nottingham, NG7 2RD, United Kingdom

ABSTRACT

Hole dominated avalanche multiplication and thus breakdown characteristics of ion implanted 4H-SiC p^+-n^--n^+ photodiodes were determined by means of photomultiplication measurements using 325 nm UV light. All the tested diodes exhibited low reverse leakage current and reasonably uniform avalanche breakdown. With avalanche widths of 0.2 μm to 1.5 μm and the capability to measure multiplication factor as low as 1.001, the room temperature impact ionization coefficients were precisely deduced from these 4H-SiC diodes using a local ionization model for electric fields ranging from 1.25 MV/cm to 2.8 MV/cm. The results agree with those reported by Ng et al. and are within the accuracy of both the C-V measurements and electric-field determinations.

INTRODUCTION

Silicon Carbide (SiC) has many favorable properties such as a wide bandgap (E_g for 4H-SiC = 3.23 eV at 300K), high thermal conductivity, large breakdown electric field strength, high saturated drift velocity, outstanding physical toughness and chemical inertness [1, 2]. All these desirable characteristics make SiC a promising candidate to overcome the performance limitations of much more readily available conventional Silicon (Si) based technology for high temperature, high frequency and high power applications [3-5]. Additionally, SiC is also an attractive alternative to the III- nitride based avalanche photodiodes (APDs) for ultraviolet (UV) detection as SiC APDs exhibit extremely low dark current and good visible blind performance [6, 7] even at elevated temperatures, owing to the low intrinsic carrier concentration and thermal stability.

An accurate knowledge of the impact ionization mechanism which governs the multiplication process is crucial in order to determine the device breakdown and thus the optimum device design. Nevertheless, most of the impact ionization results [8-11] reported to date are based on p-n diodes fabricated from heavily doped n-type and p-type 4H-SiC epitaxial structures and they show wide variations. Work by Ng et al. [10] is of particular interest as it is the only report of ionization coefficients extracted using excitations by both short (230 nm) and long (365 nm) wavelengths, thus resulting in more realistic hole and electron initiated multiplication behaviors. In Ng et al.'s work [10], both the multiplication and excess noise characteristics of submicron 4H-SiC diodes were interpreted using a non-local model which takes into account the dead space effect, non-uniform electric field and wavelength dependent carrier injection profiles. A similar experimental approach was used in current work. The avalanche multiplication characteristics were measured in 4H-SiC photodiodes with abrupt p^+-n^- junctions formed by ion implantation with avalanche widths of 0.3 μm and 1.15 μm. Hole ionization coefficients, β were extracted using a local model and compared with previously published results [8-11].

EXPERIMENTAL TECHNIQUES AND MODELING

The 4H-SiC p^+-n^--n^+ photodiodes used in this work are depicted schematically in figure 1. The diodes were formed on epi-wafer purchased from Cree Inc. The initial epitaxial structure comprised a lightly doped (1×10^{16} cm^{-3}) 4.7 μm thick layer on a 0.5 μm heavily doped n^+ buffer layer grown on 0.019 Ω·cm, 3 mp/cm^2 8° off-orientated substrate. The epilayer was thinned by Reactive Ion Etching (RIE) to obtain a set of diodes with different breakdown voltages. A p^+ anode region was then formed by Al implantation at 600°C to get a box doping profile with Al concentration of 1×10^{19} cm^{-3} for the depth of 0.4 μm. Implanted species were activated by annealing at 1600°C for 30 min in Ar ambient under graphite capping layer. Square mesa structures with side lengths of 410 μm, 310 μm and 210 μm were fabricated by RIE for a diode junctions termination. The diode surfaces were passivated using a thermally grown layer of SiO$_2$. A 100 nm thick nickel metallization was deposited to use as a bottom contact to the n^+ substrate. Semi-transparent top contacts were formed by deposition of 28 nm thick Ti/Al/Ti stacked film in previously opened windows in silicon oxide. All metals were annealed simultaneously at 1040°C in vacuum to form ohmic contacts. 600 nm thick gold layers were deposited on both sides of the samples for contacts reinforcement (blank on the back side and patterned by the lift-off procedure on the contacts to the p^+ layer).

The fabricated diodes demonstrated breakdown voltages from 50 to 270 V depending on the thickness of n^- epitaxial layers which were varied from 0.2 μm to 1.5 μm. Results from two different samples denoted as M11 and M15 with avalanche widths of 0.3 μm and 1.15 μm respectively are reported in this work as they were the most consistent and reproducible.

The photomultiplication characteristics of the devices were investigated by p-n junction illumination through the semi-transparent top contact with light from a 325 nm He-Cd laser (spot size of about 15 μm). Both DC and AC (phase sensitive detection) photomultiplication measurements were carried out to ascertain the reproducibility of the multiplication characteristics. Photocurrent normalization by baseline correction was performed to eliminate the effect of the increment in collection efficiency of the primary carriers resulting from the movement of the depletion region edge towards the top surface [12]. The photomultiplication data were reproducible and not dependent on the intensity and position of the incoming excitation, indicating that defects and variation in doping across the device had negligible effect. At least three devices of each sample were tested to ensure the reproducibility and consistency of the results. Given the fact that the avalanche region of the tested devices is thick enough to neglect any dead space effect, the multiplication characteristics were modeled using local model along with the ionization coefficients reported by Ng *et al.* [10].

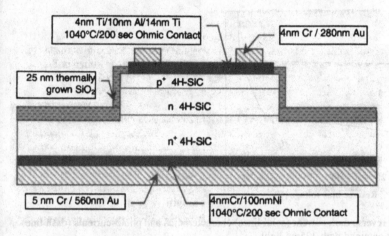

Figure 1. Schematic diagram of the 4H-SiC p^+-n^--n^+ photodiodes and its top view (mesa size is 0.41×0.41 mm).

RESULTS AND DISCUSSION

As depicted in figure 2, the diodes exhibit very low dark currents prior to avalanche breakdown and the measured photocurrents are at least two orders of magnitude greater than the dark current. The breakdown voltages of samples M11 and M15 were estimated to be 83 V and 253 V respectively by fitting the multiplication profile to Miller's multiplication expression [13] and are in close agreement with the reverse dark *I-V* results.

Given that the absorption coefficient of 4H-SiC at 325 nm is about 1350 cm^{-1} [14], most of the 325 nm light will reach the n^+ region of both samples and the arising multiplication will therefore be virtually all hole initiated (M_h). Multiplication values of 8 and in excess of 10 were obtained from samples M11 and M15 respectively. The experimental data agree well with those predicted multiplication characteristics using a local model with the ionization coefficients of Ng *et al.* [10] as illustrated in figure 3 for samples M5 and M11. The plot of ($M(V)$-1) in semi-logarithmic scale shown in figure 3(b) reveals the low multiplication values at low bias. The onset of measured avalanche multiplication for samples M11 and M15 appear to be at 40V and 190V respectively. The small discrepancies between the measured and modeled multiplication characteristics for sample M15 at low biases may be due to the errors in normalizing the measured photocurrent from the baseline correction.

Figure 4 presents the hole ionization coefficients experimentally determined in this work along with those published elsewhere [8-11]. The experimental results form a consistent set of data covering the electric fields from 1.25 to 2.8 MV/cm and match well β values measured in

4H-SIC diode with p-n junction formed by epitaxial grown on p$^+$ [8, 9] and n$^+$ substrate [10]. As seen in figure 4, the data reported by Ng *et al.* [10] gives the best agreement with the experimentally deduced hole ionization coefficient reported here.

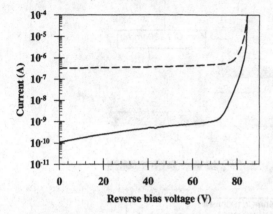

Figure 2. Typical reverse dark current (solid lines) characteristics and photo-currents (dash line) of the M11 diodes obtained with 325nm light.

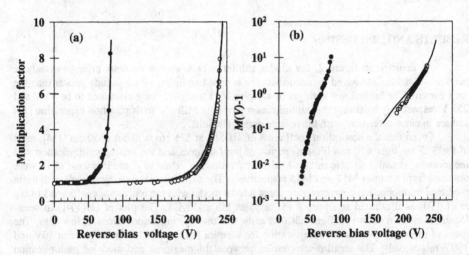

Figure 3. The experimental data (symbols) along with the modeled (lines) multiplication characteristics and (b) (*M(V)*-1) in semi-logarithmic scale using ionization coefficients of Ng *et al.* [10] for samples M11 (filled symbol) and M15 (unfilled symbol).

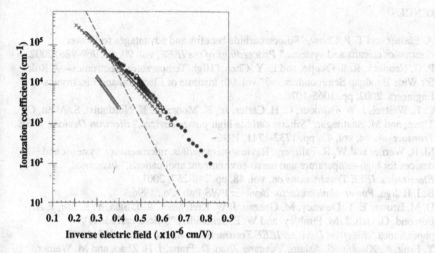

Figure 4. Hole ionization coefficients measured from samples M11 (filled circles) and M15 (open circles) in comparison with experimental data reported by Konstantinov *et al.* [9] (×), Raghunathan *et al.* [11] (Δ), Hatakeyama *et al.* [8] (grey dashed lines) and Ng *et al.* [10] (solid lines).

CONCLUSION

Hole initiated multiplication characteristics for ion-implanted p^+-n 4H-SiC diodes with avalanche widths of 0.3 µm and 1.15 µm were deduced using 325 nm light at room temperature. The locally modeled multiplication characteristics at room temperature using Ng *et al.'s* [10] ionization coefficients, agrees well with the experimental results. Furthermore, good agreement was obtained between the hole ionization coefficients extracted in this work for ion implanted *p*-*n* junctions and those of Ng *et al.'s* [10] for epitaxial diodes at electric fields in the range of 1.25 to 2.8 MV/cm. The ionization coefficients reported by Ng *et al.* [10] can thus be applied to predict the multiplication characteristics of the ion implanted 4H-SiC photodiodes used in this work where the non-local effects can be neglected.

ACKNOWLEDGEMENT

This work was funded in part by the UK Engineering and Physical Sciences Research Council (EPSRC).

REFERENCES

[1]   A. Elasser and T. P. Chow, "Silicon carbide benefits and advantages for power electronics circuits and systems," *Proceedings of the IEEE*, vol. 90, pp. 969-986, 2002.

[2]   P. G. Neudeck, R. S. Okojie, and L. Y. Chen, "High-Temperature Electronics—A Role for Wide Bandgap Semiconductors?" vol. 90: Institute of Electrical and Electronics Engineers, 2002, pp. 1065-1076.

[3]   C. E. Weitzel, J. W. Palmour, C. H. Carter, Jr., K. Moore, K. K. Nordquist, S. Allen, C. Thero, and M. Bhatnagar, "Silicon carbide high-power devices," *Electron Devices, IEEE Transactions on*, vol. 43, pp. 1732-1741, 1996.

[4]   M. R. Werner and W. R. Fahrner, "Review on materials, microsensors, systems and devices for high-temperature and harsh-environment applications," *Industrial Electronics, IEEE Transactions on*, vol. 48, pp. 249-257, 2001.

[5]   B. J. Baliga, *Power semiconductor devices*: PWS Pub. Co., 1996.

[6]   D. M. Brown, E. T. Downey, M. Ghezzo, J. W. Kretchmer, R. J. Saia, Y. S. Liu, J. A. Edmond, G. Gati, J. M. Pimbley, and W. E. Schneider, "Silicon carbide UV photodiodes," *Electron Devices, IEEE Transactions on*, vol. 40, pp. 325, 1993.

[7]   Y. Feng, X. Xiaobin, S. Aslam, Yuegang Zhao, D. Franz, J. H. Zhao, and M. Weiner, "4H-SiC UV photo detectors with large area and very high specific detectivity," *Quantum Electronics, IEEE Journal of*, vol. 40, pp. 1315, 2004.

[8]   T. Hatakeyama, T. Watanabe, T. Shinohe, K. Kojima, K. Arai, and N. Sano, "Impact ionization coefficients of 4H silicon carbide," *Applied Physics Letters*, vol. 85, pp. 1380-1382, 2004.

[9]   A. O. Konstantinov, Q. Wahab, N. Nordell, and U. Lindefelt, "Ionization rates and critical fields in 4H silicon carbide," *Appl. Phys. Lett.*, vol. 71, pp. 90-92, 1997.

[10]   B. K. Ng, J. P. R. David, R. C. Tozer, G. J. Rees, F. Yan, J. H. Zhao, and M. Weiner, "Nonlocal effects in thin 4H-SiC UV avalanche photodiodes," *IEEE Trans. on Electron Devices*, vol. 50, pp. 1724-1732, 2003.

[11]   R. Raghunathan and B. J. Baliga, "Temperature dependence of hole impact ionization coefficients in 4H and 6H-SiC," *Sol. State Electron.*, vol. 43, pp. 199-211, 1999.

[12]   M. H. Woods, W. C. Johnson, and M. A. Lampert, "Use of a Schottky barrier to measure impact ionization coefficients in semiconductors," *Solid State Electron*, vol. 16, pp. 381-384, 1973.

[13]   S. M. Sze, *Physics of Semiconductor Devices*, 2nd ed: John Wiley and Sons, 1981.

[14]   S. G. Sridhara, R. P. Devaty, and W. J. Choyke, "Absorption coefficient of 4H silicon carbide from 3900 to 3250 Å," *J. of Appl. Phys.*, vol. 84, pp. 2963, 1998.

Mater. Res. Soc. Symp. Proc. Vol. 1069 © 2008 Materials Research Society   1069-D07-17

Degradation of Majority Carrier Conductions and Blocking Capabilities in 4H-SiC High Voltage Devices due to Basal Plane Dislocations

Sei-Hyung Ryu[1], Qingchun Zhang[1], Husna Fatima[1], Sarah Haney[1], Robert Stahlbush[2], and Anant Agarwal[1]

[1]Cree, Inc, Durham, NC, 27703

[2]Naval Research Laboratory, Washington, DC, 20375

ABSTRACT

This paper presents the effect of recombination-induced stacking faults on the drift based forward conduction and leakage currents of high voltage 4H-SiC power devices. To show the effects, 10 kV 4H-SiC MPS (Merged PiN Schottky) diodes have been fabricated on a standard wafer and a low BPD (Basal Plane Dislocation) wafer, and their IV characteristics were evaluated before and after a forward bias stress, which resulted in minority carrier recombination and conductivity modulation in the drift epilayer of the diodes. After the stressing, the diode fabricated on standard wafer showed a significant increase in forward voltage drop, as well as a marked increase in leakage current, which were due to induction of stacking faults. The diode on the low BPD wafer showed very little change after the stress because the induction of stacking faults was minimized. Similar results were also observed on a 10 kV 4H-SiC DMOSFET. The results suggest that recombination-induced stacking faults are detrimental to all device types, and injection of minority carriers in majority carrier devices should be avoided at all times.

INTRODUCTION

The effects of recombination-induced stacking faults (SFs) on forward conduction of 4H-SiC PiN diodes have been extensively studied and published [1]. It was discovered that the basal plane dislocations (BPDs), mostly originated from the substrates, result in the formation of stacking faults in the drift epilayer, which behave as potential barriers that decreases the area available for the current flow [2,3].

It was also speculated that the SFs only affect the recombination-based conduction, and have negligible effects on drift based conduction or reverse leakage characteristics. However, this is not a reasonable speculation, and therefore, should be challenged, because induction of SFs in the drift layer can be equated to creation of defects in the high E-field drift region, which can cause increased leakage currents in the off-state. Also, it should be noted that a potential barrier can adversely affect majority carrier drift conduction as well. This effect of SFs on majority carrier devices was first reported on 10 kV 4H-SiC JBS diodes [4], then, on 10 kV 4H-SiC DMOSFETs [5]. However, further studies are necessary to better understand the problem. So far, this has not been a problem for typical majority carrier devices because minority carrier injection was not utilized. However, with recent advances in 4H-SiC device technologies, several majority carrier devices that utilize minority carrier injection to further improve their performances, such as MPS diodes, power MOSFETs and JFETs with body diodes, emerged.

The recombination-induced SFs may not cause serious problems for low voltage majority carrier devices (< 1200 V), but this can be a very relevant issue for high voltage devices with thick drift layers.

In this work, a 10 kV 4H-SiC MPS (Merged PiN Schottky) diode (Fig. 1) fabricated on a low BPD wafer [6], and a diode fabricated on a standard substrate were subjected to a high current stressing to induce SFs, and their IV characteristics were compared. Results obtained from high current stressing of a 10 kV 4H-SiC DMOSFET (Fig. 2) are also reported in this paper.

EXPERIMENT

A 10 kV 4H-SiC MPS diode fabricated on a low BPD 4H-SiC wafer was compared to a diode fabricated on a standard 4H-SiC wafer for this experiment. Both wafers received $6 \times 10^{14} cm^{-3}$ doped, 120 μm thick n-type epilayers as drift layers. A simplified cross-sectional view of the MPS diode is given in Fig. 1. Nickel was used as the Schottky metal, and Junction Termination Extention (JTE) structure was used as edge termination. Ohmic contacts to the p^+ implanted regions were made to ensure the minority carrier injection, which is necessary for the induction of stacking faults, during the high current stress. The device size was 4 mm x 4 mm, and the active area was 2.6 mm^2. Prior to device fabrication, the wafers were optically mapped for BPDs to check the correlation between the device degradation and BPD density [7], as shown in Fig. 3. It should be noted that the BPDs show up as continuous lines, while other types of defects show up as points. Several BPDs were observed in a standard wafer (see Fig. 3(a)), whereas no BPDs could be observed in the low BPD wafer (Fig. 3(b)).

Pulsed forward I-V characteristics of the MPS diodes were measured on wafer to minimize SF induction, using Tektronix 371 curve tracer at room temperature. The leakage current of the device was also measured in Flourinert oil up to 11 kV at room temperature. After the initial measurements, the diodes were stressed on wafer at a current density of 115 A/cm^2 for 1 hour. The temperature of the chuck was kept at 15°C during stressing. A blue light emission was observed from the junctions during the stressing, which indicates minority carrier injection

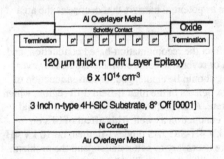

Fig. 1. Simplified cross-section of a 10 kV 4H-SiC MPS diode.

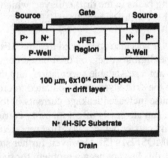

Fig. 2. Simplified cross-section of the 10 kV 4H-SiC DMOSFET [8]

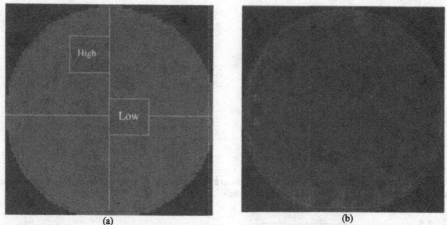

(a)   (b)

Fig. 3. Maps of Basal Plane Dislocations (BPD). (a) Standard wafer, and (b) low BPD wafer.

and recombination. After the stress, the I-V characteristics of diodes were measured again at room temperature to observe the effects of recombination induced stacking faults.

Similar experiments were also performed on a 5 A 10 kV 4H-SiC DMOSFET [8] from a standard wafer. Simplified cross-sections of these devices are given in Fig. 3. The built-in body diode of a 4H-SiC DMOSFET is reverse biased and no minority carrier is injected during the on-state. To stress the body diode and inject minority carrier into the drift layer, the MOS channel must be turned off and a negative drain bias (V_{DS} <0V) must be applied. The on-state characteristics and the leakage currents for V_{DS} up to 6 kV were measured before stressing the body diode. Then, with the MOS channel completely shut off by applying a gate bias (V_{GS}) of − 10 V, the body diode was stressed at an I_D of 5A, which is equal to the device on-state current rating, for up to 3 hours. The case temperature was kept at 50 °C during the stressing. The device was re-characterized after the stressing.

DISCUSSION

10 kV MPS diodes

Fig. 4 shows the IV characteristics of the MPS diode from the standard wafer. A significant degradation of device characteristics was observed after the high current stress on this device. After a 3 A (115 A/cm^2), 1 hour stress, the on-resistance of the diode increased by a factor of 3. It is believed that this is because the recombination induced stacking faults impede the electron drift, which has the same effect as reducing the active area of the device. When a reverse bias was applied, two orders of magnitude increase in the leakage current were observed. In addition, the avalanche voltage of the diode decreased by 1000 V. This is an indication that the recombination induced stacking faults are behaving as electrically active defects. This problem is more noticeable on high voltage devices since the stacking faults tends to extend throughout the length of the drift layer, with an offcut angle (in this case, 8°), which can result in

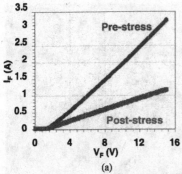
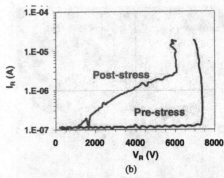

(a)

(b)

Fig. 4. IV characteristics of a 10 kV 4H-SiC MPS diode on a standard wafer, before and after the high current stress. (a) Forward characteristics, and (b) Reverse characteristics.

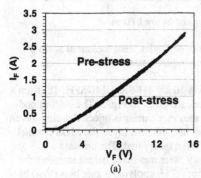
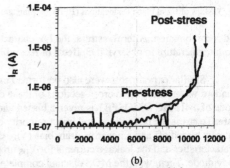

(a)

(b)

Fig. 5. IV characteristics of a 10 kV 4H-SiC MPS diode on a low BPD wafer, before and after the high current stress. (a) Forward characteristics, and (b) Reverse characteristics.

very large cross-sectional areas. On the other hand, the MPS diode from the wafer with low BPD processing showed minimal degradation, if any, as shown in Fig. 5. No noticeable changes in forward characteristics, leakage currents, or avalanche voltage were observed. Fig. 6 shows a close-up look at the BPD map for the MPS diode. Solid line shows the active area of the device. It can be seen very clearly that there are several BPD's, shown as line defects, are located within the active area. On the other hand, the device on low BPD wafer did not include any BPD's within the active area. This is a strong indication that the recombination-induced stacking faults, which grows on the BPDs, causes significant degradation of drift conductions and blocking capabilities in 4H-SiC high voltage majority carrier devices.

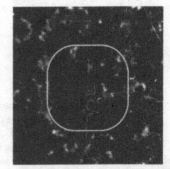

Fig. 6. Close-up BPD map for the MPS diode on standard wafer. Solid line indicates the active area of the device.

10 kV 4H-SiC DMOSFETs

The pulsed IV characteristics of the 4H-SiC 10 kV 5A DMOSFET at room temperature is shown in Fig. 7. It is clearly shown that the 4H-SiC DMOSFET built-in PiN body diode was degraded as expected due to induction of stacking faults (Fig. 7(a)). The diode initially had a V_F of 9 V at an I_A of 5 A, which increased to 13 V after 3 hr stress. Fig. 7(b) shows the DMOSFET on-state IV-characteristics. V_{GS} values up to 15 V (E_{ox} = 3 MV/cm) were used for the measurements. The forward voltage drop at an I_D of 5 A (V_{GS} = 15V) increased from 3.8 V before stressing to 6 V after 1 hr of stressing, then to 10 V after 3 hr of stressing. The DMOSFET transfer characteristics (I_D vs. V_{GS} characteristics) were also compared at the same interval to check the effect of diode stressing to the MOS channel. Fig. 8 show that there are very little changes in the transfer characteristics for V_{GS} up to 6 V, at which the MOS channel characteristics dominates the DMOSFET behavior. This suggests that there are very little, if any, changes in the MOS channel characteristics, which include the threshold voltage. The transfer characteristics for V_{GS} greater than 6 V are heavily affected by the series resistance. Fig. 8 shows a significant reduction of I_D for V_{GS} greater than 6V after stressing, which translates into a substantial increase in series resistance. In the 10 kV, 4H-SiC DMOSFET structure, the drift resistance is represented as the series resistance. From these results we can conclude that the drift

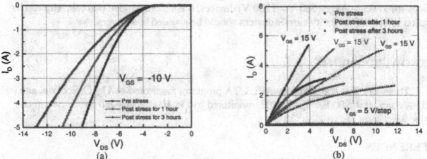

Fig. 7. IV characteristics of a 10 kV 4H-SiC DMOSFET, before and after the high current stress. (a) Characteristics of the body diode, and (b) On-state characteristics of the DMOSFET.

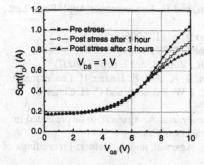

Fig. 8. DMOSFET transfer characteristics before and after the induction of stacking faults.

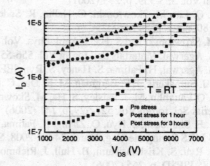

Fig. 9. Off-state IV characteristics of the DMOSFET before and after the induction of stacking faults.

resistance increased as a result of the stressing, while the MOS channel remains unchanged. Fig. 9 shows the blocking characteristics of the 4H-SiC DMOSFET. The leakage current shows a substantial increase after stressing. After 3 hr stressing, the leakage current increased by an order of magnitude, strongly suggesting that the recombination induced stacking faults are behaving as electrically active defects, which is in agreement with the results observed in the 4H-SiC MPS diodes.

CONCLUSIONS

The adverse effects of recombination-induced stacking faults on majority carrier drift conduction and leakage current in 4H-SiC power devices have been demonstrated. The experiments were performed on two types of devices, which include 10 kV 4H-SiC MPS diodes and a 10 kV 4H-SiC DMOSFET. In both case, substantial increases in on-resistance have been observed, which indicates that the stacking faults can severely affect the majority carrier drift conduction as well as the minority carrier recombination based conduction reported previously. This effect is more noticeable for very high voltage devices with very thick drift layers. In addition, it was shown that the recombination-induced stacking faults in 4H-SiC devices can cause orders of magnitude increase in the leakage currents. Although this effect may not be obvious for lower voltage (600 V – 1200 V) devices, it is recommended that minority carrier injection in 4H-SiC majority carrier devices should be avoided at all times.

ACKNOWLEDGMENTS

This work was supported by ARL CTA program, monitored by Dr. C. Scozzie, and DARPA contract # N00014-05-C-0202, monitored by Dr. H. Dietrich and Dr. S. Beerman-Curtin.

REFERENCES

1. J. P. Bergman, H. Lendenmann, P.-Å. Nilsson , U. Lindefelt, and P. Skytt: Material Sci. forum Vol. 353-359, p. 299 (2001)
2. J. Q. Liu, M. Skowronski, C. Hallin, R. Söderholm, and H. Lendenmann: Appl. Phys. Lett., Vol. 80, no. 5, p. 749 (2002)
3. M. Skowronski and S. Ha: J. Appl. Phys. Vol. 99, no. 1, p. 11 (2006)
4. A. Agarwal: Materials Sci. forum Vol. 556-557, p. 687 (2007)
5. A. Agarwal, H. Fatima, S. Haney, S. Ryu, IEEE EDL, Vol. 28, no. 7, p. 587 (2007)
6. J. J. Sumakeris, J. P. Bergman, M. K. Das, C. Hallin, B. A. Hull, E. Janzen, H. Lendenmann, M. J. O'Laughlin, M. J. Paisley, S. Ha, M. Skowronski, J. W. Palmour, and C. H. Carter, Jr.: Material Sci. forum Vol. 527-529, p. 141 (2006)
7. R. E. Stahlbush, S. Ryu, Q. Zhang, H. Fatima, S. Haney, and A. Agarwal, to be presented in 2008 MRS Spring meeting, March 24-28, 2008. San Francisco, CA.
8. S. Ryu, S. Krishnaswami, B. Hull, J. Richmond, A. Agarwal, and A. Hefner: Proceedings of the 18th ISPSD, p. 265 (2006)

Mater. Res. Soc. Symp. Proc. Vol. 1069 © 2008 Materials Research Society   1069-D07-18

Effective Channel Mobility in Epitaxial and Implanted 4H-SiC Lateral MOSFETs

Sarah Kay Haney[1,2], Sei-Hyung Ryu[1], Sarit Dhar[1], Anant Agarwal[1], and Mark Johnson[2]

[1]Power R&D, Cree, Inc, 4600 Silicon Drive, Durham, NC, 27703
[2]Department of Materials Science, North Carolina State University, Raleigh, NC, 27695

Abstract

In this paper, we investigate the effective inversion layer mobility of lateral 4H-SiC MOSFETs. Initially, lateral n-channel MOSFETs were fabricated to determine the effect of p-type epi-regrowth on a highly doped p-well surface. The negative effects of the high p-well doping are still seen with 1500 Å p-type regrowth, while growing 0.5 μm or more appears to be sufficient to grow out of the damaged area. A second experiment was performed to examine the effects of doping during epitaxial regrowth versus using ion implantation after regrowth. Comparable mobilities and threshold voltages were observed for equivalent epitaxial and implanted doping concentrations.

Introduction

The successful, commercial production of a SiC MOSFET has been a long-term goal of the silicon carbide community for the past decade. While much progress has been made, there are still many problems that must be solved before this goal can be truly achieved. The reduction of interface states in SiC MOS system has been a primary focus of research during the last 15 years. In the past decade, much work has been done through the use of NO anneals to passivate the MOS interface.[1,2] NO annealing is effective in reducing the density of interface traps, D_{it}, near the conduction band edge by an order of magnitude. However this increase in effective mobility also yields a negative shift in the threshold voltage, V_{th}, which can lead to MOSFETs with negative V_{th}. Controlling the threshold voltage is one of the most critical issues faced in current MOSFET fabrication. This difficulty stems from the fact that the threshold voltage is determined by a difference of two large numbers, namely, a large positive fixed charge and a large negative charge in the interface traps and possibly in-bulk traps.[3] So in order to effectively increase the mobility by reducing the D_{it}, one must also work to decrease the large amount of fixed charge in the oxide in order to maintain a reasonable threshold voltage. These problems have proved to be crucial in the production of DMOSFETs, where highly doped p-wells are needed. This high surface p-well doping reduces the inversion layer mobility by means of impurity scattering. In this series of initial experiments, the regrowth of lightly doped p-type layer was examined as a way to reduce impurity scattering.

Experimental

Experiment I

For experiment I, lateral n-channel MOSFETs with a W/L equal to 400 μm/ 400 μm were fabricated on 1e10^{16} cm^{-3} Al-doped p-type epilayers grown on p-type 4H-SiC substrates off-cut 4° from the (0001) Si-face vicinal surface. This initial experiment was designed to determine the effect of p-type epitaxial regrowth on the highly doped p-well surface. In this experiment, a p-well was first formed by implanting a heavy dose of p^{+} impurity (Al) with a surface concentration of $5x10^{18}$ cm^{-3}. P-well activation temperatures of 1650°C and 1800°C were investigated. After carrying out a sacrificial oxidation to remove about 20 nm of the SiC surface, 4 lightly doped ($1x10^{16}$ cm^{-3}) p-type regrowth thicknesses were investigated:

(a) No p-type regrowth. This MOSFET displayed an average μ_{neff} of 2 cm^2/V-s, but the highest V_{th} of 9V.
(b) 0.15 μm p-type regrowth. This split showed a higher average μ_{neff} of 22 cm^2/V-s but also a lower V_{th} of 5V compared to (d).
(c) 0.5 μm and 2 μm of regrowth. These MOSFETs exhibited the highest μ_{neff} of 27 cm^2/V-s, but also, as expected, the lowest V_{th} of 4V.

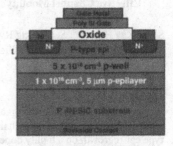

Figure 1. Experiment I Cross Sections

Phosphorus was implanted at a heavy dose to form the source and drain regions. Thermal gate oxide was grown at 1175°C Dry O$_2$, followed by 950°C Wet O$_2$ and 1175°C NO anneals. Boron doped Poly-Si films were deposited to form the gate electrode. Nickel ohmic contacts were made to the source and drain. Implants were activated at 1200°C, as it was previously determined that this activation temperature did not impact the effective channel mobility.[4] Figure 1 illustrates the detailed side view of the completed lateral MOSFETS for this first experiment.

Experiment II

A second experiment was then performed on similarly fabricated lateral MOSFETs. In this experiment, p-wells of varied concentration were formed either by doping during epitaxial regrowth or by ion implantation into the regrown layer. Epitaxial p-well doping concentrations of $1x10^{15}$ cm^{-3}, $1x10^{16}$ cm^{-3}, $5x10^{16}$ cm^{-3}, and $1x10^{17}$ cm^{-3} were investigated, while $5x10^{15}$ cm^{-3}, $1x10^{16}$ cm^{-3}, $5x10^{16}$ cm^{-3}, $1x10^{17}$ cm^{-3} and $5x10^{18}$ cm^{-3} were investigated for the implanted p-well doses. The ion-implanted doses were activated at 1650°C for all wafers. The cross section of the lateral MOSFETs fabricated for the third experiment can be seen in Figure 2. Field effect mobility and threshold voltage were measured for each experiment. Mobility was measured with a V_{ds} of 50 mV, sweeping the gate voltage from –5 up to 30 volts. Threshold voltage was extracted using linear extrapolation of the transfer characteristics to zero current.

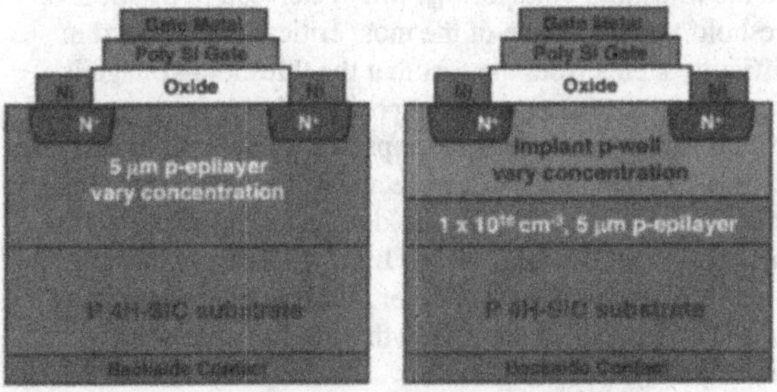

Figure 2. Experiment II Cross Sections

Results and Discussion

The results of the initial experiment are consistent with the general knowledge of the relationship between mobility and threshold voltage. Figure 3 shows the mobility versus gate voltage curves for the process splits. From these results it is clear that the 0.15 μm p-type

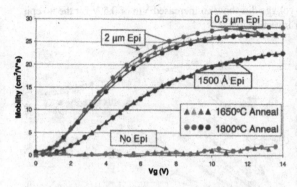

Figure 3. MOS channel mobility versus gate voltage curve for varied p-type regrowths in Experiment I.

regrowth exhibited a lower average peak μ_{neff} of 22 cm²/V-s, while growing a p-type epilayer 0.5 μm or more yielded an average μ_{neff} of 26-27 cm²/V-s. These results can be explained in two ways. First, the sample with 0.15 μm of epilayer thickness may contain many defects stemming from the high p-well dopant concentration in the implanted layer. This high density of bulk traps tends to lower the effective inversion layer electron mobility through capture of electrons and coulombic scattering.[3] In contrast, the samples with 0.5 μm or more of epitaxial regrowth may have fewer remaining defects as they may have been effectively grown out epitaxially. Another possible way to view these results is to consider the maximum depletion width in strong inversion, which can be calculated to be about 0.6 μm for 1×10^{16} cm⁻³ doping density. Thus, the depletion width extends well past the 0.15 μm re-grown epilayer into the heavily implanted p-well. This would also result in the lower average peak μ_{neff} of 22 cm²/V-s seen for the thinner regrowth as a layer of implant damage would be present in the depletion region. Whereas for the thicker regrowths the depletion region stays almost within the 0.5 μm regrown p-layer and certainly well within the 2 μm regrown p-layer. Therefore, the bulk defects in the thin re-grown epilayer case do affect the inversion layer electron mobility and threshold voltage of the MOSFET according to our new model. From Figure 3, it is also clear that little difference was seen in mobility for the varied p-well activation temperatures of 1650°C and 1800°C.

As expected the samples with the lowest mobility results exhibited the highest threshold voltage. These threshold voltage results can be seen from the I_d vs. V_g curve in Figure 4. While a V_{th} of approximately 9 V was seen on the samples that did not received any regrowth, the 0.15 μm p-type regrowth samples exhibited an average V_{th} of about 5 V. The lowest V_{th} was seen with the 0.5 μm and 2 μm of regrowth samples which exhibited a V_{th} of 3.5-4 V. While little difference is seen in Figure 4 when comparing the 1650°C and 1800°C p-well activation temperatures,

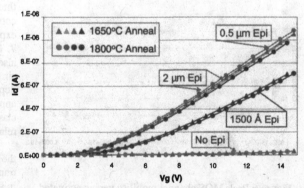

Figure 4. Drain current versus gate voltage for Experiment I.

linear extrapolation of the threshold voltage did show an increased Vth of 0.5 V for the no epi, 0.15 μm and 2 μm samples.[4] Further investigation

As scope of experiment two was quite extensive, initial results are provided here while work on the samples remains in progress. The mobility versus gate voltage results can be seen in Figure 5. The results are consistent with previous work on 6H-SiC[5] where mobility was found to decrease with increasing channel doping. Experiment II results showed that the mobility for both the epitaxially doped and ion implanted MOSFETs

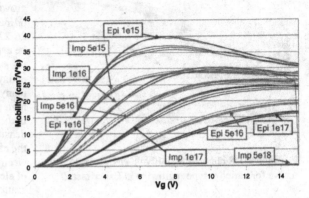

Figure 5. MOS channel mobility versus gate voltage curves for varied p-well concentrations in Experiment II.

increases as doping decreases. It was initially anticipated that the devices made on the epitaxially doped p-wells would exhibit higher mobilities. However, for equivalent epitaxial and implanted doping concentrations, comparable mobilities were observed. As can be seen in Figure 6, for similar doping concentrations the change in mobility from epitaxial to ion-implanted p-wells is only approximately 6-8 cm^2/V-s. This result suggests that the dominant mobility limiting mechanism is not related to the doping process. Rather, this is a strong indication that electron transport at the surface is limited by the oxidation process (albeit with NO annealing) that creates or activates defects at the SiO$_2$/SiC interface and possibly in the SiC sub-surface, as described in

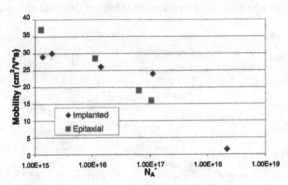

Figure 6.Peak MOS channel mobility versus activated acceptor doping for Experiment II.

the last section The relationship between the linearly extrapolated threshold voltage and doping concentration for the second experiment can be seen in Figure 7. These threshold voltage results also show comparable values for similar epitaxial and implanted doping concentrations and still appear to be limited by the classic mobility/threshold relationship.

Differentiation between the interface traps and the bulk traps will be of great importance in future research. Much work during the past decade has focused on improving the oxidation conditions to reduce the interface trap density, while very little attention has been committed to the reduction of bulk traps. If this bulk trap model is indeed true, it will encourage the researchers to optimize implantation conditions and implant activation anneals to reduce the bulk trap density, therefore improving the effective inversion layer electron mobility in MOS structures. However these results have shown that the process of fabrication of MOSFETs may not have to shift toward utilizing epitaxial layers as opposed to the implanted p-wells.

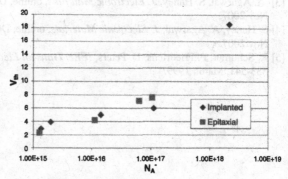

Figure 7. Threshold voltage versus activated acceptor doping for Experiment II.

Conclusion and Future Work

Our experiments revealed that p-type epitaxial regrowth up to 0.15 μm is insufficient to repair bulk damage due to the high p-well implant dose of aluminum. Thicker p-type regrowths of 0.5 μm and 2 μm resulted in higher mobilities of 26-27 cm^2/V-s. This indicates that regrowing more than 0.5 μm is adequate to either grow out of the damaged area or to insure that the heavily damaged implanted p-well is below the depletion layer. We have also shows that comparable mobilities and threshold voltages are achievable by using either ion implantation or epitaxial doping, to dope the p-well, contrary to the common belief. This should be seen as a beneficial finding, as either doping process could be used in the fabrication of MOSFETs without limiting mobility.

Future work is planned for this series of experiments. Additional DLTS measurements on the samples from both experiments #1 and #2 are being undertaken to examine the presence of the proposed bulk traps. TEM is also being used to characterize and study the SiC/SiO$_2$ interface, as well as to examine the role of excess carbon near the interface. Additional work on the third set of experiments will also include further temperature dependant analyses including D_{it}, DLTS, gated diode measurements, TEM and EELS.

Acknowledgements

This work was completed with the support the Air Force Research Laboratory and Dr. J. Scofield as well as the Army Research Laboratory and Dr. C. Scozzie.

References

[1] P. Jamet, S. Dimitrijev, *Appl. Phys. Lett.*, vol. 79, no. 3, pp. 323-325, July 2001.
[2] G. Y. Chung, C. C. Tin, J. R. Williams, K. McDonald, R. K. Chanana, R. A. Weller, S. T. Pantelides, L. C. Feldman, O. W. Holland, M. K. Das and J. W. Palmour, *IEEE Electron Dev. Lett.*, vol. 22, no. 4, pp. 176-178, April 2001.

[3] A. Agarwal, S. Haney, *J. Electronic Materials*, online, DOI 10.1007/s11664-007-0321-3, Nov 2007.

[4] S. Haney, A. Agarwal, *J. Electronic Materials*, online, DOI 10.1007/s11664-007-0310-6, Nov 2007.

[5] R. Schorner, P. Friedrichs, D. Peters, *IEEE Trans on Electron Devices.*, vol 46, no. 3, pp. 533-541, March 1999.

Mater. Res. Soc. Symp. Proc. Vol. 1069 © 2008 Materials Research Society   1069-D07-19

Spontaneous and Piezoelectric Polarization Effects on the Frequency Response of Wurtzite Aluminium Gallium Nitride / Silicon Carbide Heterojunction Bipolar Transistors

Choudhury Praharaj
Cornell University, Ithaca, NY, 14853

ABSTRACT

We present theoretical calculations for the effect of spontaneous and piezoelectric polarization on the base resistance and frequency response of wurtzite Aluminium Gallium Nitride / Silicon Carbide Heterojunction Bipolar Transistors. Heterojunction Bipolar Transistors (HBTs) built using wide band gap semiconductors with AlGaN emitter and SiC base/collector hold the promise of high-power and high-frequency operation due to lower impact ionization coefficients and higher breakdown voltages. Further, Silicon Carbide has an indirect bang gap, and a large lifetime of minority carriers compared to most other compound semiconductors, which tend to have direct band gaps. This reduces the base recombination factor when the base is made from SiC, and helps to achieve higher overall current gain. Spontaneous and piezoelectric polarizations of the order of 10^{13} electrons per cm^2 exist in wurtzite wide band-gap semiconductors. This has a non-trivial effect on band profile, charge transport and overall device characteristics since the polarization-induced charges are of the same order of magnitude as the total dopant charge content of critical device layers, and can significantly affect the amount of mobile charge depletion or accumulation in these layers. We calculate the effect of this polarization for both very thin pseudomorphic emitters and for relaxed emitter structures. We present calculations for the cases of Si-face and C-face SiC, since the signs of polarization-induced charges are different for the two cases. The intrinsic base resistance near emitter flat-band conditions is changed by about a factor of 10 depending on the alloy composition of the emitter and the polarity of growth. The maximum frequency of oscillation under emitter flat-band conditions can also be modulated by the polarization-induced charges by up to 60%. Our calculations show that the technologically less prevalent C-face SiC can give a higher advantage for frequency response, especially when the emitter thickness is larger than the critical thickness.

INTRODUCTION

The most common crystalline form for the wide band gap group III nitrides and SiC is the hexagonal wurtzite structure. These semiconductors have spontaneous polarization in the absence of externally applied electric fields [1, 2]. Further, the piezoelectric coefficients are very large [3] in these materials. The resulting polarization discontinuity at abrupt heterojunctions leads to sheet charges of the order of 10^{13} electrons per cm^2. For forward active operation of Heterojunction Bipolar transistors based on wurtzite wide band-gap semiconductors, the resistance of the base layer is modified due to this sheet charge. For base layer widths required to maintain high current gains, this affects the frequency response of the device. Modulation of base resistance in AlGaN/GaN HBTs were presented in [4] with piezoelectric effects. We study the effect of spontaneous and piezoelectric components of the polarization on the base resistance and frequency response for HBTs built using AlGaN emitters and SiC base and collector.

SPONTANEOUS AND PIEZOELECTRIC POLARIZATION

The bound sheet charge at wurtzite heterojunction interfaces is proportional to the polarization discontinuity at the interface. Figure 1 shows the sheet charges at

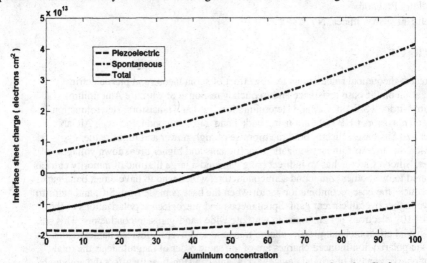

Figure 1. Polarization charges at abrupt wurtzite AlGaN/SiC interface for Ga face growth on Si-face SiC

AlGaN/SiC interfaces for Ga face growth. The contributions of the piezoelectric and spontaneous components of the polarization are shown separately. The piezoelectric component is present only for strained lattice-matched emitter layers below the critical thickness. For the AlGaN/SiC system, the critical thickness for cubic interfaces can be expected to be between the values reported for GaN and AlN in Sherwin et al [5]. Critical thickness numbers are not available for wurtzite nitride layers on wurtzite SiC. Therefore, we have assumed the critical thickness numbers for the cubic system in our calculations for the wurtzite-based HBT structures.

EFFECT OF POLARIZATION ON BASE RESISTANCE

Since the interface charges are of the order of 10^{13} electrons per cm^2, the electrostatics of the emitter-base junction is modified substantially by the presence of this charge. Figures 2 and 3 shows the intrinsic base resistance as a function of Al concentration for an AlGaN/SiC emitter base junction, near emitter flat-band conditions. All values are normalized to the value of intrinsic base resistance in the absence of polarization. The ionization factor in our calculations is assumed to be 0.01 with a base acceptor

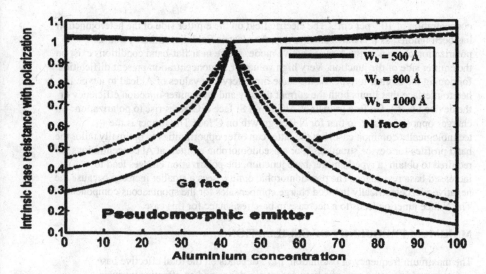

Figure 2. Intrinsic base resistance normalized to the value without polarization for an AlGaN/SiC HBT with a base acceptor concentration of 10^{20} per cm^3 and a pseudomorphic emitter for different base thicknesses W_b near emitter flat-band

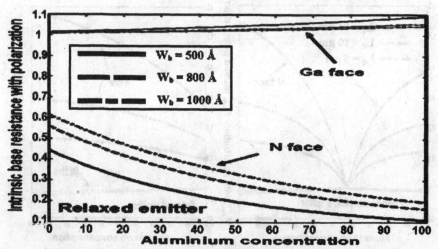

Figure 3. Intrinsic base resistance normalized to the value without polarization for an AlGaN/SiC HBT with a base acceptor concentration of 10^{20} per cm^3 and a relaxed emitter for different base thicknesses W_b near emitter flat-band

concentration of 10^{20} per cm^3. The electric field on the emitter side of the heterojunction has been ignored as it is much smaller than the electric field discontinuity due to polarization charges in the forward active mode, where near flat-band conditions exist on the emitter side of the junction. Very high values of Al concentration present difficulties for doping the emitter n- type. At the same time, very low values of Al lead to a type II heterointerface that limits both the current density and the emitter injection efficiency of the device [6, 7]. Ga face growth of AlGaN on Si face SiC gives rise to polarization charges opposite in sign to that for N face growth on C face.The former is the technologically common case, but the latter can offer opportunities to optimally tailor the band profiles for device structures. For a pseudomorphic emitter, at Al concentrations required to obtain a type I emitter-base junction, the polarization charges lead to increased base resistance. The pseudomorphic emitter sees a smaller increase because the negative piezoelectrically induced charge compensates for the spontaneous component. The N face junction leads to a decrease in base resistance for this case.

MAXIMUM FREQUENCY OF OSCILLATION

The maximum frequency of oscillation, f_{max} , depends on the total effective base resistance which is the sum of a base access resistance and an effective intrinsic resistance. The access resistance is due to the contact resistance of the base contact and the resistance of the emitter-base spacing region. The intrinsic base region contributes a resistance equal to $R_{base}/12$, where R_{base} is the undistributed base resistance, due to the

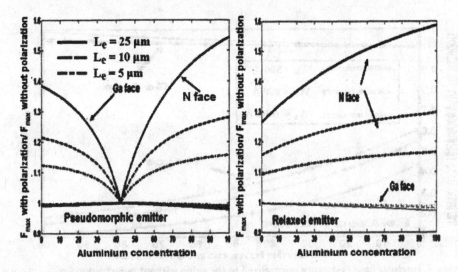

Figure 4. Fmax normalized to the value without polarization for an AlGaN/SiC HBT with a pseudomorphic emitter, near emitter flat-band. Emitter base spacing = 1 μm and base thickness = 500 A$^\circ$. L_e is the emitter width.

distributed nature of base current flow and current-crowding effects. Therefore, the effect of the polarization charges is more prominent when the total emitter width is large compared to the emitter-base spacing and when the base ohmic contact resistance is low. Figure 4 shows the effect of polarization on f_{max} for different emitter widths and for base acceptor concentration of 10^{20} per cm^3 and a fixed base width of 500 A° for the AlGaN/SiC HBT. An emitter-base spacing of 1 microns has been used in the calculations. The base ohmic contact resistance has been assumed to be zero. The effect on f_{max} is weaker than on the intrinsic base resistance due to the current crowding effects, and also due to the square root dependence of f_{max} on the base resistance. However, the modulation is upto 60% on the positive side for a base acceptor concentration of 10^{20} per cm^3.

Since the AlGaN/SiC heterojunction is of type II for low Al concentrations [6, 7], the emitter may not be close to flat-band even for high current densities. Therefore, our calculations are not completely accurate at low Al concentrations. Also, the unity gain cut-off frequency f_t has been assumed constant with aluminium concentration throughout our calculations. However, this introduces a negligible error on the calculations at high current densities since the capacitance charging times have only a small effect on f_t at high current densities. Type II heterojunction effects and the current-density dependence of polarization effects on f_{max} will be the subject of a future paper.

CONCLUSION

Polarization charges at the emitter-base junction of AlGaN/SiC heterojunction bipolar transistors change the intrinsic base resistance by factors of close to 10 near emitter flat-band conditions depending on the alloy composition of the emitter or base, and the polarity of growth. The maximum frequency of oscillation near emitter flat-band conditions is also modulated as a consequence by these charges, with a change of up to 60% for typical device dimensions. For many ranges of alloy composition, N-face devices often offer better opportunities for optimizing the frequency response due to polarization charges. Therefore, the frequency response of AlGaN/SiC heterojunction bipolar transistors can be significantly improved by using alloy compositions and growth polarities with the right sign and magnitude of polarization charges at the interface.

REFERENCES

1. F. Bernardini, F and V. Fiorentini, Nonlinear macroscopic polarization in III-V nitride alloys, Physical Review B (Condensed Matter and Materials Physics), 64, no.8 (2001)
2. A. Qteish, V. Heine and R. Needs, Polarization, band lineups and stability of SiC polytypes, Physical Review B, Vol 45, no 12, pp 6534 (1992)
3. O. Ambacher, J. Majewski, C. Miskys, A. Link, M. Hermann, M. Eickoff, M. Stutzmann, F. Bernardini, V. Fiorentini, V. Tilak, B. Schaff and L. F. Eastman , Journal of Physics: Condensed Matter, 14 , 339-434 (2002)
4. P.M. Asbeck et al, Enhancement of base conductivity via the piezoelectric effect in AlGaN/GaN HBTs, 44, Issue 2 , pp 211-219 (2000)

5. M.E. Sherwin and T.J.Drummond , Predicted elastic constants and critical layer thicknesses for cubic phase AlN, GaN and InN on β-SiC, Journal of Applied Physics, **69**, no 12 (1991)

6. E.Danielsson et al, The influence of band offsets on the IV characteristics for GaN/SiC heterojunctions, Solid State Electronics, **46**, no. 6, 827-835, (2002)

7. A.Y.Polyakov et al, Band line-up and mechanisms of current flow in n-GaN/p-SiC and n-AlgaN/p-SiC heterojunctions, Applied Physics Letters, **80**, no. 18, pp 3352, (2002)

Mater. Res. Soc. Symp. Proc. Vol. 1069 © 2008 Materials Research Society   1069-D07-20

Effect of Base Impurity Concentration on DC Characteristics of Double Ion Implanted 4H-SiC BJTs

Taku Tajima, Satoshi Uchiumi, Kenta Tsukamoto, Kazumasa Takenaka, Masataka Satoh, and Tohru Nakamura

EECE and Research Center of Ion Beam Tech., Hosei University, 3-7-2, Kajinocho, Koganei, Tokyo, 184-8584, Japan

ABSTRACT

Double ion implanted 4H-SiC bipolar junction transistors (BJTs) are fabricated by Al and N ion implantation to the base and emitter. The current gain of 3 is obtained at the base Al concentration of 1×10^{17} /cm^3. The collector current as a function of the base Gummel number suggests that double ion implanted 4H-SiC BJT operates in the intrinsic region below the emitter in the low injection level. The high base resistance restricts the base current at V_{BE} as low as 3 V.

INTRODUCTION

SiC is a promising material for ultra-low loss and high temperature devices because of high breakdown field strength and wide bandgap [1]. Many workers have studied about the device physics and processing about SiC devices such as Schottky diodes and metal-oxide-semiconductor field effect transistors (MOSFETs). Furthermore, many SiC bipolar junction transistors (BJTs), which are expected to be high temperature devices, have been fabricated using epitaxial layers, grown on SiC substrate [2,3]. In the fabrication process of SiC BJT, the ion implantation is only used to form p$^+$ contact region to the base because the ion implantation would be unsuitable to form pn junctions of base-emitter and base-collector. Indeed, it has been reported that SiC BJT with ion-implanted emitter shows the current gain less than 1, which is attributed to a low minority carrier life-time in the base-emitter space charge region [4].

In recent, however, the novel annealing system has been developed for the post ion implantation of SiC [5]. We have reported that Al ion implanted p$^+$n junction diode shows the excellent I-V characteristics with the diffusion current, which was annealed at 1900 °C using electron bombardment annealing system (EBAS, Canon-ANELVA) [6]. It is therefore expected that SiC BJT is fabricated using ion implantation techniques combined with the novel annealing technique.

In this study, we investigate the dependence of base impurity concentration on dc characteristics for the double ion implanted 4H-SiC BJT, in which ion implantation is used to form both the base and emitter regions.

EXPERIMENTAL DETAILS

Device structure

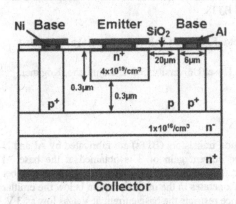

Figure 1. Structure of the fabricated
double ion implanted 4H-SiC BJT.

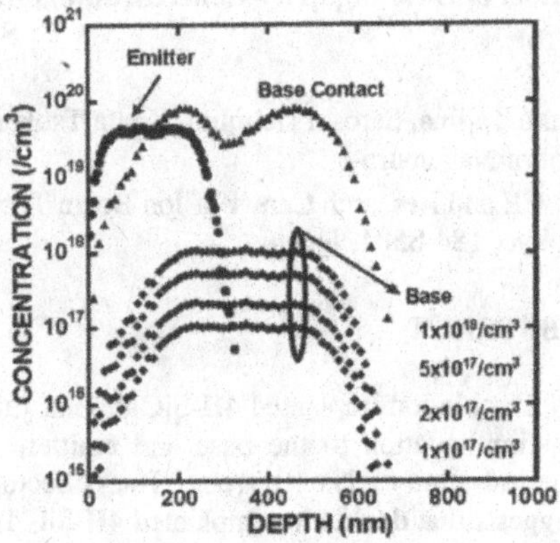

Figure 2. The doping profiles of
the base, emitter, and p^+ regions

Figure 1 schematically illustrates the structure of fabricated double ion implanted 4H-SiC BJT. The depth of i ntrinsic base is 0.3 μm. The width (W_E) and length (L_E) of the emitter were 90 and 100 μm, respectively. The p^+ contact region to the base with a width of 6 and a depth of 0.6μm is formed surrounding the emitter with a spacing of 20 μm. Figure 2 shows the do ping profiles of the base, emitter, and p^+ regions, which are calculated by using the stopping and range of ions in the matter (SRIM) simulation [7]. The base Al concentration is ranging from 1×10^{17} to 1×10^{18} /cm^3 to investigate the effect of impurity concentration on dc characteristics. The donor concentration and thickness of the emitter is 4×10^{19} /cm^3 and 0.3 μm, respectively. The averaged Al concentration in the p^+ contact region is also 5×10^{19} /cm^3.

Fabrication process

The sample is used in this study was n-type 4H-SiC(0001) epitaxial layer grown on n-type substrate with 4° off-orientation. The net donor concentration and thickness of epitaxial layer was nominally 1×10^{16} /cm^3 and 5 μm, respectively. After the sacrificial oxidation of sample, the base is fabricated by multiple-Al ion implantation with the energy ranging from 170 to 400 keV. Al ion total dose is varied from 4.05×10^{12} to 4.05×10^{13} /cm^2, which corresponds to the Al impurity concentration from 1×10^{17} to 1×10^{18} /cm^3. The p^+ contact region was also implanted by Al ions with energies of 170 and 400 keV at a total dose of 2.4×10^{15} /cm^2. Finally, the emitter is formed by multiple-N ion implantation with the energy ranging from 15 to 120 keV at a total dose of 9.2×10^{14} /cm^2. All ion implantation processes in this study were performed at room temperature. The annealing of ion implanted sample was carried out using EBAS at 1900 °C for 1 min in a vacuum less than 10^{-4} Pa.

In order to passivate the surface, the thermal oxide layer with a thickness of 20 nm is formed by dry O_2 oxidation at 1150 °C for 1.5 hr, followed by the deposition of spin-on glass (SOG)

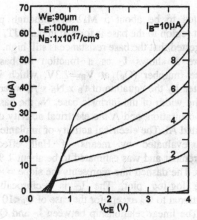

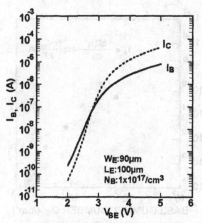

Figure 3. Static characteristics of
double ion implanted 4H-SiC BJT

Figure 4. Gummel plot of double
ion implanted 4H-SiC BJTs.

with a thickness of 0.17 μm. The ohmic contacts to the base and emitter are fabricated by depositing Ni metal layer with a thickness of 0.12 μm and subsequently annealing at 900 °C for 3 min in Ar gas flow. Finally, Al layer with a thickness of 0.2 μm is deposited onto the base and emitter Ni contact and the back surface of substrate for collector contact.

RESULTS and DISCUSSION

Figure 3 shows I_C-V_{CE} characteristics of double ion implanted 4H-SiC BJT with a base Al concentration (N_B) of 1×10^{17} /cm^3. The dimension of the emitter was W_E=90 and L_E=100 μm. The current gain is estimated to be about 3 for the base current (I_B) less than 10 μA, which is lager than that of reported devices [4]. It is confirmed that the ion implantation technique is applicable to fabricate SiC BJT. In the saturation region, however, poor linearity of I_C-V_{CE} relationship is mainly originated from the ohmic contact to the collector because the ohmic contact to the collector is formed with non-alloy Al metal layer to the C-face surface of 4H-SiC(0001) substrate. The value of V_{CEsat} is also shifted toward higher voltage with the increase of I_B, which is caused by the high emitter resistance in addition to the difference of build-in potential between the base-emitter and the base-collector junctions. In the non-saturation region, collector current is increased with V_{CE}. Since the base Al concentration in this BJT is 1×10^{17} /cm^3, the punch-through in the base leads to the lack of the linearity.

Figure 4 shows the Gummel plots of I_B and I_C for double ion implanted 4H-SiC BJT with a base Al concentration of 1×10^{17} /cm^3. In the low injection level (V_{BE}<3V), the ideality factor of I_B and I_C is estimated to be 2 and 1.5, respectively. It is suggested that I_B is dominated by the recombination current in the space charge region between the base and emitter. The ideality factor of 1.5 for I_C is larger than the case of diffusion current. As discussed above, the punch-through current flows through the base because of low base Al concentration. We think that the effective width is changed by the increase of V_{BE}, results in the increase of ideality factor of I_C. For V_{BE} above 3V, the series resistance of the base and collector restricts the increase of I_B and I_C before the diffusion current dominates I_B. The sheet resistance of Al implanted base region is

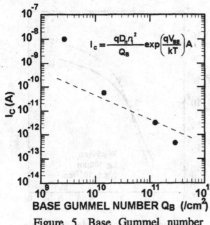

$$I_c = \frac{qD_nn^2}{Q_B}\exp\left(\frac{qV_{BE}}{kT}\right)A$$

Figure 5. Base Gummel number dependence of the collector current.

estimated to be about 5 MΩ/sq. Although p⁺ contact region to the base is formed in our BJT, it is suggested that the base resistance is still high.

Figure 5 shows I_C as a function of base Gummel number (Q_B) at V_{BE}=2.3V, which is estimated by the equation of t^i_B x N_B x A, where t^i_B is the width of the intrinsic base, N_B the base Al concentration, and A the electrical activity of implanted Al. The electrical activity of implanted Al is evaluated by means of Hall effect measurement and was estimated to be about 1 % or less. The dashed line represents the slope = -1 in the log-log plot. The I_C is reciprocally proportional to Q_B except for the case of $Q_B<10^{10}$ /cm². The linear relationship between I_C and Q_B suggests that the intrinsic BJT below the emitter operates even though in double ion implanted 4H-SiC BJT. In the case of $Q_B<10^{10}$ /cm², since the doping level in the base is the order of 10^{17}/cm³, the decrease of the effective width of the base may cause the increase of I_C through the decrease of Q_B.

CONCLUSION

We investigated the effect of base impurity concentration on dc characteristics of double ion implanted 4H-SiC BJT. The current gain of about 3 was obtained for the base Al concentration of 1 x 10^{17} /cm³. The I_C as a function of Q_B suggests that double ion implanted 4H-SiC BJT operates in the intrinsic region of BJT at the low injection level. The base resistance is still high, which might restrict the appearance of diffusion current in I_B. The optimization of device dimension and doping profile is in progress to reduce the base resistance.

REFERENCES

1. Sei-Hyung Ryu, D.Craig Capell, and John W.Palmour, *TED*, **50**, 774 (2003)
2. H.-S. Lee, M. Domeij, C.-M. Zetterling, and M. Ostling, *Mater. Sci. Forum*, **556**, 767 (2007)
3. Agarwa A.K., Sei-Hyung Ryu, Richmond J., Capell C., Palmour J.W., Balachandran S., Chow T.P., Geil B., Bayne S., Scozzie C., and Jones K.A., *ISPSD*, **24-27**, 361 (2004)
4. Agarwa A.K., Krishnaswami S., Richmond J., Capell C., Sei-Hyung Ryu, Palmour J.W., Balachandran S., Chow T.P., Geil B., Bayne S., Geil B., Scozzie C., and Jones K.A, *ISPSD*, **23-26**, 271 (2005)
5. M. Shibagaki, M. Satoh, K. Numajiri, Y. Kurematsu, F. Watanabe, S. Haga, K. Miura, S.Miyagawa, and T. Suzuki, *Mater. Sci. Forum*, **527-529**, 807(2006)
6. M. Satoh, S. Miyagawa, T. Kudoh, A. Egami, K. Numajiri, and M. Shibagaki, *Mater. Sci. Forum* (in press)
7. URL http://www.srim.org/ J.F.Ziegler

Mater. Res. Soc. Symp. Proc. Vol. 1069 © 2008 Materials Research Society

Correlation Between the V-I Characteristics of (0001) 4H-SiC PN Junctions Having Different Structural Features and Synchrotron X-ray Topography

Ryoji Kosugi[1], Toyokazu Sakata[2], Yuuki Sakuma[1], Tsutomu Yatsuo[1], Hirofumi Matsuhata[1], Hirotaka Yamaguchi[1], Ichiro Nagai[1], Kenji Fukuda[1], Hajime Okumura[1], and Kazuo Arai[1]

[1]Energy Semiconductor Electronics Research Laboratory, National Institute of Advanced Industrial Science and Technology (AIST), AIST Tsukuba Central 2, 1-1-1 Umezono, Tsukuba, Ibaraki, 305-8568, Japan
[2]R&D Associations for Future Electron Devices, AIST Tsukuba Central 2, 1-1-1 Umezono, Tsukuba, Ibaraki, 305 -8568, Japan

ABSTRACT

We have fabricated 1.5kV class four pn-type junction TEGs (Test Element Groups) of TEG_A, B, C and D having different structural features on the same wafer. The TEG_A consists of one main pn junction as a simple pn diode, while the TEG_B, C and D have p-wells in an active area. In the TEG_D, it includes the n+ region within each p-wells as well. Comparing synchrotron white beam x-ray topography (SWBXT) images with the light emission images in TEG_A, it was suggested that only a small number of TSD (threading screw dislocation) brings fatal deterioration on the breakdown voltage (V_{BD}), and the most of TSDs have less influence. The V_{BD} and fabrication yields of TEG_C and D show a strong correlation with the activation annealing condition for the ion-implanted layer.

INTRODUCTION

In a practical use of the SiC power MOSFETs, a further reduction of the channel resistance, high stability under harsh environments, and also, high production yield of large area devices are indispensable. Pn diodes with the high current rating have been already reported with high yield [1]. Meanwhile, the production yield of large area SiC power MOSFETs have not been reported in detail. Basically, power MOSFETs utilize a pn junction in the blocking states just as the pn diodes. On the other hands, there are some structural differences between the pn diodes and power MOSFETs, such as p-well structure, gate oxide between adjacent p-wells and n+ source within each p-wells. This implies that there are some specific issues in power MOSFETs. Particularly, an activation annealing for the ion-implanted region becomes a critical process in the DiMOSFETs (Double-implanted MOSFETs), since it defines an activation ratio of the ion-implanted region and the surface condition before oxide/SiC interface formation. To clarify the difference between the simple pn diodes and DiMOSFETs under blocking states, we have fabricated four pn-type junction TEGs having different structural features. Those pn junctions are close to similar structure of DiMOSFETs step-by-step from simple pn diodes. Dependence of the V-I (voltage–current) characteristics on the each structural features was surveyed over the wafer. Two activation annealing conditions have been applied for the fabrication. Prior to the fabrication, we formed grid patterns over the wafer, and then performed the SWBXT measurements to obtain the crystal defects information. This enables the direct comparison the electrical characteristics of the each pn junctions with the fingerprints of dislocations.

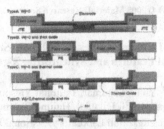

Figure 1 Schematic of the cross sectional view of four pn-type junction TEGs.

EXPERIMENTAL

Figure 1 shows a schematic figure of cross sections of each junction TEGs. The typeA TEG (hereinafter called as TEG_A) is the most simple structure similar to standard pn diodes, where the main junction was formed by a single aluminum ion implantation (I/I). JTE (Junction Termination Extension) structure was used for an edge termination for all structures. While the main junction of TEG_A consists of one p-type region, those of TEG_B, C, and D consist of p-wells in the same active area size. Difference of TEG_B and _C is an oxide thickness between the adjacent p-wells. The oxide of TEG_B consists of thick field oxide, while that of TEG_C is thermal oxide corresponding to gate oxide in DiMOSFETs (Results of TEG_B is not shown in this paper). In TEG_D, n+ region was added by phosphorus I/I as well. TEG_D is the same structure of DiMOSFETs, except that a gate and source contact metal are shorted. An active area sizes of the small and large TEGs are 150um and 990um square, respectively. The numbers of small and large TEGs are 530 and 106, respectively (The number of small TEG_C is 212). SiC substrates used in this study were 8° off-angled n+ (0001) 4H-SiC. An average effective doping density (N_d-N_a) and thickness of the epilayers were 1e16cm^{-3} and 10um, respectively. A calculated breakdown voltage (V_{BD}) of the parallel plate pn junction is estimated to be 1.66kV [2]. Before fabrication of the TEGs, grid lines were formed by an etching of SiC epilayer. Then, a SWBXT measurements were conducted at the Beam-Line 15C in Photon-Factory in High-Energy-Accelerator-Research-Organization. Three diffraction conditions, g=11-28, -1-128, and 1-108, are chosen in a grazing-incidence geometry (improved Berg-Barrett method) [3].

RESULTS AND DISCUSSION

1. Correlation between the dislocations and V-I characteristics of pn junction TEG_A

Figure 2 shows typical V-I characteristics of large TEG_A. We classified the V-I curves into five categories depending on their leakage current mode and V_{BD}. These category is not based on the origin of leakage current, but a shape of V-I characteristics.

Category I :Low V_{BD} less than 300V.
Category II :High leakage current, but high V_{BD}.
Category III :Medium leakage current, but high V_{BD}.
Category IV :Low leakage current, but slightly low V_{BD}.
Category V :Non-defective pn junctions.

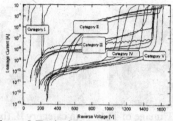

Figure 2 Typical V-I curves of large TEG_A.

Variations of V_{BD} about 100V on the TEGs in category V (hereinafter called as TEG_A_V) are attributed into those of a doping concentration and thickness of the epilayer. Figure 3(a) shows a SWBXT image for TEG_A_I. From the analysis of the SWBXT images measured at three diffraction conditions, it was assigned that the two bright circler spots (which are denoted by white open circles) in the active area correspond to threading screw dislocations (TSDs). To identify the recombination centers in the blocking states, light emission measurements were performed for TEG_A_I and _II. (Light emission measurements for category III, IV and V are under investigation.) In Fig.3(b), the light emission image at the drain voltage of 100V is superimposed on the SWBXT image. Among two TSDs, one light spot coincides with the location of one TSD and the other TSD doesn't emit light. Besides to this, it was confirmed that most TEGs in category V contain one TSD at least, but no significant deterioration is observed as shown in Fig.2. That is, only a small part of TSDs degrades the V_{BD}, and the others have less impacts on it. According to previous reports, all the pn junctions containing TSDs exhibit degraded V-I characteristics without exception[4]. One possible reason for this discrepancy is originated from the different doping concentration of drift layer, in other words, the different critical electric field (E_c). Effective doping concentration in present study is about $1 \times 10^{16} cm^{-3}$, and that of the previous work is $2.5 - 15 \times 10^{17} cm^{-3}$, where the E_c is estimated to be 2.5MV/cm and 3.8-5.5MV/cm, respectively[3]. Some unintentional particles on the metal pad is also observed in the TEGs_A_I. In those cases, the light spot doesn't coincide with the location of any dislocation, but with the location of particles on the pad. Figure 4 (a) and (b) show the same sets of the data in terms of category II. The light spot at the lower left of Fig.4(b) shows liner shape along the outer edge of JTE, which is denoted by a circular symbol in Fig.1, in different from that of

Figure 3 (a) SWBXT image of TEG_A in category I. (b) Light emission image is superimposed on the SWBXT image. Black square and white open circles in (a) correspond to the active area and TSDs in the active area, respectively.

Figure 4 (a) SWBXT image of TEG_A in category II. (b) Light emission image is superimposed on the SWBXT image. Liner bright spot as indicated by a broken circular symbol in (b) corresponds to the outer edge of JTE, which is denoted by a circular symbol in Fig.1(TypeA).

category I. In the same way, such liner light emissions were observed in all the TEGs in category II. If the amount of charges in JTE region is too high compared to the optimum one, the electric field will confine at the JTE outer edge. However, if so, the electric field must confine at their corners preferentially on the whole wafer. Other possible leakage current conduction is a surface leak path due to the local poor interface properties between the oxide and SiC on JTE region. The interface properties is significantly influenced by the surface condition before oxide formation, which relates to the activation annealing condition, as well as the gate oxide formation. Anyway, it is clear that the origin of category II is not related to dislocations, but the processing inhomogeneity. Origin of the leakage current in category III is not clear at this moment, and much confusion, since the observed dislocations in this category is not the same manner, but some patterns are observed. Figure 5 shows the (a) optical microscope image of the completed small TEGs, No89 A1, A2, A3, A4 and A5, and (b) their corresponding SWBXT image. It is clear that the square shape grid in the optical image is much strained along [1-100] direction in the SWBXT image around this region of the wafer. The white contrasts in the SWBXT image are the crowded region of a lot of TEDs (threading edge dislocations), where the each TEDs are difficult to distinguish. Similar V-I curves were observed on the crowded region with the misfit dislocations [3] and the TSDs. It is worth to say that all of the TEGs existed on the above defects don't show the leakage current in the same manner of category III, but those V-I curves are frequently observed on the crowded region of such defects.

<1-100>

Figure 5 (a) Optical micrograph image of the completed TEG_A in category III. (b) SWBXT image of the corresponding position.

2. Impact of the structural features on V-I characteristics (type A, C and D)

We fabricated two TEGs under different annealing conditions (condition 1 and 2) to understand influences of the annealing condition to V-I characteristics. Annealing temperature and period of condition 1 are 1600-1650°C and 5min, respectively, and those of condition 2 are 1650-1700°C and ~2min, respectively. Sheet resistance and surface roughness (Rms) for condition 1 obtained from the other sets of experiments are 13.7kΩ/□, 7.7-8.9nm, respectively, and those of condition 2 are 6.3kΩ/□, 0.95-2.1nm, respectively. Namely, both the activation ratio and surface roughness of condition 2 are superior to those of condition 1. Figure 6(a) and (b) show the V-I characteristics of small TEG_A and TEG_C using annealing condition2 (hereinafter called as TEG2_A and TEG2_C, respectively), respectively. The V_{BD} and fabrication yield of the small TEG2_C are almost same to those of TEG2_A except that three V-I curves in the TEG2_C show a sudden breakdown at the low voltage. This indicates that the p-well structure itself doesn't cause a significant deterioration on the V_{BD} and yield. On the other hands, it was found that the fabrication yield of type C structure was significantly influenced by the annealing condition. As shown in Fig.6(c) (Slightly lower V_{BD} in TEG1_C is attributed to the higher doping density of epilayers), a lot of V-I curves show a high leakage current at low blocking voltages. As mentioned above, pn junctions of the TEG_C consists of p-wells, and have the MOS interface on the JFET region(, that is, the region between the adjacent p-wells). As is known, the electric field in blocking states confines the bottom edge of p-wells and the gate oxide located at the center of JFET region. If the surface of JFET region is rough, the oxide will breakdown prematurely at low voltage due to a localized high electric field. Also, if an activation of p-wells is not sufficient, the depletion region will extend to inside of p-wells in the blocking states. Consequently, the shielding impact by JFET region becomes less effective, then the oxide electric field will increase. As a result, it was considered that the higher activation ratio and lower surface roughness in annealing condition 2 originate a higher fabrication yield in the TEG2_C. Despite of this, as shown in the Fig.6(d), the fabrication yield of the large TEG2_C significantly decreases compared to that of the large TEG2_A (not shown here). We guess the difference of fabrication yield between small and large TEGs is originated from a process uniformity of the activation annealing and the gate oxidation. In this work, N_2O nitridation was performed after dry thermal oxide formation. Hence, we consider that a improving the process uniformity both the activation annealing and gate oxidation is indispensable for high production yield. Figure7(a) and (b) shows the V-I curves for small TEG1_D and TEG2_D, respectively. The maximum V_{BD} of TEG2_D increases up to two times with compared to that of TEG1_D. A reach-through breakdown is known to be induced when a depletion region extends thorough p-

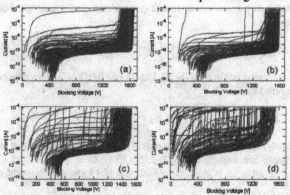

Figure 6 V-I curves of (a) small TEG2_A, (b) small TEG2_C,
(c) small TEG1_C and (d) large TEG2_C.

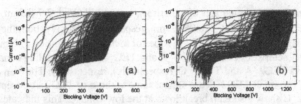

Figure 7 V-I curves of (a) small TEG1_D and (b) small TEG2_D.

well region. That is, the big improvement of V_{BD} in TEG2_D can be explained by higher activation of p-wells. It is worth to say that V_{BD} of TEG2_D, 1300V, is still lower than those of TEG2_A and C. This indicates that the amount of charges in p-wells is not sufficient to block the depletion from drift layer entirely even on the annealing condition 2.

CONCLUSIONS

Prior to the fabrication of the pn junction TEGs, the grid lines were formed over the wafer, and then the SWBXT measurements were conducted. We have fabricated four pn-type junction TEGs of TEG_A, B, C and D having different structural features on the same wafer. Those TEGs close to the similar structure of DiMOSFET from simple pn junctions step by step. From the SWBXT measurements combined with the light emission observations of TEG_A, it was suggested that, in a high blocking voltage device, only a small number of TSDs brings the fatal deterioration of the V_{BD}, and the others have less impacts on it. In the small TEG_C having p-well structures in the active region, their V_{BD} and fabrication yields are almost same to those of small TEG_A, indicating the p-well structure itself doesn't cause the fatal deterioration on those. Although it was found that the activation annealing condition for ion-implanted layer has a large impact on the V-I characteristics of TEG_C and _D, the yield in the large TEG_C significantly degrades compared to that of the large TEG_A even by utilizing the better annealing condition that we have. Consequently, we suggested that the activation annealing and gate oxidation are the critical processes to dominate a production yield of DiMOSFETs at this moment, and the improvement of their process uniformity is indispensable to obtain the high yield in large area MOS-based devices.

ACKNOWLEDGMENTS

This work is carried out in the NEDO project of "R&D of Innovative Power Electronics Inverter" supported by METI.

REFERENCES

1. B. A. Hill, M. K. Das, J. T. Richmond, B. Heath, J. J. Sumakeris, B. Geil and C. J. Scozzie, Mater. Sci. Forum, **527-529**, 1355(2006).
2. A. O. Konstantinov, Q. Wahab, N.Nordell and U. Lindefelt, Appl. Phys. Lett. **71**, 90(1997).
3. H. Matsuhata, H. Yamaguchi, I. Nagai, T. Ohno, R. Kosugi and A. Kinoshita, *Proceedings of ICSCRM2007*.
4. P. G. Neudeck and C. Fazi, *IEEE Trans Electron Devices* **46**, 478(1999).

AUTHOR INDEX

Agarwal, Anant, 195, 251, 257
Arai, Kazuo, 273
Armentrout, M.H., 115
Ayoub, Jean-Pierre, 53
Azam, S., 203

Batten, Tim, 181
Beheim, Glenn M., 209
Bergman, Peder, 141
Bockstedte, Michel, 65
Brosselard, Pierre, 141
Burk Jr., A.A., 115

Caldwell, Joshua David, 77, 169, 195
Casady, Janna, 169
Chang, Carl W., 209
Chassagne, Thierry, 163
Chen, Hui, 181
Chen, Liang-Yu, 209
Chen, Yi, 45, 77, 95
Chinthavali, Madhu S., 221
Cordier, Yvon, 163

Das, Mrinal, 195
David, John P.R., 245
De Lin, Huang, 157
Deva Reddy, Jayadeep, 109
Dhanaraj, Govindhan, 45
Dhar, Sarit, 257
Dimitrijev, Sima, 239
Dudley, Michael, 45, 77, 95, 181

Edgar, James H., 181
Ellison, Alex, 141
Evans, Laura J., 209

Fatima, Husna, 251
Ferrier, Terry L., 209
Frewin, Christopher, 109
Friedel, Bettina, 15
Friedrichs, Peter, 129
Fukuda, Kenji, 273

Gali, Ádám, 65
Glembocki, Orest J., 195

Godignon, Philippe, 141
Goh, Eric Z.J., 245
Goldsman, N., 215
Grasza, Krzysztof, 89
Green, R., 215
Greulich-Weber, Siegmund, 15
Gurfinkel, M., 215

Habersat, D., 215
Hallin, C., 115
Han, Jisheng, 239
Haney, Sarah Kay, 251, 257
Hansen, Darren, 23
Hassan, Jawad ul, 141
Hecht, Christian, 129
Henry, Anne, 141
Hens, Philip, 3
Hobart, Karl D., 195
Hock, Rainer, 3
Hold, Leonie, 239
Huang, Xianrong, 45, 77
Hull, B.A., 115

Imhoff, Eugene A., 195
Irvine, K.G., 115
Ishida, Yuuki, 123
Ito, Kazuhiro, 189

Janzén, Erik, 65, 141, 203
Johnson, C. Mark, 245
Johnson, Mark, 257
Jonsson, R., 203

Kalabukhova, E.N., 175
Kamata, Isaho, 123
Kaminski, Pawel, 33, 89
Klein, P., 169
Kojima, Kazutoshi, 123
Kong, Frederick, 239
Konias, Katja, 3
Konno, Mitsuru, 189
Koshka, Yaroslav, 157, 175
Kosugi, Ryoji, 273
Kotamraju, Siva Prasad, 157
Kozlowski, Roman, 33
Kozubal, Michal, 33, 89

Krasowski, Michael J., 209
Kravchenko, Grygoriy, 109
Kret, Slawomir, 163
Kuball, Martin, 181
Künecke, Ulrike, 3

Lancin, Maryse, 53
Lawson, Jacob, 233
Lelis, A.J., 215
Liu, Kendrick X., 77
Loboda, Mark, 23
Locke, Chris, 109
Loh, Wei Sun, 245
Lukin, S.N., 175
Lysenko, V.S., 175

MacMillan, Michael F., 45, 95
Magerl, Andreas, 3
Martin, Chris, 101
Matsuhata, Hirofumi, 273
Matsuzawa, Keiichi, 123
Mazzola, Janice, 169
Mazzola, Michael, 169
Melnychuk, Galyna, 157
Meredith, Roger D., 209
Miczuga, Marcin, 33
Momose, Kenji, 123
Müller, Ralf, 3
Murakami, Masanori, 189

Nagai, Ichiro, 273
Nakamura, Tohru, 269
Nazarov, A.N., 175
Nemoz, Maud, 163
Neudeck, Philip G., 151, 209
Nikitina, Irina, 245

Odawara, Michiya, 123
Ogunniyi, A., 215
Okojie, Robert S., 209
Okumura, Hajime, 273
O'Loughlin, M.J., 115
Onishi, Toshitake, 189
Ozpineci, Burak, 221

Pawlowski, Michal, 33

Picard, Yoosuf N., 151
Pichaud, Bernard, 53
Pons, Michel, 3
Portail, Marc, 163
Praharaj, Choudhury, 263
Prokop, Norman F., 209

Regula, Gabrielle, 53
Rengarajan, Varatharajan, 83
Rusavsky, A.V., 175
Ryu, Sei-Hyung, 251, 257

Saddow, Stephen E., 109
Sakata, Toyokazu, 273
Sakuma, Yuuki, 273
Sakwe, Sakwe Aloysius, 3
Sanchez, Edward, 45, 95
Satoh, Masataka, 269
Savchenko, D.V., 175
Scofield, James, 233
Son, Ngyen Tien, 65
Spry, David J., 209
Stahlbush, Robert E., 77, 169, 195, 251
Stein, Rene, 129
Stewart, Gray, 169
Stockmeier, Mathias, 3
Suehle, J., 215
Sumakeris, J.J., 115
Sunkari, Swapna, 169
Suzuki, Yuya, 189

Tadjer, Marko J., 195
Tajima, Taku, 269
Takahashi, Tetsuo, 123
Takeda, Hidehisa, 189
Takenaka, Kazumasa, 269
Tanner, Philip, 239
Texier, Michael, 53
Thomas, Bernd, 129
Tolbert, Leon, 221
Trunek, Andrew J., 151
Tsao, Bang-Hung, 233
Tsuchida, Hidekazu, 123
Tsukamoto, Kenta, 269
Tsukimoto, Susumu, 189

Twigg, Mark E., 151
Tymicki, Emil, 89
Tyrrell, Becky, 169

Uchiumi, Satoshi, 269

Vasin, A.V., 175
Vassilevski, Konstantin, 245
Volinsky, Alex A., 109

Wagner, Gunter, 239
Wahab, Q., 203
Wang, Guan, 181
Wang, Li, 239
Waters, Patrick, 109
Wellmann, Peter, 3
Wright, Nick G., 245

Wu, Ping, 83

Xu, Xueping, 83, 101
Xu, Zhou, 181

Yamaguchi, Hirotaka, 273
Yatsuo, Tsutomu, 273

Zelazko, Jaroslaw, 33
Zhang, Hui, 221
Zhang, Jie, 169
Zhang, Lihua, 181
Zhang, Ning, 45, 77, 95
Zhang, Qingchun, 195, 251
Zhu, Yimei, 181
Zielinski, Marcin, 163
Zwieback, Ilya, 83

SUBJECT INDEX

alloy, 269
amorphous, 175
annealing, 101
As, 181

B, 181

chemical vapor deposition (CVD)
    (chemical reaction), 129
    (deposition), 129, 141, 151,
       157, 163, 169, 239
Cl, 157
compound, 251
crystal, 95
    growth, 3, 23, 45, 83, 101, 123

deep level transient spectroscopy
    (DLTS), 89
defects, 33, 45, 53, 65, 77, 89, 175,
    195, 215, 257
devices, 195, 203, 215, 221, 239,
    251, 257, 263, 269, 273
diffusion, 189
dislocations, 3, 45, 77, 83, 123,
    151, 251
dopant, 23

electrical properties, 141, 189, 203,
    209, 215, 221, 239, 245,
    273
electron spin resonance, 65, 175
electronic material, 3, 33, 89
epitaxy, 115, 129, 141, 157, 169

film, 189

ion-implantation, 245, 257, 269

luminescence, 65

microelectro-mechanical (MEMS),
    109
microelectronics, 209, 233
morphology, 115, 169

nitride, 263

photoconductivity, 33
physical vapor deposition
    (PVD), 23
piezoelectric, 263
porosity, 15

scanning electron microscopy
    (SEM), 151
semiconducting, 15, 53, 195, 209,
    273
simulation, 83, 203, 221
sol-gel, 15
stress/strain relationship, 95
structural, 95

thermal
    conductivity, 233
    stresses, 109, 163, 233
thin film, 109, 163
transmission electron microscopy
    (TEM), 53, 181

U, 245

vapor phase epitaxy (VPE), 115,
    123

x-ray diffraction (XRD), 77, 101

Printed in the United States
By Bookmasters